Lecture Notes in Mathematics

Volume 2349

Editors-in-Chief

Jean-Michel Morel, City University of Hong Kong, Kowloon Tong, China

Bernard Teissier, IMJ-PRG, Paris, France

Series Editors

Karin Baur, University of Leeds, Leeds, UK

Michel Brion, UGA, Grenoble, France

Rupert Frank, LMU, Munich, Germany

Annette Huber, Albert Ludwig University, Freiburg, Germany

Davar Khoshnevisan, The University of Utah, Salt Lake City, UT, USA

Ioannis Kontoyiannis, University of Cambridge, Cambridge, UK

Angela Kunoth, University of Cologne, Cologne, Germany

Ariane Mézard, IMJ-PRG, Paris, France

Mark Podolskij, University of Luxembourg, Esch-sur-Alzette, Luxembourg

Mark Policott, Mathematics Institute, University of Warwick, Coventry, UK

László Székelyhidi ⓘ, MPI for Mathematics in the Sciences, Leipzig, Germany

Gabriele Vezzosi, UniFI, Florence, Italy

Anna Wienhard, MPI for Mathematics in the Sciences, Leipzig, Germany

This series reports on new developments in all areas of mathematics and their applications - quickly, informally and at a high level. Mathematical texts analysing new developments in modelling and numerical simulation are welcome. The type of material considered for publication includes:

1. Research monographs
2. Lectures on a new field or presentations of a new angle in a classical field
3. Summer schools and intensive courses on topics of current research.

Texts which are out of print but still in demand may also be considered if they fall within these categories. The timeliness of a manuscript is sometimes more important than its form, which may be preliminary or tentative. Please visit the LNM Editorial Policy (https://drive.google.com/file/d/1MOg4TbwOSokRnFJ3ZR3ciEeKs9hOnNX_/view?usp=sharing)

Titles from this series are indexed by Scopus, Web of Science, Mathematical Reviews, and zbMATH.

Frank Oertel

Upper Bounds for Grothendieck Constants, Quantum Correlation Matrices and CCP Functions

Frank Oertel
Centre for Philosophy of Natural and Social
Science (CPNSS)
London School of Economics and Political
Science (LSE)
London, UK

ISSN 0075-8434 ISSN 1617-9692 (electronic)
Lecture Notes in Mathematics
ISBN 978-3-031-57200-5 ISBN 978-3-031-57201-2 (eBook)
https://doi.org/10.1007/978-3-031-57201-2

Mathematics Subject Classification: 15A60, 62H05, 33C05, 33C45, 41A58, 05A10, 47B10, 81P45

© The Editor(s) (if applicable) and The Author(s), under exclusive license to Springer Nature Switzerland AG 2024

This work is subject to copyright. All rights are solely and exclusively licensed by the Publisher, whether the whole or part of the material is concerned, specifically the rights of translation, reprinting, reuse of illustrations, recitation, broadcasting, reproduction on microfilms or in any other physical way, and transmission or information storage and retrieval, electronic adaptation, computer software, or by similar or dissimilar methodology now known or hereafter developed.
The use of general descriptive names, registered names, trademarks, service marks, etc. in this publication does not imply, even in the absence of a specific statement, that such names are exempt from the relevant protective laws and regulations and therefore free for general use.
The publisher, the authors and the editors are safe to assume that the advice and information in this book are believed to be true and accurate at the date of publication. Neither the publisher nor the authors or the editors give a warranty, expressed or implied, with respect to the material contained herein or for any errors or omissions that may have been made. The publisher remains neutral with regard to jurisdictional claims in published maps and institutional affiliations.

This Springer imprint is published by the registered company Springer Nature Switzerland AG
The registered company address is: Gewerbestrasse 11, 6330 Cham, Switzerland

If disposing of this product, please recycle the paper.

For the maintenance of democracy

Preface

In 1956, A. Grothendieck published the celebrated "Résumé de la Théorie Métrique des Produits Tensoriels Topologiques" [57], containing a general theory of norms on tensor products of Banach spaces, describing in detail how to generate new norms and related classes of linear operators from given ones and presenting a powerful duality theory between these norms. Since 1968, the year in which J. Lindenstrauss and A. Pełczyński published their momentous description of the main results of the Résumé in the more traditional language of operators and matrices [99], there has been a veritable surge of Grothendieck's Résumé, resulting in its significant impact towards a further development of Banach space theory (cf. also [39]). In particular, Grothendieck's main result, mostly known as "the Grothendieck inequality", is now far-reaching beyond Banach space theory, including algorithmic complexity in theoretical computer science, random graphs, non-convex optimisation and semidefinite programming, foundations and philosophy of quantum mechanics, quantum information theory and even high-dimensional private data analysis. In accordance with the reformulation in [99], Grothendieck's famous inequality states the following:

Let $m, n \in \mathbb{N}$, $(H, \langle \cdot, \cdot \rangle)$ be an arbitrary Hilbert space, $A = (a_{ij})$ be an arbitrary $m \times n$ matrix with entries in $\mathbb{F}$, where $\mathbb{F} \in \{\mathbb{R}, \mathbb{C}\}$ denotes the field of scalars, and $u_1, \ldots, u_m, v_1, \ldots, v_n$ be arbitrary vectors in the closed unit ball of H, then

$$\Big|\sum_{i=1}^{m}\sum_{j=1}^{n} a_{ij}\langle u_i, v_j\rangle\Big| \leq K_G^{\mathbb{F}} \sup\Big\{\Big|\sum_{i=1}^{m}\sum_{j=1}^{n} a_{ij} p_i q_j\Big| : |p_i| \leq 1, |q_j| \leq 1 \,\forall i,j\Big\}.$$

Rounded to 3 digits, the so-called Grothendieck constant $K_G^{\mathbb{F}}$ satisfies [22, 60, 91, 93, 124, 126]

$$1.676 < K_G^{\mathbb{R}} < \frac{\pi}{2\ln(1+\sqrt{2})} \approx 1.782$$

and

$$1 < \frac{4}{\pi} < 1.338 \leq K_G^{\mathbb{C}} \leq 1.405 < K_G^{\mathbb{R}} \leq \sqrt{2}\, K_G^{\mathbb{C}}.$$

Within the framework of the search for the still unknown best possible value of the real and complex Grothendieck constant $K_G^{\mathbb{F}}$ (an open problem, unsolved since 1953), this monograph reflects our search on the smallest upper bound of both constants, $K_G^{\mathbb{R}}$ and $K_G^{\mathbb{C}}$. To this end, we establish a basic framework, built on functions which map correlation matrices to correlation matrices—entrywise—by means of the Hadamard product, such as the Krivine function in the real case or the Haagerup function in the complex case.

By making use of multivariate real and complex Gaussian analysis, higher transcendental functions, integration over spheres and combinatorics of the inversion of Maclaurin series, we provide an approach by which we also recover all famous upper bounds of Grothendieck himself ($K_G^{\mathbb{R}} \leq \sinh(\pi/2) \approx 2.301$—[57]), Krivine ($K_G^{\mathbb{R}} \leq \frac{\pi}{2\ln(1+\sqrt{2})} \approx 1.782$—[91]) and Haagerup ($K_G^{\mathbb{C}} \leq 1.405$, numerically approximated—[60]); each of them as a special case. In doing so, we aim to unify the real and complex cases as much as possible and apply our results to several concrete examples, including the Walsh-Hadamard transform ("quantum gate") and the multivariate Gaussian copula—with foundations of quantum theory and quantum information theory in mind. Furthermore, we give a slightly simplified proof of the best non computer-aided approximation known up to now, i.e., $K_G^{\mathbb{R}} < \frac{\pi}{2\ln(1+\sqrt{2})}$ ([22, 93]).

This monograph, which will be of interest to graduate students and researchers in functional analysis, complex analysis, probability, optimisation, number theory and combinatorics, in physics (particularly in relation to the foundations of quantum mechanics) and in computer science (quantum information and complexity theory), consists of nine chapters.

Chapter 1 serves as an introduction to the main topic of our contributions. In addition to providing the basic notation and terminology, we give an overview of the historical perspective and highlight the key role and strong implications of A. Grothendieck's seminal result, which lead to the introduction of a general theory of operator ideals (in the sense of A. Pietsch), an expansion of the "local" theory of Banach spaces, and which enabled a further discovery of deep connections between operator ideals, finite-dimensional subspaces of Banach spaces and various tensor products and norms in functional analysis.

In Chap. 2 (which can be studied without the subsequently following chapters), we offer a self-contained representation of complex Gaussian random vectors and unfold their rich structure. Having the Grothendieck inequality in the complex case in mind, we give special attention to the probability laws of partitioned complex Gaussian random vectors in $\mathbb{C}^{2n}$.

Chapter 3 describes the Grothendieck inequality in terms of Gram matrices and traces, and makes some connections, through a result of B. S. Tsirel'son, with quantum correlation matrices and thus with Bell's inequalities, which play a fundamental role in the foundations of quantum physics. In doing so, we show that the sign of any of the 4^m entries of the real Walsh-Hadamard transform can be determined in exactly m calculation steps. We point towards a few related open problems, unify some classical results and show how certain norms induced by Gram matrices and trace duality actually fit very well with the operator ideal formulation of the inequality.

Our task in Chap. 4 is to determine the common denominator that underlies the real and complex case. To this end, we have to consider a common source of the Grothendieck inequality, which could be viewed as an "equality in mean", induced by the classical Pearson correlation coefficient of the random signs of two suitably correlated Gaussians. We revisit uniform measures on spheres, Gaussian hypergeometric functions and even absolutely p-summing operators and reveal their connection with real and complex Gaussian variables, leading to the more general integration of expected powers of inner products of random vectors, uniformly distributed on the sphere, which are also of importance in statistical machine learning theory (in form of the so-called "kernel trick").

In Chap. 5, we introduce and investigate one of the central topics of our monograph: the class of completely correlation preserving functions (CCP), i.e., functions which map a classical correlation matrix of any size—entrywise—into a correlation matrix of the same size, such as the Grothendieck function $f(\rho) := \frac{2}{\pi} \arcsin(\rho)$, $\rho \in [-1, 1]$ (in the real case). Important connections between CCP functions, an entrywise functional calculus for positive semidefinite matrices and the famous theorem of Schoenberg, are pointed out.

In Chap. 6 (the heart of the book), we embed Grothendieck's original approach as well as Krivine's improvement into a more general framework. We show that the real Grothendieck inequality and Krivine's approach can be unified in terms of—invertible—CCP-functions and with help of Hermite polynomials. Our general approach is built on a decisive link between "suitably compressed" inverses of CCP functions and quantum correlation matrices. In general, the inverse of an invertible CCP function is not CCP. Moreover, we include a quite recent contribution of Krivine, under the form of an even more general result, again in terms of CCP-functions. We characterise the structure of CCP functions in the real case and give related examples. In doing so, we slightly extend Stein's Lemma, revisit Noise Stability, apply a few facts from the theory of distributions and test function spaces, introduce the key concept of the hyperbolic CCP transform and show that even Gaussian copulas are lurking.

Chapter 7 could be viewed as the "complex version of Chap. 6". Here, we unify the complex Grothendieck inequality and U. Haagerup's approach in similar terms, however, now under the inclusion of certain complex bivariate Hermite polynomials (originally coined by K. Itô in 1953, when he investigated complex multiple Wiener integrals—[77]).

The most important results from the previous chapters are summarised in Chap. 8. We list in detail the single steps and assumptions in the form of a "flowchart", possibly leading to a computer-aided approach regarding the implementation of an approximation to the lowest upper bound of the real and complex Grothendieck constant.

In Chap. 9, we present open problems which naturally emerge from our approach, including unsolved problems induced by the large combinatorial complexity which is subject to the implementation of Taylor series inversion.

In conclusion, we would like to thank the anonymous referees for a careful reading of the manuscript and their very helpful comments. Likewise, we express our sincere gratitude to the team members of Springer who were involved—in particular addressed to Dr. Marina Reizakis (Associate Editor, Mathematics) and Daniel Ignatius Jagadisan (Production Editor, Books). Without their indispensable and highly professional support, the present form of this monograph would not exist.

We also would like to thank Professor Feng Qi (Henan Polytechnic University, China and independent researcher, Dallas, TX 75252, USA) for a very helpful correspondence online; particularly with respect to subtleties underlying the topics of Chap. 9. Moreover, we would like to thank an anonymous helpful colleague for a very helpful correspondence by email, which in particular has expanded the list of references to include further significant sources.

London, UK Frank Oertel
February 2024

Contents

1 **Introduction and Motivation: The Outstanding Story of Grothendieck's Theorem** 1
 1.1 Historical Perspective and Theoretical Framework 1
 1.2 Preliminaries, Terminology and Notation 6

2 **Complex Gaussian Random Vectors and the Probability Law $\mathbb{C}N_{2n}(0, \Sigma_{2n}(\zeta))$** 19
 2.1 General Complex Gaussian Random Vectors in $\mathbb{C}^n$ and Their Probability Distribution 19
 2.2 Partitioned Complex Gaussian Random Vectors in $\mathbb{C}^{2n}$ and the Probability Law $\mathbb{C}N_{2n}(0, \Sigma_{2n}(\zeta))$ 31

3 **A Quantum Correlation Matrix Version of the Grothendieck Inequality** 39
 3.1 Gram Matrices, Quantum Correlation, and Beyond 39
 3.2 The Grothendieck Inequality, Correlation Matrices and the Matrix Norm $\|\cdot\|_{\infty,1}^{\mathbb{F}}$ 46
 3.3 Characterisation of $K_G^{\mathbb{F}}$ Through Operator Ideals and Violation of Bell Inequalities (A Brief Digression) 50
 3.4 $K_G^{\mathbb{R}}(2)$ and the Walsh-Hadamard Transform: Krivine's Approach Revisited 64
 3.5 The Gaussian Inner Product Splitting Property 80

4 **Powers of Inner Products of Random Vectors, Uniformly Distributed on the Sphere** 85
 4.1 Gaussian Sign-Correlation 85
 4.2 Integration Over $\mathbb{S}^{n-1}$ and the Gamma Function 87
 4.3 Integrating Powers of Inner Products of Random Vectors, Uniformly Distributed on $\mathbb{S}^{n-1}$ 97

5 Completely Correlation Preserving Functions ... 105
5.1 Completely Real Analytic Functions and the Entrywise Matrix Functional Calculus ... 105
5.2 Completely Correlation Preserving Functions and Schoenberg's Theorem ... 111

6 The Real Case: Towards Extending Krivine's Approach ... 117
6.1 Some Facts About Real Multivariate Hermite Polynomials ... 117
6.2 Real CCP Functions and Covariances: A Fourier-Hermite Analysis Approach ... 122
6.3 Examples of Real CCP Functions, Gaussian Copulas and an Extension of Stein's Lemma ... 140
6.4 Upper Bounds of $K_G^{\mathbb{R}}$ and Inversion of Real CCP Functions ... 153

7 The Complex Case: Towards Extending Haagerup's Approach ... 173
7.1 Multivariate Complex CCP Functions and Their Relation to the Real Case ... 173
7.2 On Complex Bivariate Hermite Polynomials ... 183
7.3 Upper Bounds of $K_G^{\mathbb{C}}$ and Inversion of Complex CCP Functions ... 185

8 A Summary Scheme of the Main Result ... 195

9 Concluding Remarks and Open Problems ... 199
9.1 Open Problem 1: Grothendieck Constant Versus Taylor Series Inversion ... 199
9.2 Open Problem 2: Interrelation Between the Grothendieck Inequality and Copulas ... 215
9.3 Open Problem 3: Non-commutative Dependence Structures in Quantum Mechanics and the Grothendieck Inequality ... 217

References ... 219

Index ... 227

List of Symbols

$_2F_1(a,b,c;z)$: Gaussian hypergeometric function, $c \notin -\mathbb{N}_0, \mathrm{Re}(c) > \mathrm{Re}(a+b), z \in \mathbb{T}$

$[m] := \mathbb{N} \cap [1,m] = \{1,2,\ldots,m\}, m \in \mathbb{N}$

$\mathbb{C}$: set of complex numbers

$\lceil x \rceil := \min\{v \in \mathbb{Z} : x \leq v\}, x \in \mathbb{R}$ (ceiling function)

$CS_k(S_\mathbb{F})$: set of all circularly symmetric functions, $k \in \mathbb{N}$

$C(v,k) := \{n \in \mathbb{N}_0^k : |n| = v\}$

$W_+^\omega((-r,r)) := \{\psi : \psi \in C^\omega((-r,r)) \text{ and } \left(\frac{\psi^{(n)}(0)}{n!}r^n\right)_{n \in \mathbb{N}_0} \in l_1\}, r > 0$

$h_{b,c}^\mathbb{C} : \zeta \mapsto \mathbb{E}[b(\mathbf{Z})\overline{c(\mathbf{W})}]$, $\mathrm{vec}(\mathbf{Z},\mathbf{W}) \sim \mathbb{C}N_{2k}(0,\Sigma_{2k}(\zeta)), \zeta \in \overline{\mathbb{D}}, b,c \in L^2(\mathbb{C}^k, \gamma_k^\mathbb{C}), k \in \mathbb{N}$

$H_{m,n} : (z,w) \mapsto \frac{1}{\sqrt{m!n!}} \sum_{j=0}^{m \wedge n} (-1)^j j! \binom{m}{j}\binom{n}{j} z^{m-j} w^{n-j}, z,w \in \mathbb{C}, m,n \in \mathbb{N}_0$
(complex Hermite polynomial)

$\mathrm{acx}(S)$: absolutely convex hull of S, S subset of an $\mathbb{F}$-vector space

$\mathrm{cx}(S)$: convex hull of S, S subset of an $\mathbb{F}$-vector space

$\mathrm{sign} : z \mapsto \begin{cases} \frac{z}{|z|} & \text{if } z \in \mathbb{C}^* \\ 0 & \text{if } z = 0 \end{cases}$ (complex sign function)

$\mathbb{D} := \{z \in \mathbb{C} : |z| < 1\}$ (open unit disk)

$n!! := \prod_{i=0}^{\lfloor \frac{n-1}{2} \rfloor} (n-2i), n \in \mathbb{N}_0$ (double factorial)

$E' := \mathfrak{L}(E,\mathbb{F})$, E normed (dual space of E)

$\mathbb{E}_\mathbb{P}[X] := \int_\Omega \mathrm{Re}(X)\,d\mathbb{P} + i \int_\Omega \mathrm{Im}(X)\,d\mathbb{P}$, X $\mathbb{P}$-integrable rv, where $\mathbb{P}$ is a probability measure

$\mathbf{X} \stackrel{d}{=} \mathbf{Y}$: equality of the random vectors $\mathbf{X}$ and $\mathbf{Y}$ in law

$B_{n,k}$: exponential partial Bell polynomial, $n \in \mathbb{N}_k, k \in \mathbb{N}_0$

$\mathbb{F}$: either $\mathbb{R}$ or $\mathbb{C}$, $\mathbb{F} \in \{\mathbb{R}, \mathbb{C}\}$

$n! := \prod_{i=1}^n (n-i), n \in \mathbb{N}_0$ (factorial)

$\lfloor x \rfloor := \max\{v \in \mathbb{Z} : v \leq x\}, x \in \mathbb{R}$ (floor function)

$f \otimes g$: the function $\mathbb{F}^{2k} \ni \text{vec}(x, y) \mapsto f(x)g(y)$, where $f, g : \mathbb{F}^k \longrightarrow \mathbb{F}, k \in \mathbb{N}$

γ_p : Gaussian measure on $\mathbb{R}^p$, $p \in \mathbb{N}$

$U_\rho := \int_{\mathbb{R}^k} f(\rho y + \sqrt{1-\rho^2} x) \gamma_k(\mathrm{d}x)$, $\rho \in [-1, 1]$, $f \in L^2(\gamma_k), k \in \mathbb{N}$ (Gaussian Noise Operator)

$\mathbf{Stab}_\rho[f] := \langle f, U_\rho f \rangle_{\gamma_k}$, $\rho \in [-1, 1]$, $f \in L^2(\gamma_k), k \in \mathbb{N}$ (Noise Stability)

$H_\alpha : x \mapsto \frac{(-1)^{|\alpha|}}{\sqrt{\alpha!}} \frac{D^\alpha \varphi_k(x)}{\varphi_k(x)}$, $x \in \mathbb{R}^k$, $\alpha \in \mathbb{N}_0^k$, $k \in \mathbb{N}$ (k-variate Hermite polynomial)

$\mathbb{1}_A : x \mapsto \begin{cases} 1 & \text{if } x \in A \\ 0 & \text{if } x \notin A \end{cases}$ (indicator function of a set A)

$\mathbb{N}$: set of positive integers

$\mathbb{N}_0 := \{0\} \cup \mathbb{N}$

$\mathbb{N}_m := \mathbb{N} \cap [m, \infty] = \{m, m+1, \ldots\}, m \in \mathbb{N}$

$M_\rho(x_1, x_2; n) := \frac{1}{(1-\rho^2)^{n/2}} \exp\left(\frac{2\rho\langle x_1, x_2\rangle - \rho^2(\|x_1\|^2 + \|x_2\|^2)}{2(1-\rho^2)}\right)$, $x_1, x_2 \in \mathbb{R}^n, n \in \mathbb{N}$ (Mehler kernel)

$B_{n,k}^\circ$: ordinary partial Bell polynomial, $n \in \mathbb{N}_k, k \in \mathbb{N}_0$

$\overline{\mathbb{D}} := \{z \in \mathbb{C} : |z| \leq 1\}$ (closed unit disk)

$\mathbb{R}$: set of real numbers

$C^\omega((-r, r))$: set of all real analytic functions $\psi : (-r, r) \longrightarrow \mathbb{R}, r > 0$

λ_n : real n-dimensional Lebesgue measure, also denoted as $\mathrm{d}^n x, n \in \mathbb{N}$

$\text{sign} : x \mapsto \begin{cases} 1 & \text{if } x > 0 \\ 0 & \text{if } x = 0 \\ -1 & \text{if } x < 0 \end{cases}$ (real signum function)

$\sigma_{n-1}(A) := \frac{\Gamma(n/2)}{2\pi^{n/2}} n \lambda_n(\{r\xi : 0 < r \leq 1 \text{ and } \xi \in A\})$, $A \in \mathcal{B}(\mathbb{S}^{n-1}), n \in \mathbb{N}_2$ (surface area of A)

$\mathbb{T} := \{z \in \mathbb{C} : |z| = 1\}$ (unit circle)

$B_F := \{x \in F : \|x\|_F \leq 1\}$ (closed unit ball of a normed space $(F, \|\cdot\|_F)$)

$S_F := \{x \in F : \|x\|_F = 1\}$ (unit sphere of a normed space $(F, \|\cdot\|_F)$)

$\mathbb{Z}$: set of integers

$A \otimes B$: Kronecker product of $A \in \mathbb{M}_{m,n}(\mathbb{F})$ and $B \in \mathbb{M}_{p,q}(\mathbb{F}), m, n, p, q \in \mathbb{N}$

$C_1(n; \mathbb{F})$: set of all correlation matrices of rank 1 and size n, with entries in $\mathbb{F}$

$C(n; \mathbb{F})$: set of all correlation matrices of size n, with entries in $\mathbb{F}$ (a.k.a. "elliptope")

$\Sigma_{2n}(\zeta) := \begin{pmatrix} I_n & \zeta I_n \\ \overline{\zeta} I_n & I_n \end{pmatrix}$, $\zeta \in \mathbb{F} \cap \overline{\mathbb{D}}, n \in \mathbb{N}$

$\Delta(A) \equiv \Delta^{\mathbb{F}}(A) := \frac{1}{2} \begin{pmatrix} 0 & A \\ A^* & 0 \end{pmatrix}$, $A \in \mathbb{M}_{m,n}(\mathbb{F}), m, n \in \mathbb{N}$

$e_i e_j^\top$: elementary matrix, given $(i, j) \in [m] \times [n], m, n \in \mathbb{N}$

$\mathbb{F}_2^n \equiv l_2^n(\mathbb{F}) := (\mathbb{F}^n, \langle \cdot, \cdot \rangle_2), n \in \mathbb{N}$ (n-dimensional Euclidean space)

$\langle A, B \rangle_F$: Frobenius inner product of A and $B \in \mathbb{M}_{m,n}(\mathbb{F}), m, n \in \mathbb{N}$

$\mathbb{F}_p^n \equiv l_p^n(\mathbb{F}) := (\mathbb{F}^n, \|\cdot\|_p), n \in \mathbb{N}, 1 \leq p \leq \infty$ (n-dimensional p-normed space)

List of Symbols

$G_n := \begin{pmatrix} I_n & 0 & 0 & 0 \\ 0 & 0 & I_n & 0 \\ 0 & I_n & 0 & 0 \\ 0 & 0 & 0 & I_n \end{pmatrix}, n \in \mathbb{N}$ (swap gate)

$\Gamma_H(u, v)$: Gram matrix $((u, v) \in H^m \times H^n, H \, \mathbb{F}$-inner product space, $m, n \in \mathbb{N})$

$H_1 := \frac{1}{\sqrt{2}} \begin{pmatrix} 1 & 1 \\ 1 & -1 \end{pmatrix}$ (Walsh-Hadamard transform)

I_n : identity matrix, $n \in \mathbb{N}$

$J_2(w) \equiv J_2^{[n]}(w) := \text{vec}(\text{Re}(w), \text{Im}(w)), w \in \mathbb{C}^n, n \in \mathbb{N}$

$K_G^{\mathbb{F}}$: real and complex Grothendieck constant, $\mathbb{F} \in \{\mathbb{R}, \mathbb{C}\}$

$K_G^{\mathbb{F}}(d)$: real and complex Grothendieck constant of order d, $\mathbb{F} \in \{\mathbb{R}, \mathbb{C}\}$

$K_G^{\mathbb{F}}(m, n)$: real and complex Grothendieck constant on $\mathbb{M}_{m,n}(\mathbb{F})$, $\mathbb{F} \in \{\mathbb{R}, \mathbb{C}\}$

mat := vec^{-1} ("matrixation operator")

$\mathbb{M}_{m,n}(S)$: set of $m \times n$ matrices with entries in $\emptyset \neq S \subseteq \mathbb{F}$, $m, n \in \mathbb{N}$

$\mathbb{M}_n(\mathbb{F}) := \mathbb{M}_{n,n}(\mathbb{F}), n \in \mathbb{N}$

$\mathbb{M}_n(S)^+$: set of all positive semidefinite matrices in $\mathbb{M}_n(\mathbb{F})$ with entries in $\emptyset \neq S \subseteq \mathbb{F}$, $n \in \mathbb{N}$

$O(n)$: orthogonal group, $n \in \mathbb{N}$

$\mathcal{Q}_{m,n}(\mathbb{F}) := \{S : S = \Gamma_H(u, v) \text{ for some } \mathbb{F} - \text{Hilbert space } H \text{ and } (u, v) \in S_H^m \times S_H^n\}, m, n \in \mathbb{N}$

$R_2(A) := \begin{pmatrix} \text{Re}(A) & -\text{Im}(A) \\ \text{Im}(A) & \text{Re}(A) \end{pmatrix}, A \in \mathbb{M}_{m,n}(\mathbb{C}), m, n \in \mathbb{N}$

$f[A] := (f(a_{ij})), A \equiv (a_{ij}) \in \mathbb{M}_{m,n}(\mathbb{F})), f : U \longrightarrow \mathbb{F}, \emptyset \neq U \subseteq \mathbb{F}, (i, j) \in [m] \times [n], m, n \in \mathbb{N}$

$A * B$: Hadamard produkt, $A, B \in \mathbb{M}_{m,n}(\mathbb{F}), m, n \in \mathbb{N}$ (a.k.a. Schur product)

$\mathbb{S}^{n-1}$: unit sphere in $\mathbb{R}_2^n$, $n \in \mathbb{N}_2$

$SO(n)$: special orthogonal group, $n \in \mathbb{N}$

$SU(n)$: special unitary group, $n \in \mathbb{N}$

$\text{tr}(C)$: trace of matrix $C \in \mathbb{M}_n(\mathbb{F}), n \in \mathbb{N}$

$U(n)$: unitary group, $n \in \mathbb{N}$

$\text{vec}(A) := (a_1^\top | \ldots | a_n^\top)^\top$, where $A = (a_1 | a_2 | \cdots | a_n) \in \mathbb{M}_{m,n}(\mathbb{F}), m, n \in \mathbb{N}$

$\Gamma(z) := \int_0^\infty e^{-t} t^{z-1} \, dt, \text{Re}(z) > 0$ (Gamma function)

$\mathfrak{S}_2(H, K)$: class of Hilbert-Schmidt operators, H, K $\mathbb{F}$-Hilbert spaces

$\langle S, T \rangle_{HS} := \text{tr}(T^*S)$ (Hilbert-Schmidt inner product of S and $T \in \mathfrak{S}_2(H, K)$)

$s(b) := \text{Im}(b) \circ \frac{1}{\sqrt{2}} J_2^{-1}, b : \mathbb{C}^n \longrightarrow \mathbb{C}, n \in \mathbb{N}$

$\mathfrak{L}(E, F) := \{T : E \longrightarrow F : T$ is bounded and linear$\}$, E and F normed

$r(b) := \text{Re}(b) \circ \frac{1}{\sqrt{2}} J_2^{-1}, b : \mathbb{C}^n \longrightarrow \mathbb{C}, n \in \mathbb{N}$

$\phi_\mathbf{X}$: characteristic function of the real-valued random vector $\mathbf{X}$

$\phi_\mathbf{Z}$: characteristic function of the complex-valued random vector $\mathbf{Z}$

$\gamma_n^\mathbb{C}$: Gaussian measure on $\mathbb{C}^n$, $n \in \mathbb{N}$

$\lambda_n^\mathbb{C}$: complex Lebesgue measure, $n \in \mathbb{N}$

$\mathbf{X} \sim N_n(\mu, \Sigma)$: $\mathbf{X}$ is an $N_n(\mu, \Sigma)$-distributed random vector, $n \in \mathbb{N}$ (real Gaussian law)

$\mathbf{Z} \sim \mathbb{C}N_n(\mu, \Gamma)$: $\mathbf{Z}$ is an $\mathbb{C}N_n(\mu, \Gamma)$-distributed random vector, $n \in \mathbb{N}$ (complex Gaussian law)

$\|A\|_H^G := \max\limits_{(u,v) \in S_H^m \times S_H^n} |\mathrm{tr}(A^* \Gamma_H(u, v))|$, H $\mathbb{F}$-Hilbert space, $A \in \mathbb{M}_{m,n}(\mathbb{F})$, $m, n \in \mathbb{N}$

$\|A\|_{\infty,1}^{\mathbb{F}} := \|A\|_{\mathcal{L}(l_\infty^n, l_1^m)} = \|A\|_{\mathbb{F}}^G$, $A \in \mathbb{M}_{m,n}(\mathbb{F})$, $m, n \in \mathbb{N}$

$(\mathfrak{D}_2, \|\cdot\|_{\mathfrak{D}_2})$: 1-Banach ideal of all 2-dominated operators

$(\mathfrak{L}_2, \|\cdot\|_{\mathfrak{L}_2})$: 1-Banach ideal of all 2-factorable operators

$(\mathfrak{N}, \|\cdot\|_{\mathfrak{N}})$: 1-Banach ideal of all nuclear operators

$(\mathfrak{P}_p, \|\cdot\|_{\mathfrak{P}_p})$: 1-Banach ideal of all absolutely p-summing operators, $1 \le p < \infty$

$A(\mathbb{D}) := \{g : g \in C(\overline{\mathbb{D}})$ such that $g\big|_{\mathbb{D}}$ is holomorphic on $\mathbb{D}\}$ (Disc Algebra)

$h_{f,f}^{\mathrm{hyp}} := ((h_{f,f}^{-1}|_{(-1,1)})_{\mathrm{abs}})^{-1}$, $h_{f,f}^{-1}|_{(-1,1)} \in W_+^\omega((-1, 1))$, $f \in S_{L^2(\gamma_k)}$, $k \in \mathbb{N}$

$H_{f,g} := h_{f,f} + h_{g,g}$, $f, g \in L^2(\mathbb{R}^k, \gamma_k)$, $k \in \mathbb{N}$

$h_{f,g} : \rho \mapsto \sum\limits_{n \in \mathbb{N}_0^k} \langle f, H_n \rangle_{\gamma_k} \langle g, H_n \rangle_{\gamma_k} \rho^{|n|}$, $|\rho| \le 1$, $f, g \in L^2(\mathbb{R}^k, \gamma_k)$, $k \in \mathbb{N}$

$\psi_{\mathrm{abs}} : x \mapsto \sum\limits_{n=0}^{\infty} \frac{|\psi^{(n)}(0)|}{n!} x^n$, if $\psi \in W_+^\omega((-r, r))$, $|x| \le r$, $r > 0$

$W^+(\mathbb{D}) := \{f : f(z) = \sum\limits_{n=0}^{\infty} b_n z^n$ is analytic on $\mathbb{D}$ and $\sum\limits_{n=0}^{\infty} |b_n| < \infty\}$ (Wiener Algebra)

Chapter 1
Introduction and Motivation: The Outstanding Story of Grothendieck's Theorem

1.1 Historical Perspective and Theoretical Framework

In order to better categorise the primary subject of our monograph, we briefly examine the history of an essential part of the classical theory of Banach spaces in functional analysis, which emerged from A. Grothendieck's seminal article "Résumé de la Théorie Métrique des Produits Tensoriels Topologiques" [57]. In this regard, we would like to highlight the article [118], which provides a detailed overview of this remarkable development.

In the late thirties, tensor products entered the area of functional analysis, due to the works of F. J. Murray, J. von Neumann and R. Schatten. However, it was Grothendieck who revealed the structural richness of tensor products of Banach spaces and who used their various norms to construct related classes of bounded linear operators. In this context, he actually established the origin of the "local" theory of Banach spaces, i.e., the study of the structure of Banach spaces in terms of their finite-dimensional subspaces. Here, it should be noted that Grothendieck also *introduced* and characterised the famous and highly consequential approximation property of Banach spaces (cf. [56, Section 1.5]).

Despite its emergence more than six decades ago, the techniques and results of the pioneering work of Grothendieck in [57] are still not widely known or appreciated, including his main result, mostly known as "the Grothendieck inequality". Grothendieck himself called it "the fundamental theorem of the metric theory of tensor products". It is likely that his work was rejected because he provided almost no proofs and relied on the (duality) theory of the rather abstract (yet very powerful) notion of tensor products of Banach spaces. (cf. [3, 33, 39, 57]). [39] gives a very readable and comprehensive account of the tensor product theory, developed in [57] while maintaining the symbolic language of Grothendieck. The culmination in [39] is Chap. 4, where the Grothendieck inequality and its consequences are considered in detail. Actually, [57] appeared in 1956.

It was not until the end of the sixties when the scientific community gave [57] more recognition. The interest in Grothendieck's work namely revived when J. Lindenstrauss and A. Pełczyński recast its main results in the more traditional language of operators and matrices, including the Grothendieck inequality (Theorem 1.1), on which our monograph is based (cf. [39, Theorem A.3.1] and [99]). They presented important applications to the theory of absolutely p-summing operators and translated results, which were written in terms of tensor products by Grothendieck, into properties of linear operators and operator ideals.

Almost at the same time, a general theory of operator ideals on the class of Banach spaces was developed by A. Pietsch and his academic school in Jena, yet without the use and the abstract language of Grothendieck's tensor norms. Due to Pietsch's seminal book "Operator Ideals" [123], that theory became a central theme in Banach space theory. Particularly during this time, theory and applications of operator ideals had a greater prevalence, as opposed to the tensor norm theory of Grothendieck. A comprehensive overview of Pietsch's theory and application of operator ideals—including a corresponding reformulation of Grothendieck's seminal inequality—is given in [38].

In 1993, A. Defant and K. Floret published their pathbreaking and comprehensive monograph "Tensor Norms and Operator Ideals" [33]. Here, deep interconnections between operator ideals, the "local" theory of Banach spaces and tensor norms are revealed with a high level of attention to detail. They made very clear that tensor products and operator ideals are closely connected and showed in detail how to transform tensor products to operator ideals and conversely, revealing that normed tensor products of Banach spaces (in the sense of Grothendieck) and Banach operator ideals (in the sense of Pietsch) are "two sides of the same coin"! Nowadays, many researchers follow the approach of Defant and Floret and make use of both languages simultaneously, just like we do (cf. [112–115]). The monographs [33, 39, 79, 134] are very valuable sources which strongly help to make Grothendieck's approach accessible to a wider community.

In conclusion, the Grothendieck inequality had a profound influence on the geometry of Banach spaces and operator theory; particularly between 1970 and 1990. We highly recommend the readers who have a solid knowledge of functional analysis to study Chap. 8 of the superb monograph [3]. Here is worked out in great clarity, step-by-step (even without the use of tensor products of Banach spaces, and without the use of operator ideal theory), how the Grothendieck inequality can be equivalently characterised, including Grothendieck's key result, that the inequality is equivalent to the deep fact that *any* bounded linear operator $T \in \mathfrak{L}(L^1(\mu), l_2)$ (where the measure μ lives on a σ-finite measure space) already is absolutely 1-summing and satisfies the norm inequality $\|T\|_{\mathfrak{P}_1} \leq K_G^{\mathbb{F}} \|T\|$ (cf. [3, Remark 8.3.2 (b)], [33, Theorem 23.10], [34, 38], [79, Theorem 10.7] and Remark 4.2). An exceptional proof of the latter result (which is built on a factorisation of $T \in \mathfrak{L}(l_1, l_2)$ over the disc algebra $A(\mathbb{D})$) is given in [157, Theorem III.F.7] (cf. also Remark 5.5 below).

Meanwhile, in addition to this impact, the Grothendieck inequality exhibits deep applications in different fields (such as algorithmic complexity in theoretical computer science, analysis of Boolean functions, random graphs (including the mathematics of the systemic risk in financial networks, analysis of nearest-neighbour interactions in a crystal structure (Ising model), correlation clustering and image segmentation in the field of computer vision), NP-hard combinatorial optimisation, non-convex optimisation and semidefinite programming (cf. [53]), foundations and philosophy of quantum mechanics, quantum information theory, quantum correlations (cf. Sect. 3.1), quantum cryptography, communication complexity protocols and even high-dimensional private data analysis (cf. [42])! Also in these fields there exist many challenging related open problems.

Theorem 1.1 (Grothendieck Inequality in Matrix Form) *Let $\mathbb{F} \in \{\mathbb{R}, \mathbb{C}\}$. There is an absolute constant $K > 0$ such that for any $m, n \in \mathbb{N}$, for any $A = (a_{ij}) \in \mathbb{M}_{m,n}(\mathbb{F})$, any $\mathbb{F}$-Hilbert space H, and any $(u_1, \ldots, u_m) \in B_H^m$, $(v_1, \ldots, v_n) \in B_H^n$, the following inequality is satisfied:*

$$\left| \sum_{i=1}^{m} \sum_{j=1}^{n} a_{ij} \langle u_i, v_j \rangle_H \right| \leq K \sup\left\{ \left| \sum_{i=1}^{m} \sum_{j=1}^{n} a_{ij} p_i q_j \right| : |p_i| \leq 1, |q_j| \leq 1 \, \forall i, j \right\}.$$

The smallest possible value of the corresponding absolute constant K is called the *Grothendieck constant* $K_G^{\mathbb{F}}$ (cf. also Theorem 3.1). The superscripts $\mathbb{R}$ and $\mathbb{C}$ are used to indicate the different values in the real and complex cases. Regarding functional analytic key reformulations of the Grothendieck inequality, involving the infinite-dimensional Banach spaces of type $C(K)$, $C(L)'$ and $L^1(\mu)$, we highly recommend the readers to study [126, Section 2], including the detailed and very helpful proof of the equivalence of [126, Theorem 2.3] and the Grothendieck inequality in matrix form (on which our monograph is based). Observe that in the case $m = n = 1$, already $K = 1$ satisfies the Grothendieck inequality. However, it is well-known that $K_G^{\mathbb{R}} > K_G^{\mathbb{C}} > 1$ (cf. also Corollary 3.4). In his seminal paper [57], Grothendieck proved that $K_G^{\mathbb{R}} \leq \sinh(\frac{\pi}{2}) \approx 2.301$ (within our framework recovered as special case in Example 6.7). In 1974, Grothendieck's result could be improved by R. E. Rietz, who showed that $K_G^{\mathbb{R}} < 2.261$ (cf. [128]). Until present (rounded to three digits) the following encapsulation of $K_G^{\mathbb{R}}$ holds (cf. [22, 93], Example 6.7 and Example 6.10):

$$1.676 < K_G^{\mathbb{R}} \stackrel{(!)}{<} \frac{1}{\frac{2}{\pi} \sinh^{-1}(1)} = \frac{\pi}{2 \ln(1 + \sqrt{2})} \approx 1.782.$$

The complex constant is strictly smaller than the real one. Namely, if we merge the values of the upper bounds of $K_G^{\mathbb{C}}$ achieved to date (cf. [60, 91, 124, 126], Theorem 1.2 and our approximative calculation of the number $\frac{1}{c^*} \approx 1.40449$ at the end of Example 7.1), we obtain (rounded to three digits):

$$1 < \frac{4}{\pi} < 1.338 \leq K_G^{\mathbb{C}} \leq 1.405 < \sqrt{2} < e^{1-\gamma} < \frac{\pi}{2} < K_G^{\mathbb{R}} \leq \sqrt{2} K_G^{\mathbb{C}}, \quad (1.1.1)$$

where $\gamma := \sum_{n=2}^{\infty} (-1)^n \frac{\zeta(n)}{n} = -\Gamma'(1) \approx 0.577$ denotes the Euler-Mascheroni constant. Until present, the best-known lowest upper bound of $K_G^{\mathbb{C}}$ is given by $K_G^{\mathbb{C}} \leq 1.40491$, carried out by U. Haagerup in [60] (approximatively achieved again in Example 7.1).

Regarding apparently surprising equivalent formulations of Theorem 1.1 (including their detailed verifications), revealing the depth of the structure beneath the "surface of the inequality", we refer to [79, Equivalent formulations, p. 109 ff].

Computing the best possible numerical value of the constants $K_G^{\mathbb{R}}$ and $K_G^{\mathbb{C}}$ is still an open problem (unsolved since 1953). This is where our own research continues. We look for a general framework (primarily build on methods originating from (block) matrix analysis (cf. [70]), multivariate statistics with real and complex Gaussian random vectors, theory of special functions, modelling of statistical dependence with copulas and combinatorics, whose complexity increases rapidly in dimension, though) which allows either to give the value of $K_G^{\mathbb{R}}$, respectively $K_G^{\mathbb{C}}$ explicitly or to approximate these values from above and from below at least. However, our approach—which in particular allows a short proof of the real and complex Grothendieck inequality, even with J.-L. Krivine's upper bound of $K_G^{\mathbb{R}}$—confronts us strongly with the question whether the seemingly non-avoidable combinatorial complexity actually allows us to determine the values of $K_G^{\mathbb{R}}$, respectively $K_G^{\mathbb{F}}$ explicitly, or not. A detailed description of this research problem can be studied in Sect. 9.1 of our monograph.

If either the size $m \times n$ of the arbitrarily chosen matrix $A \in \mathbb{M}_{m,n}(\mathbb{F})$ or the dimension d of the finite-dimensional Hilbert space $\mathbb{F}_2^d$ is predefined, we obtain the corresponding two weakened forms of Theorem 1.1:

Proposition 1.1 *Let* $\mathbb{F} \in \{\mathbb{R}, \mathbb{C}\}$.

(i) For any $d \in \mathbb{N}$ *there is a constant* $K^{\mathbb{F}}(d) > 1$ *such that*

$$\Big|\sum_{i=1}^{m}\sum_{j=1}^{n} a_{ij} \langle u_i, v_j \rangle_{\mathbb{F}_2^d}\Big|$$

$$\leq K^{\mathbb{F}}(d) \sup\Big\{\Big|\sum_{i=1}^{m}\sum_{j=1}^{n} a_{ij} p_i q_j\Big| : |p_i| \leq 1, |q_j| \leq 1 \,\forall i, j\Big\} \quad (1.1.2)$$

for any $m, n \in \mathbb{N}$, *for any* $A \in \mathbb{M}_{m,n}(\mathbb{F})$, *for any* $(u_1, \ldots, u_m) \in B_d^m$, *and for any* $(v_1, \ldots, v_n) \in B_d^n$.

(ii) For any $(m, n) \in \mathbb{N}^2$ *there is a constant* $K^{\mathbb{F}}(m, n) > 1$ *such that*

$$\Big|\sum_{i=1}^{m}\sum_{j=1}^{n} a_{ij} \langle u_i, v_j \rangle_H\Big|$$

$$\leq K^{\mathbb{F}}(m, n) \sup\Big\{\Big|\sum_{i=1}^{m}\sum_{j=1}^{n} a_{ij} p_i q_j\Big| : |p_i| \leq 1, |q_j| \leq 1 \,\forall i, j\Big\} \quad (1.1.3)$$

1.1 Historical Perspective and Theoretical Framework

for any Hilbert space H, for any $A \in \mathbb{M}_{m,n}(\mathbb{F})$, for any $(u_1, \ldots, u_m) \in B_H^m$, and for any $(v_1, \ldots, v_n) \in B_H^n$.

Let $K_G^{\mathbb{F}}(d)$ denote the smallest possible value of the corresponding constant $K^{\mathbb{F}}(d)$, introduced by Krivine (cf. [33, Proposition 20.17]), and let $K_G^{\mathbb{F}}(m, n)$ be the smallest possible value of the constant $K^{\mathbb{F}}(m, n)$, introduced by B. S. Tsirel'son for $\mathbb{F} = \mathbb{R}$ (cf. [148] and the detailed elaboration in [95, 96]). Consequently, $K_G^{\mathbb{F}}(d) \leq K_G^{\mathbb{F}}$ for all $d \in \mathbb{N}$, whence $\sup_{d \in \mathbb{N}} K_G^{\mathbb{F}}(d) \leq K_G^{\mathbb{F}}$. Similarly, it follows that $\sup_{(m,n) \in \mathbb{N}^2} K_G^{\mathbb{F}}(m, n) \leq K_G^{\mathbb{F}}$. It seems to us that the numbers $K_G^{\mathbb{F}}(m, n)$ and $K_G^{\mathbb{F}}(d)$ in general do not stand in relation to each other. Hence, to avoid any risk of confusion, it is important to understand whether authors refer to $K_G^{\mathbb{F}}(m, n)$ or to $K_G^{\mathbb{F}}(d)$ (or even to $K_G^{\mathbb{F}}$) in their work, when they talk about "the Grothendieck constant" (such as it is the case in [14, 23, 45, 46, 84]). For any $d \in \mathbb{N}_3$, explicit *lower* bounds of $K_G^{\mathbb{R}}(d)$ in closed, analytic form are provided in [23, Theorem 1] and [45, Theorem 2.2]. Very recently, the lower bound of $K_G^{\mathbb{R}}(3)$ (which is precisely the threshold value for the nonlocality of the two-qubit Werner state for projective measurements in quantum information theory (cf. [2], [45, Section 3] and Example 3.2)) could be improved. [14] namely reveals that $1.4367 \leq K_G^{\mathbb{R}}(3) \leq 1.4546$. To achieve this result, however, a high computing power was required. In [84], an application of duality in semidefinite programming (implemented via the so-called "convex hull algorithm" in MATLAB) lead to the following values of $K_G^{\mathbb{R}}(m, n)$: $K_G^{\mathbb{R}}(5, 5) = K_G^{\mathbb{R}}(4, n) = \sqrt{2}$, where $n \in \{4, 5, 6, 7\}$.

Note that $K_G^{\mathbb{F}}(1) = 1$. Since the sequence $(K_G^{\mathbb{F}}(d))_{d \in \mathbb{N}}$ is non-decreasing it even follows that $K_G^{\mathbb{F}} = \lim_{d \to \infty} K_G^{\mathbb{F}}(d) = \sup\{K_G^{\mathbb{F}}(d) : d \in \mathbb{N}\}$ (see Proposition 3.4). Moreover, we may add (see Corollary 3.5):

$$K_G^{\mathbb{R}}(2d) \leq K_G^{\mathbb{R}}(2) \, K_G^{\mathbb{C}}(d) = \sqrt{2} \, K_G^{\mathbb{C}}(d) \text{ for all } d \in \mathbb{N}.$$

In particular, by taking the limit $d \to \infty$, we reobtain $K_G^{\mathbb{R}} \leq \sqrt{2} \, K_G^{\mathbb{C}}$.

Another important special case of the Grothendieck inequality (known as *the little Grothendieck inequality*) appears if just positive semidefinite matrices A are considered. Let $k_G^{\mathbb{F}}$ denote the Grothendieck constant, derived from the Grothendieck inequality restricted to the set of all positive semidefinite $n \times n$ matrices, with entries in $\mathbb{F}$. Then (cf. [39, Theorem 3.5.9], [57, Théorème 4, p. 41], [109, Theorem II and Remark, p. 179], [122] and Remark 4.2):

Theorem 1.2

$$k_G^{\mathbb{R}} = \frac{\pi}{2} \text{ and } k_G^{\mathbb{C}} = \frac{4}{\pi}.$$

An approximation of the largest lower bounds of both Grothendieck constants (which is not the subject of our current work) can be found in (Davie, A.M., 1984, Lower bound for K_G (Based on ideas of U. Haagerup) (unpublished)). The real case is studied in (Reeds, J.A., 1992, A new lower bound on the real Grothendieck constant. AT&T Bell Labs, Murray Hill (unpublished research report)) as well.

1.2 Preliminaries, Terminology and Notation

This section serves to provide the foundation upon which our whole work is built. To this end, we list the basic notation and symbolic abbreviations used throughout the book. More specific terminology, including terms introduced for the first time and related symbolic shortcuts will be introduced on the spot. The few remaining symbolic shortcuts which are not explicitly described, are either self-explanatory or can be found in any well-established and relevant undergraduate textbook in mathematics.

Numbers and Sets As is usual, we denote the set of complex numbers by $\mathbb{C}$ and the set of real numbers by $\mathbb{R}$. $\mathbb{Z}$ represents the set of all integers and $\mathbb{N}$ stands for the subset of positive integers. We will use the symbol $\mathbb{F}$ to denote either the real field $\mathbb{R}$ or the complex field $\mathbb{C}$. If we wish to state a definition or a result that is satisfied for either real or complex numbers (i.e., if $\mathbb{F} \in \{\mathbb{R}, \mathbb{C}\}$), we simply will make use of the letter symbol $\mathbb{F}$. Where there is no risk of confusion, we suppress the symbol $\mathbb{F}$. In order to save unnecessary case distinctions, we constantly view the set $\mathbb{R}$ as a subset of $\mathbb{C}$, so that $\mathbb{R} = \{z \in \mathbb{F} : z = \bar{z}\} = \{z \in \mathbb{F} : \text{Im}(z) = 0\}$. $\mathbb{T} := \{z \in \mathbb{C} : |z| = 1\}$ denotes the unit circle ("one-dimensional torus"), $\mathbb{D}$ the open unit disk and $\overline{\mathbb{D}}$ the closed unit disk. If F is an arbitrary normed space, then S_F denotes its unit sphere and B_F its closed unit ball. Thus, $S_{\mathbb{F}} = \mathbb{F} \cap \mathbb{T}$. In particular, $S_{\mathbb{R}} = \{-1, 1\}$ and $S_{\mathbb{C}} = \mathbb{T}$. $\mathbb{N}_0$ denotes the set of all non-negative integers (often also somewhat unhappily denoted as $\mathbb{Z}_+$). If $m \in \mathbb{N}$, we put $[m] := \mathbb{N} \cap [1, m] = \{1, 2, \ldots, m\}$ and $\mathbb{N}_m := \mathbb{N} \cap [m, \infty) = \{m, m+1, m+2, \ldots\}$. Fix $n \in \mathbb{N}_0$. In addition to the factorial $n!$, the double factorial $n!! \in \mathbb{N}$ will play a dominant role. The latter is defined as

$$n!! := \prod_{i=0}^{\lfloor \frac{n-1}{2} \rfloor} (n - 2i),$$

where $\mathbb{R} \ni x \mapsto \lfloor x \rfloor := \max\{\nu \in \mathbb{Z} : \nu \leq x\}$ denotes the floor function (and $\mathbb{R} \ni x \mapsto \lceil x \rceil := \min\{\nu \in \mathbb{Z} : x \leq \nu\}$ the ceiling function). We adopt the usual approach to include $(-1)!! := 1$ as well. A straightforward proof by induction on $n \in \mathbb{N}_0$, including the well-known fact that $\Gamma(\frac{1}{2}) = \sqrt{\pi}$ shows that

$$n!! = \begin{cases} 2^{\frac{n}{2}} \Gamma(\frac{n}{2} + 1) & \text{if } n \text{ is even} \\ \sqrt{\frac{2}{\pi}} 2^{\frac{n}{2}} \Gamma(\frac{n}{2} + 1) & \text{if } n \text{ is odd} \end{cases},$$

where $\{z \in \mathbb{F} : \text{Re}(z) > 0\} \ni z \mapsto \Gamma(z) := \int_0^\infty e^{-t} t^{z-1} \, dt$ denotes the Gamma function which will play an important role in our monograph (cf. [142, Chapter 6.1] and Lemma 4.1).

Vectors, Matrices, Norms and Linear Operators in General Fix $m, n \in \mathbb{N}$. The set of all $m \times n$-matrices with entries in a given non-empty subset $S \subseteq \mathbb{F}$ is

1.2 Preliminaries, Terminology and Notation

denoted by $\mathbb{M}_{m,n}(S)$. The matrix algebra $\mathbb{M}_{n,n}(\mathbb{F})$ is abbreviated as $\mathbb{M}_n(\mathbb{F})$. As usual, $e_i \in \mathbb{F}^n$ denotes the column vector having a 1 in the ith place and zeros elsewhere. If we wish to emphasize the dependence on the dimension n of the vector space $\mathbb{F}^n$, then we speak of the set $\{e_1^{(n)}, e_2^{(n)}, \ldots, e_n^{(n)}\} \subseteq \mathbb{F}^n$ (cf., e.g., (3.4.6)). $I_n := (e_1 | e_2 | \cdots | e_n) \in \mathbb{M}_n(\mathbb{F})$ describes the identity matrix. Initially, if not indicated otherwise, any vector (deterministic or random) $x \in \mathbb{F}^n$ is set as column vector, so that the allocated row vector is described by transposition $(x \mapsto x^\top)$. Translated into Dirac's bra-ket language, which is also used in quantum information theory, it holds that $e_i = |i-1\rangle$ and $e_i^\top = \langle i-1|$ $(i \in [n])$. In particular, $|0\rangle = e_1$, $|1\rangle = e_2$ and $|n-1\rangle\langle 1| = e_n e_2^\top \in \mathbb{M}_{n,2}(\mathbb{F})$ (cf. [108, 141] and (3.4.7)). If $A \in \mathbb{M}_{m,n}(S)$ is given, it is sometimes very fruitful to represent the entries of A as $A_{ij} := e_i^\top A e_j = e_j^\top A^\top e_i = (A^\top)_{ji}$, so that $A = (a_{ij})$, where $a_{ij} := A_{ij}$. $\overline{A} \in \mathbb{M}_{m,n}(\mathbb{F})$ is defined as $\overline{A}_{ij} := \overline{A_{ij}}$, implying that $A^* := \overline{A}^\top = \overline{A^\top}$ and $x^* := \overline{x}^\top = \overline{x^\top}$. Recall that the Euclidean norm is given by $\|x\|_2 := \sqrt{x^* x} = \sqrt{\sum_{i=1}^{n} |x_i|^2}$ for any $x = (x_1, \ldots, x_n)^\top \in \mathbb{F}^n$. If we equip the n-dimensional vector space $\mathbb{F}^n$ with the Euclidean inner product, we obtain the n-dimensional Hilbert space $\mathbb{F}_2^n \equiv l_2^n(\mathbb{F}) := (\mathbb{F}^n, \langle \cdot, \cdot \rangle_2)$, where the inner product on $\mathbb{F}^n$ is given by $\langle x, y \rangle_2 := \langle x, y \rangle_{\mathbb{F}_2^n} := y^* x = \sum_{i=1}^{n} x_i \overline{y_i}$. In particular, $\langle z, w \rangle_{\mathbb{F}_2^1} = z \cdot \overline{w}$ for all $z, w \in \mathbb{F}$. As usual in mathematics, we adopt the convention that any inner product $\langle \cdot, \cdot \rangle_H$, defined on an arbitrary Hilbert space H, is linear in the first argument and conjugate linear in the second one, implying that $\langle x, y \rangle_H^{[P]} := \overline{\langle x, y \rangle_H} = \langle y, x \rangle_H$ $(x, y \in H)$ is conjugate linear in the first argument and linear in the second one; a rather common approach in (quantum) physics. An orthonormal basis in $\mathbb{F}_2^n$ is given by the set of vectors $\{e_1, e_2, \ldots, e_n\}$; i.e., by the standard basis of $\mathbb{F}^n$. Occasionally, if $1 \leq p \leq \infty$, we put $\mathbb{F}_p^n \equiv l_p^n(\mathbb{F}) := (\mathbb{F}^n, \|\cdot\|_p)$, where

$$\|x\|_p := \begin{cases} (\sum_{i=1}^{n} |x_i|^p)^{1/p} & \text{if } 1 \leq p < \infty \\ \max\{|x_i| : i \in [n]\} & \text{if } p = \infty \end{cases}$$

denotes the p-norm of $x = (x_1, \ldots, x_m)^\top \in \mathbb{F}^n$. If there is no risk of confusion regarding $\mathbb{F}$, we simply speak of the space l_p^n (as usual). As usual, if $n \in \mathbb{N}_2$, then $\mathbb{S}^{n-1}$ denotes the unit sphere in $\mathbb{R}_2^n$. Throughout the book, we also identify any linear operator $T : \mathbb{F}^n \longrightarrow \mathbb{F}^m$ with its representing matrix with respect to the respective standard bases: $T \equiv (T_{ij})_{(i,j) \in [m] \times [n]}$. In particular, we have

$$T_{ij} \equiv e_i^\top T e_j = \langle T e_j, e_i \rangle_{\mathbb{F}_2^m} = \langle e_j, T^* e_i \rangle_{\mathbb{F}_2^n} \equiv (T^*)_{ji} \text{ for all } i, j \in [m] \times [n],$$

where $T^* : \mathbb{F}_2^m \longrightarrow \mathbb{F}_2^n$ is the adjoint operator. Furthermore, in the case $n = 1$, $\mathbb{F}$ is considered throughout as the one-dimensional Hilbert space $(\mathbb{F}_2^1, \langle \cdot, \cdot \rangle_2)$, where $\langle z, w \rangle_2 := z \overline{w}$ for all $z, w \in \mathbb{F}$. As usual, $O(n)$ denotes the orthogonal group, consisting of all invertible matrices $A \in \mathbb{M}_n(\mathbb{R})$ such that $A^{-1} = A^\top$. $U(n)$ describes the unitary group, consisting of all invertible matrices $A \in \mathbb{M}_n(\mathbb{C})$ such that $A^{-1} = A^*$. $SO(n) := \{A \in O(n) : \det(A) = 1\}$ is the special orthogonal group, and $SU(n) := \{A \in U(n) : \det(A) = 1\}$ describes the special unitary group.

An important inner product on the $\mathbb{F}$-vector space $\mathbb{M}_{m,n}(\mathbb{F})$ of all $m \times n$-matrices with entries in $\mathbb{F}$, which turns $\mathbb{M}_{m,n}(\mathbb{F})$ into an mn-dimensional Hilbert space, is the Frobenius inner product, which is defined as follows: if $A = (a_{ij}) \in \mathbb{M}_{m,n}(\mathbb{F})$ and $B = (b_{ij}) \in \mathbb{M}_{m,n}(\mathbb{F})$, then

$$\langle A, B \rangle_F := \mathrm{tr}(AB^*) = \mathrm{tr}(B^*A) = \sum_{i=1}^{m}\sum_{j=1}^{n} a_{ij}\overline{b_{ij}} = \overline{\langle B, A \rangle_F},$$

where

$$\mathrm{tr}(C) := \sum_{i=1}^{n}\langle Ce_i, e_i \rangle_2 = \sum_{i=1}^{n} c_{ii} = \mathrm{tr}(C^\top) = \overline{\mathrm{tr}(C^*)} = \overline{\mathrm{tr}(\overline{C})}$$

denotes the trace of a given (quadratic) matrix $C = (c_{ij}) \in \mathbb{M}_{n,n}(\mathbb{F})$. One can easily verify the well-known fact that the set of all elementary matrices $\{e_i e_j^\top : (i, j) \in [m] \times [n]\}$ is an orthonormal basis in the mn-dimensional $\mathbb{F}$-Hilbert space $(\mathbb{M}_{m,n}(\mathbb{F}), \|\cdot\|_F)$ (since $(e_i e_j^\top)_{\alpha\beta} = \delta_{i\alpha}\delta_{j\beta}$ for all $(\alpha, \beta) \in [m] \times [n]$ and $\mathrm{tr}(xy^\top) = y^\top x$ for all $x, y \in \mathbb{F}^n$). We adopt the symbolic notation of the "$\mathfrak{L}$-community" to represent the set of all bounded linear operators between two normed spaces $(E, \|\cdot\|_E)$ and $(F, \|\cdot\|_F)$ by $\mathfrak{L}(E, F)$. As usual, the Banach space $E' := \mathfrak{L}(E, \mathbb{F})$ denotes the dual space of E. (The "$\mathfrak{B}$-community", often encountered among researchers in the field of C^*-algebras, uses $\mathfrak{B}(E, F)$ instead, so that for example, $\mathfrak{B}(H) = \mathfrak{L}(H, H)$, if H is a given Hilbert space (cf., e.g., [66, Chapter 1.3])). Remember that every linear operator $T : E_0 \longrightarrow F$ from a finite-dimensional normed space E_0 to an arbitrary normed space F already is bounded. It should also be noted that actually $\langle A, B \rangle_F = \langle A, B \rangle_{HS}$ coincides with the Hilbert-Schmidt inner product, defined on the Hilbert space $\mathfrak{L}(\mathbb{F}_2^n, \mathbb{F}_2^m)$. To this end, recall that if H and K are arbitrarily given $\mathbb{F}$-Hilbert spaces, $T \in \mathfrak{L}(H, K)$ is a Hilbert-Schmidt operator if and only if $(\|Te_\iota\|_K)_{\iota \in J} \in l^2(J)$ for some orthonormal basis $(e_\iota)_{\iota \in J}$ in H. Here, J denotes an arbitrary index set which must be neither finite nor at most countable (cf., e.g., [80, Proposition 20.2.7]). Hence, if $S, T \in \mathfrak{S}_2(H, K)$ are two Hilbert-Schmidt operators, the Cauchy-Schwarz inequality implies that

$$\langle S, T \rangle_{HS} := \sum_{\iota \in J} \langle Te_\iota, Se_\iota \rangle_K = \sum_{\iota \in J} \langle e_\iota, T^*Se_\iota \rangle_K = \mathrm{tr}(T^*S),$$

is a well-defined inner product on the $\mathbb{F}$-vector space $\mathfrak{S}_2(H, K)$ of all Hilbert-Schmidt operators. In fact, it turns $\mathfrak{S}_2(H, K)$ into a Hilbert space itself (cf. [30,

1.2 Preliminaries, Terminology and Notation

Exercises IX.2.19, IX.2.20] and [80, Proposition 20.2.7]). Let us also note the easy-to-prove fact that

$$\Pi : (\mathbb{M}_{m,n}(\mathbb{F}), \|\cdot\|_F) \xrightarrow{\cong} (\mathbb{M}_{m,n}(\mathbb{F}), \|\cdot\|_F)', A \mapsto (B \mapsto \mathrm{tr}(BA^\top))$$

is an isometric isomorphism, whose inverse is given by $\Pi^{-1} = \Psi$, where

$$\mathbb{M}_{m,n}(\mathbb{F}) \ni \Psi(t) := (t(e_i e_j^\top))_{i,j} \text{ for all } t \in (\mathbb{M}_{m,n}(\mathbb{F}), \|\cdot\|_F)'.$$

Obviously, also the canonically defined mapping

$$\Theta : (\mathbb{M}_{m,n}(\mathbb{F}), \|\cdot\|_F)' \xrightarrow{\cong} (\mathbb{M}_{n,m}(\mathbb{F}), \|\cdot\|_F)', t \mapsto (M \mapsto \langle M^\top, t \rangle)$$

is an isometric isomorphism, implying that the composition of these two isometric isomorphisms lead to the finite-dimensional version of *trace duality* with respect to the norm $\|\cdot\|_F = \|\cdot\|_{HS}$ (cf. [37, Theorem 6.4]):

$$\Pi \circ \Theta : (\mathbb{M}_{m,n}(\mathbb{F}), \|\cdot\|_F) \xrightarrow{\cong} (\mathbb{M}_{n,m}(\mathbb{F}), \|\cdot\|_F)', A \mapsto (B \mapsto \mathrm{tr}(BA)).$$

Although it is our intention that the main ideas developed in our monograph can be captured without having any knowledge of advanced functional analysis and related operator theory, we will add a few text passages which should show how also our approach extends into the area of functional analysis and operator ideal theory. Related references will be listed, of course. In particular—despite its elegance and power—we intentionally avoid the explicit use of the language of abstract tensor products of Banach spaces and related tensor norms (originally coined by A. Grothendieck in his seminal paper [57]) as far as possible. Of course, any attentive reader will recognise that tensor products occasionally also are lurking in our framework (primarily in the form of concrete Kronecker products of matrices). Remarks in this regard could be skipped at the first reading. However, for particularly stubborn readers and authors, we strongly refer to [30, 33, 37, 66, 79, 123, 154].

Since (symmetrically) partitioned random vectors and block matrices play a key role in our analysis, it is sometimes very useful to transform matrices into column vectors by making use of a technique known as matrix vectorisation (cf. [1, Chapter 10]). If $A = (a_1 | a_2 | \cdots | a_n) \in \mathbb{M}_{m,n}(\mathbb{F})$, with columns $a_j \in \mathbb{F}^m$ ($j \in [n]$), then

$$\mathrm{vec}(A) := \mathrm{vec}(a_1, \ldots, a_n) := (a_1^\top | \ldots | a_n^\top)^\top \in \mathbb{F}^{mn}$$

denotes the *column* vector constructed by stacking the columns of A on top of each other. A concise *entrywise* implementable construction (built on Euclidean division with remainder) of $\mathrm{vec}(A)$ will be studied at the beginning of Sect. 3.4 (cf. (3.4.4)). Obviously,

$$\mathrm{vec} : (\mathbb{M}_{m,n}(\mathbb{F}), \langle \cdot, \cdot \rangle_F) \xrightarrow{\cong} \mathbb{F}_2^{mn}, A \mapsto \mathrm{vec}(A)$$

is an isometric isomorphism (between finite-dimensional Hilbert spaces). In particular,

$$\langle \text{vec}(A), \text{vec}(B) \rangle_2 = \text{tr}(B^*A) \text{ and } \|\text{vec}(A)\|_2 = \sqrt{\text{tr}(A^*A)} \tag{1.2.1}$$

for all $A, B \in \mathbb{M}_{m,n}(\mathbb{F})$. We also need vec's cousin, the Kronecker product of matrices (cf. [1, Chapter 10]), on which the construction of a matrix is based which delivers $\sqrt{2}$ as a lower bound of the real Grothendieck constant $K_G^{\mathbb{R}}$ and plays a key role in the foundations of quantum mechanics, quantum information and even in evolutionary biology: the Walsh-Hadamard transform (cf. Example 3.3, Remark 3.10 and Remark 3.11)! The Kronecker product is constructed as follows: if $A \in \mathbb{M}_{m,n}(\mathbb{F})$ and $B \in \mathbb{M}_{p,q}(\mathbb{F})$, then

$$\mathbb{M}_{mp,nq}(\mathbb{F}) \ni A \otimes B = \begin{pmatrix} M_{11} & M_{12} & \cdots & M_{1n} \\ M_{21} & M_{22} & \cdots & M_{2n} \\ \vdots & \vdots & \vdots & \vdots \\ M_{m1} & M_{m2} & \cdots & M_{mn} \end{pmatrix},$$

where $M_{ij} := a_{ij}B \in \mathbb{M}_{p,q}(\mathbb{F})$. If $C \in \mathbb{M}_{n,r}(\mathbb{F})$ and $D \in \mathbb{M}_{q,s}(\mathbb{F})$ are two further matrices, then elementary block matrix multiplication instantly results into the well-known fact that

$$(A \otimes B)(C \otimes D) = AC \otimes BD. \tag{1.2.2}$$

We will also provide a rigorous *entrywise* implementable construction of the Kronecker product at the beginning of Sect. 3.4 (again built on Euclidean division with remainder). In the context of partitioned random vectors, we will apply the vec operator in the following sense: if $x = (x_1, \ldots, x_n, x_{n+1}, \ldots, x_{2n})^\top \in \mathbb{F}^{2n}$, then $x = \text{vec}(a_1, a_2)$, where $a_1 := (x_1, \ldots, x_n)^\top \in \mathbb{F}^n$ and $a_2 := (x_{n+1}, \ldots, x_{2n})^\top \in \mathbb{F}^n$.

Fix $v = (v_1, \ldots, v_l)^\top \in \mathbb{F}^l$ and $A = (a_{ij}) \in \mathbb{M}_{m,n}(\mathbb{F})$. Let $\Psi : [m] \times [n] \longrightarrow [l]$ and $\Lambda : [l] \longrightarrow [m] \times [n]$ be given. In the context of our analysis, Ψ should be viewed as a mapping which maps an index $(i, j) \in [m] \times [n]$ of the matrix element $a_{ij} \in \mathbb{F}$ to the index $\Psi(i, j) \in [l]$ of an allocated vector element. Conversely, Λ should be regarded as a mapping which maps the index $\alpha \in [l]$ of the vector element $v_\alpha \in \mathbb{F}$ to an index $\Lambda(\alpha) = (\Lambda_1(\alpha), \Lambda_2(\alpha)) \in [m] \times [n]$ of an allocated matrix element. More precisely formulated, if we consider the (linear) composition operator $C_\Psi : \mathbb{F}^l \longrightarrow \mathbb{M}_{m,n}(\mathbb{F})$, we map the given vector $v \in \mathbb{F}^l$ to a matrix $C_\Psi(v) := v_\Psi \in \mathbb{M}_{m,n}(\mathbb{F})$ as follows:

$$(v_\Psi)_{i,j} := v_{\Psi(i,j)} \text{ for all } (i, j) \in [m] \times [n].$$

1.2 Preliminaries, Terminology and Notation

Analogously, we map the matrix $A = (a_{ij}) \in \mathbb{M}_{m,n}(\mathbb{F})$ to a vector $A_\Lambda \in \mathbb{F}^l$, according to the rule

$$(C_\Lambda(A))_\alpha := (A_\Lambda)_\alpha := a_{\Lambda(\alpha)} = a_{(\Lambda_1(\alpha), \Lambda_2(\alpha))} \text{ for all } \alpha \in [l],$$

where $C_\Lambda : \mathbb{M}_{m,n}(\mathbb{F}) \longrightarrow \mathbb{F}^l$ denotes the related linear composition operator. Similarly, the matrix $A \in \mathbb{M}_{m,n}(\mathbb{F})$ can be mapped to the matrix $A_\sigma \in \mathbb{M}_{r,s}(\mathbb{F})$, where σ is now a given mapping of type $\sigma : [r] \times [s] \longrightarrow [m] \times [n]$. Observe that A_σ consists of entries of the originally given matrix A, so that we could view A_σ as a subordinated matrix of A (cf. Remark 5.2). For example, $A^\top = A_\tau = C_\tau(A)$, where the mapping $\tau : [n] \times [m] \longrightarrow [m] \times [n]$ is defined as transposition: $\tau(\nu, \mu) := (\mu, \nu)$ (cf. (3.4.4)). A very recent application of vectorisation within the framework of single quantum systems (where entangled states do not play a role), including a related application of the Grothendieck inequality can be found in [152, 153].

Partitioning, $\mathbb{C}^{2n}$ Versus $\mathbb{R}^{4n}$, Matrices with Special Form and Positive Semidefiniteness With regard to the study of a class of crucially important partitioned multivariate complex Gaussian random vectors for our analysis (cf. Chap. 2), we firstly have to shed some light on the structure of the following two important mappings, which we will encounter many times in our book. Similar constructions and particular cases are listed in [5, Chapter 1.2] and [70, Problem 1.3.P20 and Problem 1.3.P21]. To this end, fix $m, n \in \mathbb{N}$ and $w \in \mathbb{C}^n$. Put

$$\mathbb{C}^n \ni w = \text{Re}(w) + i\,\text{Im}(w) \mapsto J_2(w) \equiv J_2^{[n]}(w)$$
$$\equiv J_2^{[n]}(w) := \text{vec}(\text{Re}(w), \text{Im}(w)) \in \mathbb{R}^{2n} \tag{1.2.3}$$

and

$$\mathbb{M}_{m,n}(\mathbb{C}) \ni A \mapsto R_2(A) := \begin{pmatrix} \text{Re}(A) & -\text{Im}(A) \\ \text{Im}(A) & \text{Re}(A) \end{pmatrix} \in \mathbb{M}_{2m,2n}(\mathbb{R}) \tag{1.2.4}$$

Observe, that if $n \geq 2$, $J_2(w) = \text{vec}(\text{Re}(w), \text{Im}(w)) = \text{vec}(\text{Re}(w_1), \ldots, \text{Re}(w_n), \text{Im}(w_1), \ldots, \text{Im}(w_n))$ in general does not coincide with $\text{vec}(J_2(w_1), J_2(w_2), \ldots, J_2(w_n))$. However, given arbitrary $z, w \in \mathbb{C}^n$, we obtain an important equality which will be applied several times in our monograph; namely:

$$J_2^{[2n]}(\text{vec}(z, w)) = \text{vec}(\text{Re}(z), \text{Re}(w), \text{Im}(z), \text{Im}(w))$$
$$= G\,\text{vec}(\text{Re}(z), \text{Im}(z), \text{Re}(w), \text{Im}(w)) = G\,\text{vec}(J_2(z), J_2(w)), \tag{1.2.5}$$

where

$$G \equiv G_n := \begin{pmatrix} I_n & 0 & 0 & 0 \\ 0 & 0 & I_n & 0 \\ 0 & I_n & 0 & 0 \\ 0 & 0 & 0 & I_n \end{pmatrix} = G^\top = G^{-1} \in O(4n) \text{ is orthogonal}. \quad (1.2.6)$$

Observe that the matrix $G_1 \in O(4)$ precisely coincides with the "swap gate" (also known as "flip operator"), used in quantum information theory (cf. [141, Problem 28] and Example 3.2). In general, if $\text{vec}(x_1, x_2, x_3, x_4) \in \mathbb{R}^{4n}$ is given, then $G \equiv G_n$ swaps the column vectors $x_2 \in \mathbb{R}^n$ and $x_3 \in \mathbb{R}^n$ and maps $\text{vec}(x_1, x_2, x_3, x_4)$ to $\text{vec}(x_1, x_3, x_2, x_4)$. Also the matrix G will be needed repeatedly; for example, in the proofs of Lemma 2.2 and Corollary 2.2.

Not only in this context, we occasionally need the following construction. Let $f, g : \mathbb{F}^k \longrightarrow \mathbb{F}$ be two arbitrary functions ($k \in \mathbb{N}$). Then the function $f \otimes g : \mathbb{F}^{2k} \longrightarrow \mathbb{F}$ is (purely symbolic) defined as

$$(f \otimes g)(\text{vec}(x, y)) := f(x)g(y) \text{ for all } x, y \in \mathbb{F}^k.$$

Clearly the mapping $J_2 : \mathbb{C}^n \longrightarrow \mathbb{R}^{2n}$ is bijective, and $\|J_2(a)\|_{\mathbb{R}^{2n}_2} = \|a\|_{\mathbb{C}^n_2}$ for any $a \in \mathbb{C}^n$. Moreover,

$$J_2(\bar{a}) = \begin{pmatrix} I_n & 0 \\ 0 & -I_n \end{pmatrix} J_2(a) \text{ for all } a \in \mathbb{C}^n. \quad (1.2.7)$$

Recall that $\mathbb{S}^{\nu-1}$ denotes the unit sphere in $\mathbb{R}^\nu_2$ ($\nu \in \mathbb{N}_2$). Thus, $S_{\mathbb{C}^n_2} = J_2^{-1}(\mathbb{S}^{2n-1})$ describes the unit sphere in $\mathbb{C}^n_2$. Let $z, w \in \mathbb{C}^n_2$. Then,

$$\text{Re}(\langle w, z \rangle_{\mathbb{C}^n_2}) = \text{Re}(z^* w) = J_2(z)^\top J_2(w) = \langle J_2(w), J_2(z) \rangle_{\mathbb{R}^{2n}_2} \quad (1.2.8)$$

induces an inner product which turns $\mathbb{C}^n$ into a *real* finite-dimensional Hilbert space and J_2 into an isometric isomorphism between the *real* Hilbert spaces $(\mathbb{C}^n, \text{Re}(\langle \cdot, \cdot \rangle_{\mathbb{C}^n_2}))$ and $\mathbb{R}^{2n}_2$. Clearly, J_2 cannot be extended to a linear mapping between $\mathbb{C}^n$ and $\mathbb{C}^{2n} \supseteq \mathbb{R}^{2n}$. However, by construction $J_2^{-1} : \mathbb{R}^{2n} \longrightarrow \mathbb{C}^n$ clearly satisfies

$$x_1 + i x_2 = J_2^{-1} x = (I_n \mid i I_n) x \text{ for all } x = \text{vec}(x_1, x_2) \in \mathbb{R}^{2n} \cong \mathbb{R}^n \times \mathbb{R}^n, \quad (1.2.9)$$

1.2 Preliminaries, Terminology and Notation

implying that the linear and non-injective mapping between the complex vector spaces $\mathbb{C}^{2n}$ and $\mathbb{C}^n$, induced by the matrix $(I_n \mid i I_n) \in \mathbb{M}_{n,2n}(\mathbb{C})$ actually is a linear extension of J_2^{-1}. Moreover, it follows that

$$\operatorname{Im}(\langle w, z\rangle_{\mathbb{C}_2^n}) = \operatorname{Im}(z^* w) = \operatorname{Re}(z^*(-i w)) = J_2(z)^\top J_2(-i w)$$
$$= \langle J_2(-i w), J_2(z)\rangle_{\mathbb{R}_2^{2n}} \qquad (1.2.10)$$

and $J_2(A w) = R_2(A) J_2(w)$ and $R_2(r A) = r R_2(A)$ for all $r \in \mathbb{R}$, $z, w \in \mathbb{C}^n$ and $A \in \mathbb{M}_{m,n}(\mathbb{C})$.
Note also that $R_2(I_n) = I_{2n}$ and

$$R_2(G H) = R_2(G) R_2(H) \text{ for all } (G, H) \in \mathbb{M}_{k,m}(\mathbb{C}) \times \mathbb{M}_{m,n}(\mathbb{C}).$$

Moreover,

$$R_2(A^*) = R_2(A)^\top \text{ for any } A \in \mathbb{M}_{m,n}(\mathbb{C}). \qquad (1.2.11)$$

In particular,

$$R_2(A^\top)^\top = R_2(\overline{A}) = \begin{pmatrix} I_n & 0 \\ 0 & -I_n \end{pmatrix} R_2(A) \begin{pmatrix} I_n & 0 \\ 0 & -I_n \end{pmatrix} \text{ for any } A \in \mathbb{M}_{m,n}(\mathbb{C}).$$

Thus, from the algebraic viewpoint, $R_2 : M_n(\mathbb{C}) \longrightarrow \mathbb{M}_{2n}(\mathbb{R})$ is an injective unitary *-ring homomorphism. In particular, $A \in M_n(\mathbb{C})$ is invertible if and only if $R_2(A) \in \mathbb{M}_{2n}(\mathbb{R})$ is invertible (since $R_2(A) R_2(A^{-1}) = I_{2n} = R_2(A^{-1}) R_2(A)$ for any $A \in GL(n; \mathbb{C})$). In the case $n = 1$, we reobtain the well-known Abelian group isomorphism $R_2 : \mathbb{T} \xrightarrow{\cong} SO(2)$. Consequently, it follows that

$$\operatorname{Re}(\langle A w, z\rangle_{\mathbb{C}_2^m}) = \operatorname{Re}(z^* A w) = J_2(z)^\top R_2(A) J_2(w)$$
$$= \langle R_2(A) J_2(w), J_2(z)\rangle_{\mathbb{R}_2^{2m}} \text{ and}$$
$$\operatorname{Im}(\langle A w, z\rangle_{\mathbb{C}_2^m}) = \operatorname{Im}(z^* A w) = -J_2(z)^\top R_2(i A) J_2(w)$$
$$= -\langle R_2(i A) J_2(w), J_2(z)\rangle_{\mathbb{R}_2^{2m}} \qquad (1.2.12)$$

for all $(z, w) \in \mathbb{C}^m \times \mathbb{C}^n$ and $A \in \mathbb{M}_{m,n}(\mathbb{C})$.

Lemma 1.1 *Let $n \in \mathbb{N}$ and $C \in \mathbb{M}_n(\mathbb{R})$. Then the following statements are equivalent:*

(i) C *is skew symmetric.*
(ii) $x^\top C y = -y^\top C x$ *for all* $x, y \in \mathbb{R}^n$.
(iii) $z^\top C z = 0$ *for all* $z \in \mathbb{C}^n$.
(iv) $x^\top C x = 0$ *for all* $x \in \mathbb{R}^n$.

Proof Since C is skew symmetric, it follows that $x^\top C y = (x^\top C y)^\top = y^\top(-C)x$ for all $x, y \in \mathbb{R}^n$, wherefrom the implication $(i) \Rightarrow (ii)$ follows. (ii) obviously implies condition (iii): we just have to develop $(x^\top + i y^\top) C (x + i y)$ for arbitrary $x, y \in \mathbb{R}^n$ and apply (ii) to the four consecutive factors. $(iii) \Rightarrow (iv)$ is trivial. Assume that the hypothesis (iv) holds. Let $x, y \in \mathbb{R}^n$ be given. Then $x^\top C x = 0$, $y^\top C y = 0$ and $(x^\top + y^\top) C (x + y) = 0$. Consequently,

$$0 = (x^\top + y^\top) C (x + y) = x^\top C y + y^\top C x = x^\top (C + C^\top) y,$$

and (i) follows. □

Combining Lemma 1.1 with (1.2.11) and (1.2.12), we immediately obtain another neat result, including a full characterisation of Hermitian matrices $A = A^* \in \mathbb{M}_n(\mathbb{C})$ by their symmetric real representation $R_2(A) = R_2(A)^\top \in \mathbb{M}_{2n}(\mathbb{R})$.

Proposition 1.2 *Let $n \in \mathbb{N}$ and $\Gamma = \mathrm{Re}(\Gamma) + i\,\mathrm{Im}(\Gamma) \in \mathbb{M}_n(\mathbb{C})$. Then the following statements are equivalent:*

(i) Γ *is Hermitian.*
(ii) $i\Gamma$ *is skew Hermitian.*
(iii) $\mathrm{Re}(\Gamma) \in \mathbb{M}_n(\mathbb{R})$ *is symmetric and* $\mathrm{Im}(\Gamma) \in \mathbb{M}_n(\mathbb{R})$ *is skew symmetric.*
(iv) $R_2(\Gamma)$ *is symmetric.*
(v) $z^* \Gamma z \in \mathbb{R}$ *for all* $z \in \mathbb{C}^n$.
(vi) $R_2(i\Gamma)$ *is skew symmetric.*

In particular, if $\Sigma = \begin{pmatrix} A & B \\ C & D \end{pmatrix} \in \mathbb{M}_2(\mathbb{M}_n(\mathbb{R}))$, *then the following applies:*

$\Sigma = R_2(\Gamma)$ *for some Hermitian matrix* $\Gamma \in \mathbb{M}_n(\mathbb{C})$ *if and only if* $A = D$

and $B + C = 0$ *and* Σ *is symmetric.* (1.2.13)

Thereby, the uniquely defined Hermitian matrix is given by $\Gamma = A + iC$.

Another important implication refers to the role of the matrix $R_2(A)$ regarding a full clarification of the reason for the difference between the structure of positive semidefinite matrices in $\mathbb{M}_n(\mathbb{C})$ and the structure of positive semidefinite matrices in $\mathbb{M}_n(\mathbb{R})$ (cf. e.g. [70, Theorem 4.1.10] and the field-independent definition in the form of Lemma 1.2 below):

1.2 Preliminaries, Terminology and Notation

Corollary 1.1 *Let $n \in \mathbb{N}$ and $A = \mathrm{Re}(A) + i\,\mathrm{Im}(A) \in M_n(\mathbb{C})$. Then the following statements are equivalent:*

(i) $z^*Az \geq 0$ for all $z \in \mathbb{C}^n$.
(ii) $z^*Az \geq 0$ for all $z \in \mathbb{C}^n$ and A is Hermitian.
(iii) $x^\top R_2(A)\,x \geq 0$ for all $x \in \mathbb{R}^{2n}$ and $R_2(A)$ is symmetric.

If in addition $\mathrm{Im}(A) = 0$, then (i) is equivalent to

(i)' $x^\top Ax \geq 0$ for all $x \in \mathbb{R}^n$ and A is symmetric.

Corollary 1.1 reveals the role of symmetry in the established definition of a positive semidefinite real matrix. For example, if we consider the non-symmetric real matrix $A := \begin{pmatrix} 0 & 1 \\ -1 & 0 \end{pmatrix}$ and $z := (1, i)^\top \in \mathbb{C}^2$, then $x^\top Ax = 0$ for all $x \in \mathbb{R}^2$, but $z^*Az = 2i$. Thus, throughout the book, we apply the following characterisation of positive semidefinite (respectively positive definite) matrices, which does not depend on the choice of the field $\mathbb{F} \in \{\mathbb{R}, \mathbb{C}\}$:

Lemma 1.2 *Let $\mathbb{F} \in \{\mathbb{R}, \mathbb{C}\}$, $n \in \mathbb{N}$ and $A \in M_n(\mathbb{F})$. A is positive semidefinite (respectively positive definite) in $M_n(\mathbb{F})$ if the following two conditions are satisfied:*

(i) $A = A^*$.
(ii) $z^*Az \geq 0$ (respectively $z^*Az > 0$) for all $z \in \mathbb{F}^n \setminus \{0\}$.

In particular, $A \in M_n(\mathbb{R})$ is positive semidefinite in $M_n(\mathbb{R})$, if and only if $A \in M_n(\mathbb{R}) \subseteq M_n(\mathbb{C})$ is positive semidefinite in $M_n(\mathbb{C})$.

Remark 1.1 If we identify (bounded) linear operators $A \in \mathcal{L}(\mathbb{F}^n, \mathbb{F}^n)$ and matrices $A \in M_n(\mathbb{F})$, then $A \in \mathcal{L}(\mathbb{F}^n, \mathbb{F}^n)$ is positive semidefinite if and only if A is a positive self-adjoint operator. Since any positive operator $A \in \mathcal{L}(\mathbb{C}^n, \mathbb{C}^n)$ is self-adjoint, positivity coincides with positive semidefiniteness on $\mathcal{L}(\mathbb{C}^n, \mathbb{C}^n)$, in contrast to positivity on $\mathcal{L}(\mathbb{R}^n, \mathbb{R}^n)$; i.e., there are positive non-symmetric operators $B \in \mathcal{L}(\mathbb{R}^n, \mathbb{R}^n)$ (such as $B := \begin{pmatrix} 0 & 1 \\ -1 & 0 \end{pmatrix}$), implying that these operators cannot be positive semidefinite in $M_n(\mathbb{C})$.

Given $\emptyset \neq S \subseteq \mathbb{F}$, we put (cf. [59])

$$\mathbb{M}_n(S)^+ := \{A : A \in \mathbb{M}_n(S) \text{ and } A \text{ is psd in } \mathbb{M}_n(\mathbb{F})\}.$$

Thus, $\mathbb{M}_n(\mathbb{C})^+ = \{A \in \mathbb{M}_n(\mathbb{C}) : z^*Az \geq 0 \text{ for all } z \in \mathbb{C}^n\}$ and $\mathbb{M}_n(\mathbb{R})^+ = \{A \in \mathbb{M}_n(\mathbb{R}) : A = A^\top \text{ and } x^\top Ax \geq 0 \text{ for all } x \in \mathbb{R}^n\}$. Moreover, $A \in \mathbb{M}_n(\mathbb{C})^+$ if and only if $R_2(A) \in \mathbb{M}_{2n}(\mathbb{R})^+$. Here, the subclass of all correlation matrices, i.e., of all psd matrices with ones on their diagonal (cf. [83, Definition 2.14.] and Lemma 3.1) plays the main role in our work. Only through their structure, including the deep impact of correlation-preserving mappings (cf. Definition 5.4) our main results could be developed. We actually work with exactly those correlation matrices that are used in statistics. So, our approach could also be interesting for the statistical community; especially for those researchers who are working in spatio-temporal modelling and functional data analysis (FDA).

All basic properties of positive semidefinite (respectively, positive definite) matrices including the "striking if not almost magical" structure of related 2×2 block matrices, used and listed throughout the book (without giving any proof) can be found in [16, Chapter 1]. A further, very detailed analysis of the convex psd cone and its geometry, considered from the point of view of convex optimisation is listed in [32, Chapter 2.9]. (Note also that in addition to the symbol "$\mathbb{M}_n(\mathbb{R})^+$", the terms "$\mathbb{S}_n^+$" and "$\mathbb{P}_n$" are often found in the literature.)

Measurability, Probability, Random Vectors If not specified differently, $(\Omega, \mathscr{F})$ always denotes a measurable space which is not specified in more detail. However, we have to make use of different probability spaces, including $(\mathbb{F}^k, \mathcal{B}(\mathbb{F}^k), \gamma_k^\mathbb{F})$, where $\gamma_k^\mathbb{F}$ denotes the real or complex Gaussian measure, described in detail in Sect. 2.1. As usual, if $\mathbb{P}$ is a given probability measure on some $(\Omega, \mathscr{F})$ and $X : \Omega \longrightarrow \mathbb{F}$ a $\mathbb{P}$-integrable $\mathbb{F}$-valued random variable, then (cf., e.g., [11])

$$\mathbb{E}_\mathbb{P}[X] := \mathbb{E}_\mathbb{P}[\operatorname{Re}(X)] + i\,\mathbb{E}_\mathbb{P}[\operatorname{Im}(X)] := \int_\Omega \operatorname{Re}(X)\,d\mathbb{P} + i \int_\Omega \operatorname{Im}(X)\,d\mathbb{P}.$$

In order not to unnecessarily complicate readability, we use the symbols $d^n x$ and λ_n interchangeably to denote the real n-dimensional Lebesgue measure (e.g., $\int_{\mathbb{R}^n} f\,d\lambda_n = \int_{\mathbb{R}^n} f(x)\lambda_n(dx) = \int_{\mathbb{R}^n} f(x)\,d^n x$). Unless otherwise stated, random *variables* will be denoted by capital letters (such as $X : \Omega \longrightarrow \mathbb{R}$, or $Z : \Omega \longrightarrow \mathbb{C}$), whereas random *vectors* will be denoted by bold capital letters (such as $\mathbf{X} : \Omega \longrightarrow \mathbb{R}^n$ or $\operatorname{vec}(\mathbf{Z}, \mathbf{W}) : \Omega \longrightarrow \mathbb{C}^{2n}$). $\mathbf{X} \stackrel{d}{=} \mathbf{Y}$ stands for the equality $\mathbb{P}_\mathbf{X} = \mathbb{P}_\mathbf{Y}$ of the respective probability laws.

Within the framework of standard measure theory (including classical L^p-spaces), we will tacitly assume that we always are working with equivalence classes of almost everywhere coinciding $\mathbb{F}$-valued functions, respectively vector valued measurable mappings on some underlying measure space $(\Omega, \mathscr{F}, \mu)$. However, since paths of stochastic processes will not play a role in this book, we do not have to pay special attention to the structure of null sets. In this regard, a typical example is the real-valued signum function:

$$\operatorname{sign}(x) := \begin{cases} 1 & \text{if } x > 0 \\ 0 & \text{if } x = 0 \\ -1 & \text{if } x < 0 \end{cases}.$$

If we namely view sign as an element of $L^\infty(\mathbb{R})$ (with $\|\operatorname{sign}\|_\infty = 1$), it follows that

$$\operatorname{sign} = 2\,\mathbb{1}_{[0,\infty)} - 1 \text{ in } L^\infty(\mathbb{R})$$

1.2 Preliminaries, Terminology and Notation

(since $\{0\}$ is a Lebesgue null set), where $\mathbb{1}_A$ denotes the indicator function of A. Observe that $H := \mathbb{1}_{[0,\infty)}$ is also well-known as Heaviside step function, which is especially used for applications of Fourier analysis in electricity engineering. This perspective will become an important part of our approach (cf. Example 6.6).

Finally, let us remark, that we also make use of the purely symbolic notation $x \equiv y$ to indicate that x can be canonically identified with the quantity y (such as $\mathbb{M}_{n,1}(\mathbb{F}) \equiv \mathbb{F}^n$) or that it is just a shortcut for the previously rigorously defined quantity y (cf., e.g., (2.1.2)).

Chapter 2
Complex Gaussian Random Vectors and the Probability Law $\mathbb{C}N_{2n}(0, \Sigma_{2n}(\zeta))$

2.1 General Complex Gaussian Random Vectors in $\mathbb{C}^n$ and Their Probability Distribution

Regarding a deeper analysis of the underlying structure of the Grothendieck inequality in the complex case, including the Haagerup equality, it is very helpful to work with centred random vectors whose probability law follows the multivariate *complex* Gaussian distribution, fully characterised through certain correlation matrices, whose entries are elements of $\overline{\mathbb{D}}$. This approach allows us to generalise the Haagerup equality by *substituting* the complex sign function, chosen by Haagerup (see [60]), through arbitrary "circularly odd" functions $b : \mathbb{C}^k \longrightarrow \mathbb{T}$, where $k \in \mathbb{N}$ (see Corollary 7.1).

It is far beyond the scope of the present contribution, to recall the rich structure of the multivariate complex Gaussian distribution in detail. However, for the convenience of the readers we list and describe the whole properties of that probability law which are implemented in some of our following proofs in relation to the complex version of the Grothendieck inequality and beyond in a self-contained way. We highly recommend the readers to study related chapters in the references [5, 55], respectively [71, Chapter 2.1] and [74, Appendix E.2], where that class of random vectors and their distribution functions is comprehensively and rigorously introduced including the related symbolic (mostly self-explaining) notation. Significant facts about real Gaussian random vectors are also listed and discussed thoroughly in [97, Chapter 5.II.1] and [135, Chapter 1.10]. In [11, Chapter 30], a real Gaussian random vector is viewed and studied as a special case of a measurable mapping between a probability space and a measurable space. Recall the powerful general characterisation of the Gaussian law of random vectors with values in $\mathbb{R}^n$ (cf. e.g. [11, Theorem 30.2]):

Proposition 2.1 *Let* $\mathbf{X} \sim N_n(\mu, \Sigma)$ *and* $\mathbf{X} = vec(X_1, X_2, \ldots, X_n)$ *be a random vector in* $\mathbb{R}^n$. *Then the following statements are equivalent:*

(i)
$$\mathbf{X} \sim N_n(\mu, \Sigma);$$

(ii) *For all* $a \in \mathbb{R}^n$,
$$a^\top \mathbf{X} = \sum_{i=1}^n a_i X_i \sim N_1(a^\top \mu, a^\top \Sigma a);$$

(iii) *The characteristic function of* $\mathbf{X}$ *is given by*
$$\mathbb{R}^n \ni a \mapsto \phi_\mathbf{X}(a) := \mathbb{E}[\exp(ia^\top \mathbf{X})] = \exp(ia^\top \mu - \tfrac{1}{2}a^\top \Sigma a).$$

Consequently, the following important fact (which we frequently apply in this book) follows at once:

If $\mathbf{X} \sim N_n(\mu, \Sigma)$, then $A\mathbf{X} + b \sim N_n(A\mu + b, A\Sigma A^\top)$

for all $m \in \mathbb{N}$ and $(b, A) \in \mathbb{R}^m \times \mathbb{M}_{m,n}(\mathbb{R})$. (2.1.1)

Furthermore, recall that a random vector $\mathbf{Z} = vec(Z_1, Z_2, \ldots, Z_n)$ which maps into $\mathbb{C}^n$ is a complex random vector if for all $\nu \in [n]$ $Z_\nu = X_\nu + i Y_\nu$, where $X_\nu = \text{Re}(Z_\nu)$ and $Y_\nu = \text{Im}(Z_\nu)$ both are real random variables (each one defined on the same probability space). Along the lines of the notation for (deterministic) vectors in $\mathbb{C}^n$ one puts

$$\mathbf{X} \equiv \text{Re}(\mathbf{Z}) := vec(\text{Re}(Z_1), \text{Re}(Z_2), \ldots, \text{Re}(Z_n)) = vec(X_1, X_2, \ldots, X_n)$$

and

$$\mathbf{Y} \equiv \text{Im}(\mathbf{Z}) := vec(\text{Im}(Z_1), \text{Im}(Z_2), \ldots, \text{Im}(Z_n)) = vec(Y_1, Y_2, \ldots, Y_n),$$

implying that $\mathbf{Z} = \mathbf{X} + i\mathbf{Y} = \text{Re}(\mathbf{Z}) + i\,\text{Im}(\mathbf{Z})$. Let

$$\lambda_n^\mathbb{C} := (J_2^{-1})_*\lambda_{2n}$$

be the Lebesgue measure on $\mathbb{C}^n$ (i.e., the image measure of the real Lebesgue measure λ_{2n}). Fix $0 < p < \infty$. If $z = x + iy \in \mathbb{C}$, then

$$|z|^p = (x^2 + y^2)^{p/2} \leq \max\{2^{(p/2)-1}, 1\}(|x|^p + |y|^p) \leq \max\{2^{p/2}, 2\}|z|^p$$

2.1 General Complex Gaussian Random Vectors in $\mathbb{C}^n$ and Their Probability... 21

(see [80, 2.10.E]). Consequently, the change-of-variables formula (cf. e.g. [12, Chapter 19]), applied to the image measure $\lambda_n^{\mathbb{C}}$, implies that

$$h \in L^p(\mathbb{C}^n, \lambda_n^{\mathbb{C}}) \text{ if and only if } \operatorname{Re}(h) \circ J_2^{-1} \in L^p(\mathbb{R}^{2n}, \lambda_{2n})$$

$$\text{and } \operatorname{Im}(h) \circ J_2^{-1} \in L^p(\mathbb{R}^{2n}, \lambda_{2n}).$$

The construction of $\lambda_n^{\mathbb{C}}$ namely implies that

$$\int_{\mathbb{C}^n} |h(z)|^p \, \lambda_n^{\mathbb{C}}(\mathrm{d}z) \equiv \int_{\mathbb{C}^n} |h|^p \, \mathrm{d}\lambda_n^{\mathbb{C}} = \int_{\mathbb{R}^{2n}} |h|^p \circ J_2^{-1} \, \mathrm{d}\lambda_{2n}$$

$$\equiv \int_{\mathbb{R}^{2n}} |h(x+iy)|^p \, \lambda_{2n}(\mathrm{d}(x,y)). \tag{2.1.2}$$

Equipped with these basic, well-known facts about the Lebesgue measure on $\mathbb{C}^n$, we reintroduce complex Gaussian random vectors in the following, seemingly elementary way:

Definition 2.1 (Complex Gaussian Random Vector) An n-dimensional complex random vector $\mathbf{Z}$ is a complex Gaussian random vector if the real $2n$-dimensional random vector $J_2(\mathbf{Z}) = \operatorname{vec}(\operatorname{Re}(\mathbf{Z}), \operatorname{Im}(\mathbf{Z}))$ is a real Gaussian random vector.

Although that definition of complex Gaussian random vectors seems to be a quite inconspicuous one, it encapsulates a rich underlying structure which strongly differs from that one of real Gaussian random vectors. Firstly, without having to know any further details about the structure of complex Gaussian random vectors, the change-of-variables formula (cf. e.g. [12, Chapter 19]) implies that

$$\mathbb{E}[g(\mathbf{Z})] \equiv \mathbb{E}[g \circ \mathbf{Z}] = \int_{\Omega} g \circ \mathbf{Z} \, \mathrm{d}\mathbb{P} = \int_{\mathbb{C}^n} g \, \mathrm{d}\mathbb{P}_{\mathbf{Z}} = \int_{\mathbb{C}^n} g \, \mathrm{d}(J_2^{-1})_* \mathbb{P}_{J_2(\mathbf{Z})}$$

can be written as

$$\mathbb{E}[g(\mathbf{Z})] = \mathbb{E}_{\mathbb{P}_{\mathbf{X}}}[g \circ J_2^{-1}] = \mathbb{E}[\operatorname{Re}(g(J_2^{-1}(\mathbf{X})))] + i\, \mathbb{E}[\operatorname{Im}(g(J_2^{-1}(\mathbf{X})))]$$

for any $\mathbb{P}_{\mathbf{Z}}$-integrable function $g = \operatorname{Re}(g) + i \operatorname{Im}(g)$ and any complex Gaussian random vector $\mathbf{Z}$, where $\mathbf{X} \stackrel{d}{=} J_2(\mathbf{Z})$ is a real $2n$-dimensional Gaussian random vector. Consequently, the expectation vector $\mu \equiv \mathbb{E}[\mathbf{Z}] := \operatorname{vec}(\mathbb{E}[Z_1], \mathbb{E}[Z_2], \ldots, \mathbb{E}[Z_n])$ as well as the variance matrix $\Gamma \equiv \operatorname{var}(\mathbf{Z}) := \mathbb{E}[(\mathbf{Z}-\mu)(\mathbf{Z}-\mu)^*] = (\mathbb{E}[(Z_i - \mu_i)(\overline{Z_j} - \overline{\mu_j})])_{1 \le i,j \le n} \in \mathbb{M}_n(\mathbb{C})^+$ and the cross-covariance matrix $C \equiv \operatorname{cov}(\mathbf{Z}, \overline{\mathbf{Z}}) := \mathbb{E}[(\mathbf{Z}-\mu)(\overline{\mathbf{Z}-\mu})^*] = \mathbb{E}[(\mathbf{Z}-\mu)(\mathbf{Z}-\mu)^\top] = (\mathbb{E}[(Z_i - \mu_i)(Z_j - \mu_j)])_{1 \le i,j \le n} \in M_n(\mathbb{C})$ are well-defined. Let $S \in \mathbb{M}_{2n}(\mathbb{R})^+$ be the variance matrix of $J_2(\mathbf{Z})$. Then

$$J_2(\mathbf{Z}) \sim N_{2n}(J_2(\mu), S),$$

where $S = \mathbb{E}[J_2(\mathbf{Z} - \mu) J_2(\mathbf{Z} - \mu)^\top]$. A straightforward computation of $C + \Gamma = 2\mathbb{E}[(\mathbf{Z} - \mu)(\text{Re}(\mathbf{Z} - \mu))^\top]$ and $C - \Gamma = 2i\,\mathbb{E}[(\mathbf{Z} - \mu)(\text{Im}(\mathbf{Z} - \mu))^\top]$ implies that

$$2S = \begin{pmatrix} \text{Re}(C + \Gamma) & \text{Im}(C - \Gamma) \\ \text{Im}(C + \Gamma) & -\text{Re}(C - \Gamma) \end{pmatrix} = R_2(\Gamma) + \begin{pmatrix} \text{Re}(C) & \text{Im}(C) \\ \text{Im}(C) & -\text{Re}(C) \end{pmatrix}$$

$$= \Lambda_{2n}^* \begin{pmatrix} \Gamma & C \\ \overline{C} & \overline{\Gamma} \end{pmatrix} \Lambda_{2n}, \tag{2.1.3}$$

where $\Lambda_{2n} := \frac{1}{\sqrt{2}} \begin{pmatrix} I_n & i\, I_n \\ I_n & -i\, I_n \end{pmatrix} \in U(2n)$ is an unitary matrix (with $\det(\Lambda_{2n}) = (-i)^n$). Observe that

$$\begin{pmatrix} \Gamma & C \\ \overline{C} & \overline{\Gamma} \end{pmatrix} = \mathbb{E}[\mathbf{W}\mathbf{W}^*],$$

where the complex random vector $\mathbf{W} := \text{vec}(\mathbf{Z} - \mu, \overline{\mathbf{Z} - \mu}) = \text{vec}(\mathbf{Z}, \overline{\mathbf{Z}}) - \tilde{\mu}$, with $\tilde{\mu} := \text{vec}(\mu, \overline{\mu})$, maps into $\mathbb{C}^{2n}$. Consequently, $\begin{pmatrix} \Gamma & C \\ \overline{C} & \overline{\Gamma} \end{pmatrix}$ is the variance matrix of the random vector $\text{vec}(\mathbf{Z}, \overline{\mathbf{Z}})$. Observe that

$$\mathbb{R}^{4n} \ni J_2(\text{vec}(\mathbf{Z}, \overline{\mathbf{Z}})) = A J_2(\mathbf{Z}) \sim N_{4n}(A J_2(\mu), A S A^\top), \tag{2.1.4}$$

where $A := \begin{pmatrix} I_n & 0 \\ I_n & 0 \\ 0 & I_n \\ 0 & -I_n \end{pmatrix} \in \mathbb{M}_{4n,2n}(\mathbb{R})$. Thus, since $a^\top S a = \frac{1}{2}(\Lambda_{2n} a)^* \begin{pmatrix} \Gamma & C \\ \overline{C} & \overline{\Gamma} \end{pmatrix} \Lambda_{2n}$
$a = \frac{1}{2}\mathbb{E}[(\Lambda_{2n} a)^* \mathbf{W}\mathbf{W}^* \Lambda_{2n} a] = \mathbb{E}[(\mathbf{W}^* \Lambda_{2n} a)^* \mathbf{W}^* \Lambda_{2n} a]$ for all $a \in \mathbb{R}^{2n}$, it follows that the real matrix S is always positive semidefinite (cf. Corollary 1.1); i.e., $S \in \mathbb{M}_{2n}(\mathbb{R})^+$. If in addition $C\Gamma = \Gamma C$, then (see [139, Theorem 3])

$$\det(S) = \frac{1}{4^n} \det\left(R_2(\Gamma) + \begin{pmatrix} \text{Re}(C) & \text{Im}(C) \\ \text{Im}(C) & -\text{Re}(C) \end{pmatrix} \right) = \frac{1}{4^n} \det(\Gamma \overline{\Gamma} - C\overline{C}).$$

In particular, if $C = 0$, then $\frac{1}{2} R_2(\Gamma) = S$ is positive semidefinite, implying that

$$\text{Re}(\Gamma) = \text{Re}(\Gamma)^\top \quad \text{and} \quad -\text{Im}(\Gamma) = \text{Im}(\Gamma)^\top \tag{2.1.5}$$

(cf. Proposition 1.2) and

$$\det(R_2(\Gamma)) = |\det(\Gamma)|^2. \tag{2.1.6}$$

Hence, $\det(\sqrt{R_2(\Gamma)}) = |\det(\Gamma)|$. In particular, $\text{Re}(\mathbf{Z}) \sim N_n(\text{Re}(\mu), \frac{1}{2}\text{Re}(\Gamma))$, $\text{Im}(\mathbf{Z}) \sim N_n(\text{Im}(\mu), \frac{1}{2}\text{Re}(\Gamma))$ and $\mathbb{E}[\text{Re}(Z_i)\text{Im}(Z_i)] = \text{Im}(\Gamma_{ii}) = 0$ for all $i \in [n]$. However, because of Proposition 2.2, the random *vectors* $\text{Re}(\mathbf{Z})$ and $\text{Im}(\mathbf{Z})$ in general are not independent!

Since the distribution of $J_2(\mathbf{Z})$ is fully specified by μ, Γ and C (due to (2.1.3)), we write $\mathbf{Z} \sim \mathbb{C}N_n(\mu, \Gamma, C)$ if $\mathbf{Z}$ is an n-dimensional complex Gaussian random vector. Thus, if $\mathbf{X}$ and $\mathbf{Y}$ are n-dimensional real random vectors, then

$$\tfrac{1}{\sqrt{2}}\mathbf{X} + i\tfrac{1}{\sqrt{2}}\mathbf{Y} \sim \mathbb{C}N_n(\mu, \Gamma, C) \text{ if and only if } \text{vec}(\mathbf{X}, \mathbf{Y})$$

$$\sim N_{2n}\left(J_2(\mu), R_2(\Gamma) + \begin{pmatrix} \text{Re}(C) & \text{Im}(C) \\ \text{Im}(C) & -\text{Re}(C) \end{pmatrix}\right),$$

or equivalently that

$$\mathbf{Z} \sim \mathbb{C}N_n(\mu, \Gamma, C) \text{ if and only if } J_2(\sqrt{2}\mathbf{Z})$$

$$\sim N_{2n}\left(J_2(\mu), R_2(\Gamma) + \begin{pmatrix} \text{Re}(C) & \text{Im}(C) \\ \text{Im}(C) & -\text{Re}(C) \end{pmatrix}\right).$$

Regarding the main topic of our monograph, we only need to work with $C = 0$, what will happen from now on. In this case, we just write $\mathbf{Z} \sim \mathbb{C}N_n(\mu, \Gamma)$. Thus, $\mathbf{Z} = \tfrac{1}{\sqrt{2}}\mathbf{X} + i\tfrac{1}{\sqrt{2}}\mathbf{Y} \sim \mathbb{C}N_n(\mu, \Gamma)$, if and only if

$$\text{vec}(\mathbf{X}, \mathbf{Y}) = \sqrt{2}J_2(\mathbf{Z}) = \sqrt{2}\text{vec}(\text{Re}(Z_1), \ldots, \text{Re}(Z_n),$$
$$\text{Im}(Z_1), \ldots, \text{Im}(Z_n)) \sim N_{2n}(J_2(\mu), R_2(\Gamma)),$$

implying that (2.1.1) carries over to the complex case:

$$\text{if } \mathbf{Z} \sim \mathbb{C}N_n(\mu, \Gamma), \text{ then } A\mathbf{Z} + b \sim \mathbb{C}N_n(A\mu + b, A\Gamma A^*)$$
$$\text{for all } m \in \mathbb{N} \text{ and } (b, A) \in \mathbb{C}^m \times \mathbb{M}_{m,n}(\mathbb{C}). \quad (2.1.7)$$

Remark 2.1 Let $n \in \mathbb{N}$ and $0 \neq \Sigma \in \mathbb{M}_n(\mathbb{R})^+$ be given. Fix some $\mathbf{X} \sim N_n(0, \Sigma)$. A natural question would be, to ask whether $\mathbf{Z} := \tfrac{1}{\sqrt{2}}\mathbf{X} + i\mathbf{0} \sim \mathbb{C}N_n(0, \Sigma)$ in particular is a complex Gaussian random vector? However, if this were the case, it would follow that

$$\text{vec}(\mathbf{X}, \mathbf{0}) = J_2(\sqrt{2}\mathbf{Z}) \sim N_{2n}(0, R_2(\Sigma)) = N_{2n}\left(0, \begin{pmatrix} \Sigma & 0 \\ 0 & \Sigma \end{pmatrix}\right),$$

implying that $0 \neq \Sigma = \mathbb{E}[\mathbf{X}\mathbf{X}^\top] = \mathbb{E}[\mathbf{0}\mathbf{0}^\top] = 0$, which is absurd.

Remark 2.2 In general, the random vector $\text{vec}(\mathbf{Z}, \overline{\mathbf{Z}})$ is not a complex Gaussian one, even if $\mathbf{Z}$ is. In order to recognise this, let e.g. $Z \sim \mathbb{C}N_1(0, 1)$ be given. Then $J_2(\sqrt{2}Z) \sim N_2(0, I_2)$. Thus, (2.1.4) implies that

$$J_2(\sqrt{2}\,\text{vec}(Z, \overline{Z})) \sim N_4(0, AA^\top).$$

However, since the matrix

$$AA^\top = \begin{pmatrix} 1 & 1 & 0 & 0 \\ 1 & 1 & 0 & 0 \\ 0 & 0 & 1 & -1 \\ 0 & 0 & -1 & 1 \end{pmatrix}$$

does not coincide with a block matrix of type $\begin{pmatrix} A & -B \\ B & A \end{pmatrix} \in M_2(M_2(\mathbb{R}))$, $\text{vec}(Z, \overline{Z})$ is not a complex Gaussian random vector. That observation also holds in the multi-dimensional case (see Lemma 2.1).

Occasionally, in view of embedding both, the complex and the real case into a single statement, we also unambiguously say that $\mathbf{Z} \sim \mathbb{F}N_n(\mu, \Gamma)$ if the random vector $\mathbf{Z}$ maps into $\mathbb{F}^n$, where $\mathbb{F} \in \{\mathbb{R}, \mathbb{C}\}$. Note that for both fields, we explicitly include the case of variance matrices $\Gamma \in \mathbb{M}_n(\mathbb{F})^+$ which are not invertible, so that a probability *density* function of $\mathbf{Z} \sim \mathbb{F}N_n(\mu, \Gamma)$ would not have to exist; as opposed to the characteristic function of $\mathbf{Z}$ which completely determines the probability law $\mathbb{P}_\mathbf{Z}$. The characteristic function of $\mathbf{Z} \sim \mathbb{C}N_n(\mu, \Gamma)$ can be reduced to the well-known characteristic function of the $2n$-dimensional real Gaussian random vector $J_2(\mathbf{Z}) \sim N_{2n}(J_2(\mu), \frac{1}{2}R_2(\Gamma))$. This follows from

$$\mathbb{C}^n \ni c \mapsto \phi_\mathbf{Z}(c) := \mathbb{E}[\exp(i\,\text{Re}(c^*\mathbf{Z}))] = \mathbb{E}[\exp(i\,J_2^\top(c)J_2(\mathbf{Z}))]$$

$$= \exp(i\,J_2(c)^\top J_2(\mu))\exp(-\tfrac{1}{4}J_2(c)^\top R_2(\Gamma)J_2(c))$$

$$= \exp(i\,\text{Re}(c^*\mu))\exp(-\tfrac{1}{4}\text{Re}(c^*\Gamma c))$$

$$= \exp(i\,\text{Re}(c^*\mu))\exp(-\tfrac{1}{4}c^*\Gamma c)$$

(cf. [5, Theorem 2.7], [11, Theorem 30.2], [74, Definition E.1.13 and Theorem E.1.16] and Lemma 2.1 below).

Proposition 2.2 *Let $n \in \mathbb{N}$, $\mu \in \mathbb{C}^n$, $\Gamma \in \mathbb{M}_n(\mathbb{C})^+$ and $\mathbf{Z} \sim \mathbb{C}N_n(\mu, \Gamma)$. Then $\text{Re}(\mathbf{Z})$ and $\text{Im}(\mathbf{Z})$ are independent if and only if $\text{Im}(\Gamma) = 0$.*

Proof (2.1.5) and Lemma 1.1 imply that

$$\phi_{\text{vec}(\text{Re}(\mathbf{Z}),\text{Im}(\mathbf{Z}))}(\text{vec}(\text{Re}(c),\text{Im}(c))) = \phi_{\mathbf{Z}}(c)$$
$$= \phi_{\text{Re}(\mathbf{Z})}(\text{Re}(c))\,\phi_{\text{Im}(\mathbf{Z})}(\text{Re}(c))$$
$$\exp(-\tfrac{1}{2}\text{Im}(c)^\top \text{Im}(\Gamma)\text{Re}(c))$$

for all $c \in \mathbb{C}^n$. Hence, $\text{Re}(\mathbf{Z})$ and $\text{Im}(\mathbf{Z})$ are independent if and only if

$$\phi_{\text{Re}(\mathbf{Z})}(\text{Re}(c))\,\phi_{\text{Im}(\mathbf{Z})}(\text{Re}(c))(1 - \exp(-\tfrac{1}{2}\text{Im}(c)^\top \text{Im}(\Gamma)\text{Re}(c))) = 0$$

for all $c \in \mathbb{C}^n$. Consequently, if we apply the absolute value of the latter equality to any vector $e_l + i e_k \in \mathbb{C}^n$, where $k, l \in [n]$, it follows that

$$|1 - \exp(-\tfrac{1}{2}\text{Im}(\Gamma_{kl}))| = 0$$

for all $k, l \in [n]$, and the claim follows. □

Similarly, if Γ (respectively $R_2(\Gamma)$) is invertible, the complex density function under the Lebesgue measure $\lambda_n^\mathbb{C}$ on $\mathbb{C}^n$ can be constructed as

$$\mathbb{C}^n \ni a \mapsto \varphi_{\mu,\Gamma}(a) := \varphi_{J_2(\mu), \frac{1}{2} R_2(\Gamma)}(J_2(a))$$
$$= \frac{1}{\pi^n \sqrt{\det(R_2(\Gamma))}} \exp(-(J_2(a-\mu))^* R_2(\Gamma^{-1})(J_2(a-\mu)))$$
$$\stackrel{(2.1.6)}{=} \frac{1}{\pi^n |\det(\Gamma)|} \exp(-(a-\mu)^* \Gamma^{-1}(a-\mu)).$$

These facts, including (2.1.1), (2.1.5), (2.1.7) and Proposition 2.1, immediately imply the following comprehensive characterisation of the probability law $\mathbb{C}N_n(\mu, \Gamma)$.

Proposition 2.3 *Let $n \in \mathbb{N}$, $\mu \in \mathbb{C}^n$, $\Gamma \in \mathbb{M}_n(\mathbb{C})^+$ and $\mathbf{Z}$ a complex n-dimensional random vector. Then the following statements are equivalent:*

(i) *For all $c \in \mathbb{C}^n$, $\phi_{\mathbf{Z}}(c) = \mathbb{E}[\exp(i\,\text{Re}(c^*\mathbf{Z}))] = \exp(i\,\text{Re}(c^*\mu))\exp(-\tfrac{1}{4}c^*\Gamma c)$.*
(ii) *$\sqrt{2}\,J_2(\mathbf{Z}) = \text{vec}(\sqrt{2}\,\text{Re}(Z_1),\dots,\sqrt{2}\,\text{Re}(Z_n),\sqrt{2}\,\text{Im}(Z_1),\dots,\sqrt{2}\,\text{Im}(Z_n))$*
 $\sim N_{2n}(\sqrt{2}\,J_2(\mu), R_2(\Gamma))$.
(iii) *$\mathbf{Z} \sim \mathbb{C}N_n(\mu, \Gamma)$.*
(iv) *For all $\alpha \in \mathbb{T}$, $\alpha \mathbf{Z} \sim \mathbb{C}N_n(\alpha\mu, \Gamma)$.*
(v) *For all $c \in \mathbb{C}^n$, $c^*\mathbf{Z} \sim \mathbb{C}N_1(c^*\mu, c^*\Gamma c)$.*
(vi) *For all $c \in \mathbb{C}^n$, $\sqrt{2}\,J_2(c^*\mathbf{Z}) \sim N_2(\sqrt{2}\,J_2(c^*\mu), R_2(c^*\Gamma c))$.*
(vii) *For all $c \in \mathbb{C}^n$, $\sqrt{2}\,\text{Re}(c^*\mathbf{Z}) \sim N_1(\sqrt{2}\,\text{Re}(c^*\mu), c^*\Gamma c)$.*

In particular, if $\mathbf{Z} \sim \mathbb{C}N_n(\mu, \Gamma)$, then $\text{Re}(\mathbf{Z}) \stackrel{d}{=} \text{Im}(\mathbf{Z}) \sim N_n(0, \text{Re}(\Gamma))$, and $\text{Re}(Z_i)$ and $\text{Im}(Z_i)$ are independent for all $i \in \mathbb{N}$. $\mathbf{Z} \sim \mathbb{C}N_n(\mu, \Gamma)$ if and only if $\overline{\mathbf{Z}} \sim \mathbb{C}N_n(\overline{\mu}, \overline{\Gamma})$. Moreover, if $\mu = 0$ and $\Gamma = I_n$, then $\{\text{Re}(Z_1), \ldots, \text{Re}(Z_n), \text{Im}(Z_1), \ldots, \text{Im}(Z_n)\}$ are pairwise independent.

In relation to a further investigation of the complex Grothendieck constant (built on complex Hermite polynomials), we need a further analysis of the structure of the random vector $\text{vec}(\mathbf{Z}, \overline{\mathbf{Z}})$ (cf. Theorem 7.3). That analysis, encapsulated in the next lemma, includes a short proof of a generalisation of a result of L. J. Halliwell (see [61, Appendix B]), built on a "change of mean trick", that allows us to pull both, the characteristic function and the moment generating function of a real Gaussian random vector out of a single formula, without having to assume the existence of a density function. In doing so, we will recognise again that in general, the complex random vector $\text{vec}(\mathbf{Z}, \overline{\mathbf{Z}})$ in $\mathbb{C}^{2n}$ is not Gaussian, even if the random vector $\mathbf{Z}$ in $\mathbb{C}^n$ were a complex Gaussian one.

Lemma 2.1 *Let $n \in \mathbb{N}$, $\Sigma \in \mathbb{M}_n(\mathbb{R})^+$ and $\Gamma \in \mathbb{M}_n(\mathbb{C})^+$. Let $\mathbf{X} \sim N_n(0, \Sigma)$ and $\mathbf{Z} \sim \mathbb{C}N_n(0, \Gamma)$. Then*

(i) $\mathbb{E}[\exp(c^\top \mathbf{X})] = \exp(\frac{1}{2} c^\top \Sigma c)$ *for all $c \in \mathbb{C}^n$.*
(ii) $\mathbb{E}[\exp(a^* \mathbf{Z} + b^* \overline{\mathbf{Z}})] = \exp(a^* \Gamma \overline{b})$ *for all $a, b \in \mathbb{C}^n$.*
(iii) *For all $a, b \in \mathbb{C}^n$,*

$$\phi_{\text{vec}(\mathbf{Z},\overline{\mathbf{Z}})}(\text{vec}(a, b)) = \exp(-\tfrac{1}{4} a^* \Gamma a) \exp(-\tfrac{1}{4} b^* \overline{\Gamma} b) \exp(-\tfrac{1}{2} \text{Re}(a^* \Gamma \overline{b}))$$

$$= \exp\left(-\tfrac{1}{4} \text{vec}(a,b)^\top \begin{pmatrix} \Gamma & 0 \\ 0 & \overline{\Gamma} \end{pmatrix} \text{vec}(a,b)\right)$$

$$\exp(-\tfrac{1}{2} \text{Re}(a^* \Gamma \overline{b})).$$

In particular, $\text{vec}(\mathbf{Z}, \overline{\mathbf{Z}}) \sim \mathbb{C}N_{2n}\left(0, \begin{pmatrix} \Gamma & 0 \\ 0 & \overline{\Gamma} \end{pmatrix}\right)$ if and only if $\Gamma = 0$. Moreover, $\mathbb{E}[\exp(a^* \mathbf{Z})] = 1$ for all $a \in \mathbb{C}^n$.

Proof

(i) Let $c = \alpha + i\beta \in \mathbb{C}^n$ be given, where $\alpha, \beta \in \mathbb{R}^n$. Fix an arbitrary $\mathbf{Y} \sim N_n(0, I_n)$. Since $\Sigma = R^2$, for some (uniquely determined and not necessarily invertible) matrix $R \in \mathbb{M}_n(\mathbb{R})^+$, (2.1.1) implies that $\mathbf{X} \stackrel{d}{=} R\mathbf{Y}$. Hence,

$$\mathbb{E}[\exp(c^\top \mathbf{X})] = \mathbb{E}[\exp(\alpha^\top R\mathbf{Y}) \exp(i(R\beta)^\top \mathbf{Y})]$$

$$= (2\pi)^{-\frac{n}{2}} \int_{\mathbb{R}^n} \exp(\alpha^\top Ry - \tfrac{1}{2}\|y\|_2^2) \exp(i(R\beta)^\top y) \, d^n y.$$

2.1 General Complex Gaussian Random Vectors in $\mathbb{C}^n$ and Their Probability... 27

An easy calculation shows that

$$\exp(\alpha^\top Ry - \tfrac{1}{2}\|y\|_2^2) = \exp(-\tfrac{1}{2}\|y - R\alpha\|_2^2)\exp(\tfrac{1}{2}\alpha^\top R^2 \alpha)$$
$$= (2\pi)^{\frac{n}{2}} \varphi_{R\alpha, I_n}(y) \exp(\tfrac{1}{2}\alpha^\top \Sigma \alpha).$$

Consequently, it follows that

$$\mathbb{E}[\exp(c^\top X)] = \mathbb{E}[\exp(i(R\beta)^\top \widetilde{Y})]\exp(\tfrac{1}{2}\alpha^\top \Sigma \alpha) = \phi_{\widetilde{Y}}(R\beta)\exp(\tfrac{1}{2}\alpha^\top \Sigma \alpha)$$

is the product of the value of the characteristic function of the random vector $\widetilde{Y} \stackrel{d}{=} Y + R\alpha \sim N_n(R\alpha, I_n)$ at $R\beta$ and the real number $\exp(\tfrac{1}{2}\alpha^\top \Sigma \alpha)$. Thus, since $\mathbb{E}[\exp(i(R\beta)^\top \widetilde{Y})] = \exp(i\beta^\top \Sigma \alpha)\exp(-\tfrac{1}{2}\beta^\top \Sigma \beta)$, we finally obtain

$$\mathbb{E}[\exp(c^\top X)] = \exp(i\beta^\top \Sigma \alpha)\exp(-\tfrac{1}{2}\beta^\top \Sigma \beta)\exp(\tfrac{1}{2}\alpha^\top \Sigma \alpha) = \exp(\tfrac{1}{2}c^\top \Sigma c).$$

(ii) Put $\mathbb{C}^{2n} \ni c := \mathrm{vec}(\overline{a} + \overline{b}, i\overline{a} - i\overline{b})$. Then $c = \begin{pmatrix} I_n & I_n \\ iI_n & -iI_n \end{pmatrix} \mathrm{vec}(\overline{a}, \overline{b}) = \sqrt{2}\, \Lambda_{2n}\mathrm{vec}(\overline{a}, \overline{b})$ (cf. (2.1.3)) and $a^*Z + b^*\overline{Z} = c^\top J_2(Z)$. Since $X := J_2(Z) \sim N_{2n}(0, \tfrac{1}{2}R_2(\Gamma))$ (due to Proposition 2.3), we may consequently apply (i) to X and c, and it follows that

$$\mathbb{E}[\exp(a^*Z + b^*\overline{Z})] = \exp(\tfrac{1}{4}c^\top R_2(\Gamma)c).$$

A straightforward calculation shows that

$$\Lambda_{2n}^\top R_2(\Gamma) = \begin{pmatrix} 0 & \Gamma \\ \overline{\Gamma} & 0 \end{pmatrix} \Lambda_{2n}^*.$$

Consequently, since $\Lambda_{2n} = (\Lambda_{2n}^*)^{-1}$ is unitary, the construction of the vector c implies that

$$c^\top R_2(\Gamma)c = 2\mathrm{vec}(\overline{a},\overline{b})^\top (\Lambda_{2n}^\top R_2(\Gamma)\Lambda_{2n})\mathrm{vec}(\overline{a},\overline{b})$$
$$= 2\mathrm{vec}(\overline{a},\overline{b})^\top \begin{pmatrix} 0 & \Gamma \\ \overline{\Gamma} & 0 \end{pmatrix}\mathrm{vec}(\overline{a},\overline{b}) = 2(a^*\Gamma \overline{b} + b^*\overline{\Gamma}\overline{a})$$
$$= 2(a^*\Gamma \overline{b} + a^*\Gamma^* \overline{b}).$$

Since $\Gamma = \Gamma^*$, the equality (ii) follows.

(iii) If we apply (2.1.4) and equality (i) to $c := i\, A^\top J_2(\text{vec}(a,b)) = i\, J_2(a+\overline{b}) \in \mathbb{C}^{2n}$ and $\mathbf{X} := J_2(\mathbf{Z}) \sim N_{2n}(0, \tfrac{1}{2} R_2(\Gamma))$, it follows that

$$\phi_{\text{vec}(\mathbf{Z},\overline{\mathbf{Z}})}(\text{vec}(a,b)) = \mathbb{E}[\exp(i\, J_2(\text{vec}(a,b))^\top A J_2(\mathbf{Z}))] = \mathbb{E}[c^\top J_2(\mathbf{Z})]$$
$$= \exp(\tfrac{1}{4} c^\top R_2(\Gamma) c)$$
$$= \exp(-\tfrac{1}{4}\operatorname{Re}((a^* + b^\top)\Gamma(a+\overline{b})))$$

However, since Γ is Hermitian, it follows that $b^\top \Gamma \overline{b} = (b^\top \Gamma \overline{b})^\top = b^* \overline{\Gamma} b$. In the same way, we obtain $b^\top \Gamma a = a^* \overline{\Gamma b}$, which completes the proof of (iii). $\square$

Remark 2.3 The main difficulty in the proof of Lemma 2.1 arises from the fact that the normal random variables $\alpha^\top \mathbf{X}$ and $\beta^\top \mathbf{X}$ are correlated, so that we cannot simply represent $\mathbb{E}_{\mathbb{P}_\mathbf{X}}[\exp(\alpha^\top \mathbf{X})\exp(i\beta^\top \mathbf{X})]$ as a product of two expectations. However, it is possible to construct a completely different proof of Lemma 2.1, which is built on (an application of the one-dimensional case of) Theorem 6.2. We strongly encourage the readers to work out the details.

Let $p \in \mathbb{N}$. It is well-known that the image measure $\mathbb{P}_\mathbf{X}$ of a real Gaussian random vector $\mathbf{X} \sim N_p(0, I_p)$ actually coincides with the Gaussian measure γ_p on $\mathbb{R}^p$ (cf. e.g. [20, Proposition 1.2.2.]), constructed via

$$\mathcal{B}(\mathbb{R}^p) \ni B \mapsto \gamma_p(B) := (2\pi)^{-p/2} \int_B \exp(-\tfrac{1}{2}\|x\|_2^2)\lambda_p(\mathrm{d}x) = \mathbb{P}(\mathbf{X} \in B).$$

Due to Proposition 2.3 this fact can be easily transferred to the complex case. To this end, let $n \in \mathbb{N}$, $\mathbf{Z} \sim \mathbb{C}N_n(0, I_n)$ and $b = \operatorname{Re}(b) + i\operatorname{Im}(b) : \mathbb{C}^n \longrightarrow \mathbb{C}$. Consider the two mappings

$$r(b) := \operatorname{Re}(b) \circ \tfrac{1}{\sqrt{2}} J_2^{-1} : \mathbb{R}^{2n} \longrightarrow \mathbb{R} \text{ and } s(b) := \operatorname{Im}(b) \circ \tfrac{1}{\sqrt{2}} J_2^{-1} : \mathbb{R}^{2n} \longrightarrow \mathbb{R}.$$

By construction, it follows that for any $x, y \in \mathbb{R}^n$,

$$r(b)(\text{vec}(x,y)) = \operatorname{Re}(b(\tfrac{1}{\sqrt{2}} x + i\, \tfrac{1}{\sqrt{2}} y)) \text{ and } s(b)(\text{vec}(x,y))$$
$$= \operatorname{Im}(b(\tfrac{1}{\sqrt{2}} x + i\, \tfrac{1}{\sqrt{2}} y)). \qquad (2.1.8)$$

Obviously, $s(b) = r(-i\,b)$, $b \circ \tfrac{1}{\sqrt{2}} J_2^{-1} = r(b) + i s(b)$, $\operatorname{Re}(b) = r(b) \circ \sqrt{2} J_2$, $\operatorname{Im}(b) = s(b) \circ \sqrt{2} J_2$ and $r(\alpha b) = \alpha r(b)$ for any $\alpha \in \mathbb{R}$.

Let $p, q \in [1, \infty)$, such that $\frac{1}{p} + \frac{1}{q} = 1$. A direct application of Hölder's inequality (to the vectors $(r(b), s(b))^\top \in \mathbb{R}^2$ and $(1, 1)^\top \in \mathbb{R}^2$) implies that

$$\max\{|r(b)|^p, |s(b)|^p\} \leq r(|b|)^p = (|b|^p \circ \frac{1}{\sqrt{2}} J_2^{-1}) \leq (|r(b)| \cdot 1 + |s(b)| \cdot 1)^p$$
$$\leq 2^{p/q} (|r(b)|^p + |s(b)|^p).$$

Hence, $\max\{|r(b)|^p, |s(b)|^p\} \in L^p(\mathbb{R}^{2n}, \gamma_{2n})$, if and only if $b \in L^p(\mathbb{C}^n, \mathbb{P}_\mathbf{Z})$, and

$$\mathbb{E}[b(\mathbf{Z})] = \int_{\mathbb{C}^n} b \, d\mathbb{P}_\mathbf{Z} = \int_{\mathbb{C}^n} b \, d(\frac{1}{\sqrt{2}} J_2^{-1})_* \mathbb{P}_{\sqrt{2} J_2(\mathbf{Z})}$$
$$= \int_{\mathbb{R}^{2n}} b(\frac{1}{\sqrt{2}}(x + iy)) \gamma_{2n}(d(x, y))$$
$$= \int_{\mathbb{R}^{2n}} r(b) \, d\gamma_{2n} + i \int_{\mathbb{R}^{2n}} s(b) \, d\gamma_{2n} = \mathbb{E}[r(b)(\mathbf{X})] + i \, \mathbb{E}[s(b)(\mathbf{X})], \quad (2.1.9)$$

where $\mathbf{X} \stackrel{d}{=} \sqrt{2} J_2(\mathbf{Z}) \sim N_{2n}(0, I_{2n})$. In particular,

$$\mathcal{B}(\mathbb{C}^n) \ni A \mapsto \gamma_n^\mathbb{C}(A) := \mathbb{P}(\mathbf{Z} \in A) = \mathbb{E}[\mathbb{1}_A(\mathbf{Z})] \stackrel{(2.1.9)}{=} \gamma_{2n}(J_2(\sqrt{2}A))$$
$$= \pi^{-n} \int_{J_2(A)} \exp(-\|x\|_2^2) \lambda_{2n}(dx)$$
$$\stackrel{(2.1.2)}{=} \pi^{-n} \int_A \exp(-\|J_2(z)\|_2^2) \lambda_n^\mathbb{C}(dz)$$

emerges as the Gaussian measure on $\mathbb{C}^n$, implying that

$$\gamma_n^\mathbb{C} = (\frac{1}{\sqrt{2}} J_2^{-1})_* \gamma_{2n} = \mathbb{P}_\mathbf{Z}$$

is absolutely continuous with respect to $\lambda_n^\mathbb{C}$, with Radon-Nikodým derivative $\frac{d\gamma_n^\mathbb{C}}{d\lambda_n^\mathbb{C}} = \pi^{-n} \exp(-\|J_2(z)\|_2^2)$.

Remark 2.4 In [33, Section 8.7], the complex Gaussian measure on $\mathbb{C}^n$ is defined in such a manner that it coincides with the real Gaussian measure γ_{2n} on $\mathbb{R}^{2n}$. Given that construction, the important factor $\sqrt{2}$ - which actually emerges from the underlying structure of the probability law of a complex Gaussian random vector—is ignored. In our view, that approach creates a bit of dissonance. For example, Corollary 4.4, which shows us that for both fields, $\mathbb{F} = \mathbb{R}$ and $\mathbb{F} = \mathbb{C}$, the little

Grothendieck constant $k_G^{\mathbb{F}}$ actually emerges from a common source, can no longer be maintained.

Hence, $b \in L^p(\mathbb{C}^n, \gamma_n^{\mathbb{C}})$ if and only if $\max\{|r(b)|^p, |s(b)|^p\} \in L^p(\mathbb{R}^{2n}, \gamma_{2n}^{\mathbb{R}})$, and

$$\int_{\mathbb{C}^n} \mathrm{Re}(b)\, d\gamma_n^{\mathbb{C}} + i \int_{\mathbb{C}^n} \mathrm{Im}(b)\, d\gamma_n^{\mathbb{C}} = \int_{\mathbb{C}^n} b\, d\gamma_n^{\mathbb{C}} = \mathbb{E}[b(\mathbf{Z})]$$

$$\stackrel{(2.1.9)}{=} \int_{\mathbb{R}^{2n}} r(b)\, d\gamma_{2n} + i \int_{\mathbb{R}^{2n}} s(b)\, d\gamma_{2n}.$$
(2.1.10)

In particular, $b \in L^2(\mathbb{C}^n, \gamma_n^{\mathbb{C}})$ if and only if $r(b) \in L^2(\mathbb{R}^{2n}, \gamma_{2n})$ and $s(b) \in L^2(\mathbb{R}^{2n}, \gamma_{2n})$, so that (in either case)

$$\mathbb{E}[b(\mathbf{Z})\overline{b}(\mathbf{Z})] = \|b\|_{\gamma_n^{\mathbb{C}}}^2 = \int_{\mathbb{C}^n} |b|^2 d\gamma_n^{\mathbb{C}} \stackrel{(2.1.10)}{=} \int_{\mathbb{R}^{2n}} r(|b|^2)\, d\gamma_{2n}$$

$$= \|r(b)\|_{\gamma_{2n}}^2 + \|s(b)\|_{\gamma_{2n}}^2 \qquad (2.1.11)$$

(since $r(|b|^2) = r(b)^2 + s(b)^2$). In particular, for any function $f : \mathbb{R}^{2n} \longrightarrow \mathbb{C}$, it follows that $f \circ \sqrt{2}J_2 \in L^2(\mathbb{C}^n, \gamma_n^{\mathbb{C}})$ if and only if $\mathrm{Re}(f) \in L^2(\mathbb{R}^{2n}, \gamma_{2n})$ and $\mathrm{Im}(f) \in L^2(\mathbb{R}^{2n}, \gamma_{2n})$, whence

$$\|f \circ \sqrt{2}J_2\|_{\gamma_n^{\mathbb{C}}}^2 = \|\mathrm{Re}(f)\|_{\gamma_{2n}}^2 + \|\mathrm{Im}(f)\|_{\gamma_{2n}}^2. \qquad (2.1.12)$$

in either case. Consequently,

$$\|g^{\mathbb{C}}\|_{\gamma_n^{\mathbb{C}}} = \|g\|_{\gamma_n} \qquad (2.1.13)$$

for any real-valued function $g : \mathbb{R}^n \longrightarrow \mathbb{R}$, where $g^{\mathbb{C}} := (g \otimes 1) \circ \sqrt{2}\, J_2$ (since $\gamma_{2n} = \gamma_n \otimes \gamma_n$). Moreover, if $b, c \in L^2(\mathbb{C}^n, \gamma_n^{\mathbb{C}})$, then $r(b\overline{c}) = r(b)r(c) + s(b)s(c)$ and $s(b\overline{c}) = s(b)r(c) - r(b)s(c)$. (2.1.10) therefore implies that

$$\langle b, c \rangle_{\gamma_n^{\mathbb{C}}} = \langle r(b), r(c) \rangle_{\gamma_{2n}} + \langle s(b), s(c) \rangle_{\gamma_{2n}}$$

$$+ i \langle s(b), r(c) \rangle_{\gamma_{2n}} - i \langle r(b), s(c) \rangle_{\gamma_{2n}}. \qquad (2.1.14)$$

Remark 2.5 If in addition the function $b : \mathbb{C}^n \longrightarrow \mathbb{C}$ is holomorphic, then b actually is an element of the Segal-Bergmann space (cf. [73, Chapter 3.10])

$$\mathcal{H}L^2(\mathbb{C}^n) := \{c : \mathbb{C}^d \longrightarrow \mathbb{C} : c \text{ is holomorphic and } \|c\|_{\gamma_n^{\mathbb{C}}} < \infty\}.$$

In a similar vein, one can now prove easily the more comprehensive

Corollary 2.1 *Let* $n \in \mathbb{N}$, $p \in [1, \infty)$, $\mu \in \mathbb{C}^n$, $\Gamma \in M_n(\mathbb{C})^+$ *and* $\mathbf{Z} \sim \mathbb{C}N_n(\mu, \Gamma)$. *Let* $b = \mathrm{Re}(b) + i\,\mathrm{Im}(b) : \mathbb{C}^n \longrightarrow \mathbb{C}$, *such that*

$$\mathbb{R}^{2n} \ni y \mapsto r(b)(y + \sqrt{2}\,J_2(\mu)) \in L^p(\mathbb{R}^{2n}, \gamma_{2n})$$

and

$$\mathbb{R}^{2n} \ni y \mapsto s(b)(y + \sqrt{2}\,J_2(\mu)) \in L^p(\mathbb{R}^{2n}, \gamma_{2n}).$$

Then $b \in L^p(\mathbb{C}^n, \mathbb{P}_{\mathbf{Z}})$, *and*

$$\mathbb{E}[b(\mathbf{Z})] = |\det(\Gamma)|(\int_{\mathbb{R}^{2n}} r(b)(y + \sqrt{2}\,J_2(\mu))\gamma_{2n}(\mathrm{d}y)$$

$$+ i \int_{\mathbb{R}^{2n}} s(b)(y + \sqrt{2}\,J_2(\mu))\gamma_{2n}(\mathrm{d}y)).$$

Proof We just have to observe that

$$\mathbb{E}[b(\mathbf{Z})] = \mathbb{E}[r(b)(\mathbf{Y})] + i\,\mathbb{E}[s(b)(\mathbf{Y})]$$

$$= \int_{\mathbb{R}^{2n}} r(b)(\sqrt{R_2(\Gamma)}x + \sqrt{2}\,J_2(\mu))\gamma_{2n}(\mathrm{d}x)$$

$$+ i \int_{\mathbb{R}^{2n}} s(b)(\sqrt{R_2(\Gamma)}x + \sqrt{2}\,J_2(\mu))\gamma_{2n}(\mathrm{d}x)$$

$$\stackrel{(2.1.6)}{=} |\det(\Gamma)|(\int_{\mathbb{R}^{2n}} r(b)(y + \sqrt{2}\,J_2(\mu))\gamma_{2n}(\mathrm{d}y)$$

$$+ i \int_{\mathbb{R}^{2n}} s(b)(y + \sqrt{2}\,J_2(\mu))\gamma_{2n}(\mathrm{d}y)),$$

where $\mathbf{Y} \stackrel{d}{=} \sqrt{2}\,J_2(\mathbf{Z}) \sim N_{2n}(\sqrt{2}\,J_2(\mu), R_2(\Gamma))$. □

2.2 Partitioned Complex Gaussian Random Vectors in $\mathbb{C}^{2n}$ and the Probability Law $\mathbb{C}N_{2n}(0, \Sigma_{2n}(\zeta))$

For the remainder of the monograph we put, without loss of generality, $\mu = 0$, so that we are working with centred Gaussian random vectors (with respect to both fields, $\mathbb{R}$ and $\mathbb{C}$). Moreover, we make use of a specific class of partitioned correlation matrices, which turns out to be of crucial importance regarding the topic of our monograph. To this end, let $\mathbb{F} \in \{\mathbb{R}, \mathbb{C}\}$, $n \in \mathbb{N}$ and $\zeta \in \mathbb{F} \cap \overline{\mathbb{D}}$. Put

$$\Sigma_{2n}(\zeta) := \begin{pmatrix} I_n & \zeta I_n \\ \overline{\zeta} I_n & I_n \end{pmatrix} = \Sigma_2(\zeta) \otimes I_n = \begin{pmatrix} 1 & 0 & \cdots & 0 & \zeta & 0 & \cdots & 0 \\ 0 & 1 & \cdots & 0 & 0 & \zeta & \cdots & 0 \\ \vdots & \vdots & \ddots & \vdots & \vdots & \vdots & \ddots & \vdots \\ 0 & 0 & \cdots & 1 & 0 & 0 & \cdots & \zeta \\ \overline{\zeta} & 0 & \cdots & 0 & 1 & 0 & \cdots & 0 \\ 0 & \overline{\zeta} & \cdots & 0 & 0 & 1 & \cdots & 0 \\ \vdots & \vdots & \ddots & \vdots & \vdots & \vdots & \ddots & \vdots \\ 0 & 0 & \cdots & \overline{\zeta} & 0 & 0 & \cdots & 1 \end{pmatrix}$$

$$= \Sigma_{2n}(\mathrm{Re}(\zeta)) + R_2(-i\,\mathrm{Im}(\zeta) I_n). \tag{2.2.1}$$

Since $|\zeta| \leq 1$, it follows that

$$\left\langle \begin{pmatrix} I_n & \zeta I_n \\ \overline{\zeta} I_n & I_n \end{pmatrix} \begin{pmatrix} a \\ b \end{pmatrix}, \begin{pmatrix} a \\ b \end{pmatrix} \right\rangle_{\mathbb{F}^{2n}} = \|a + \zeta b\|_{\mathbb{F}^n}^2 + (1 - |\zeta|^2) \|b\|_{\mathbb{F}^n}^2 \geq 0$$

for all $a, b \in \mathbb{F}^n$, implying that $\Sigma_{2n}(\zeta)$ in fact is positive semidefinite and hence a correlation matrix (due to Lemma 3.1). Lemma 3.1 also clearly implies that

$$C(2; \mathbb{F}) = \left\{ \Sigma_2(\zeta) = \begin{pmatrix} 1 & \zeta \\ \overline{\zeta} & 1 \end{pmatrix} : \zeta \in \mathbb{F} \cap \overline{\mathbb{D}} \right\}. \tag{2.2.2}$$

Moreover, the determinant of the Kronecker product $\Sigma_{2n}(\zeta) = \Sigma_2(\zeta) \otimes I_n$ is calculated as (cf. [69, Problem 4.2.1]):

$$\det(\Sigma_{2n}(\zeta)) = \det(I_n)^2 \det(\Sigma_2(\zeta))^n = (1 - |\zeta|^2)^n.$$

If in addition $|\zeta| < 1$, then $\|a + \zeta b\|_{\mathbb{F}^n}^2 + (1 - |\zeta|^2) \|b\|_{\mathbb{F}^n}^2 > 0$ for all $(a, b) \in (\mathbb{F}^n \times \mathbb{F}^n) \setminus \{(0, 0)\}$, implying that in this case $\Sigma_{2n}(\zeta)$ even is positive definite and hence invertible, with inverse $\Sigma_{2n}(\zeta)^{-1} = \frac{1}{1-|\zeta|^2} \Sigma_{2n}(-\zeta) = \frac{1}{1-|\zeta|^2} \begin{pmatrix} I_n & -\zeta I_n \\ -\overline{\zeta} I_n & I_n \end{pmatrix}$, implying that also $(1 - |\zeta|^2)\, \Sigma_{2n}(\zeta)^{-1} = \Sigma_{2n}(-\zeta)$ is a correlation matrix of rank $2n$.

In particular, for any $\rho \in (-1, 1)$, the density function of the $2n$-dimensional random vector $\mathrm{vec}(\mathbf{X}_1, \mathbf{X}_2) \sim N_{2n}(0, \Sigma_{2n}(\rho))$, where both, $\mathbf{X}_1$ and $\mathbf{X}_2$ are n-dimensional random vectors, exists. It is given by

$$\varphi_{0, \Sigma_{2n}(\rho)}(x_1, x_2) = \frac{1}{(2\pi)^n (1 - \rho^2)^{n/2}} \exp\left(-\frac{1}{2(1-\rho^2)} \langle \Sigma_{2n}(-\rho) x, x \rangle_{\mathbb{R}_2^{2n}}\right)$$

$$= \frac{1}{(2\pi)^n (1 - \rho^2)^{n/2}} \exp\left(-\frac{\|x_1\|^2 + \|x_2\|^2 - 2\rho\langle x_1, x_2\rangle}{2(1-\rho^2)}\right)$$

2.2 Partitioned Complex Gaussian Random Vectors in $\mathbb{C}^{2n}$ and the Probability... 33

$$\begin{aligned}
&= M_\rho(x_1, x_2; n)\, \varphi_{0, I_{2n}}(x_1, x_2) \\
&= M_\rho(x_1, x_2; n)\, \varphi_{0, I_n}(x_1)\, \varphi_{0, I_n}(x_2) \\
&= \varphi_{0, \Sigma_{2n}(\rho)}(x_2, x_1),
\end{aligned} \quad (2.2.3)$$

where $x_1, x_2 \in \mathbb{R}^n$, $x := \mathrm{vec}(x_1, x_2)$ and

$$M_\rho(x_1, x_2; n) := \frac{1}{(1-\rho^2)^{n/2}} \exp\left(\frac{2\rho\langle x_1, x_2\rangle - \rho^2(\|x_1\|^2 + \|x_2\|^2)}{2(1-\rho^2)}\right)$$

$$= M_{-\rho}(x_1, -x_2; n) \quad (2.2.4)$$

denotes the n-dimensional Mehler kernel (cf. [62]). Moreover, since $\det(\Sigma_{2n}(\zeta)) > 0$ if and only if $\zeta \in \mathbb{D}$, we achieve the following result:

Proposition 2.4 *Let* $\mathbb{F} \in \{\mathbb{R}, \mathbb{C}\}$, $n \in \mathbb{N}$ *and* $\zeta \in \overline{\mathbb{D}} \cap \mathbb{F}$. *Then*

$$\Sigma_{2n}(\zeta) := \begin{pmatrix} I_n & \zeta I_n \\ \overline{\zeta} I_n & I_n \end{pmatrix} \in C(2n; \mathbb{F}).$$

Moreover, the following statements are equivalent

(i) $\Sigma_{2n}(\zeta)$ *is a correlation matrix of rank* $2n$.
(ii) $\Sigma_{2n}(\zeta)$ *is invertible.*
(iii) $\zeta \in \mathbb{D}$.

If one of these equivalent statements is given, then $\Sigma_{2n}(\zeta)^{-1} = \frac{1}{1-|\zeta|^2} \Sigma_{2n}(-\zeta)$. *In particular, also the $2n \times 2n$-matrix $(1 - |\zeta|^2)\Sigma_{2n}(\zeta)^{-1}$ is a correlation matrix of rank $2n$.*

It is not obvious that the probability law $N_{2n}(0, \Sigma_{2n}(\rho))$ can also be described as follows (cf. [111, Definition 11.6 and Definition 11.10]):

Proposition 2.5 *Let* $n \in \mathbb{N}$ *and* $\rho \in [-1, 1]$. *Let* $\mathbf{X} = (X_1, \ldots, X_n)^\top$ *and* $\mathbf{Y} = (Y_1, \ldots, Y_n)^\top$ *be two $\mathbb{R}^n$-valued random vectors. Then the following statements are equivalent:*

(i) $\mathrm{vec}(\mathbf{X}, \mathbf{Y}) \sim N_{2n}(0, \Sigma_{2n}(\rho))$.
(ii) $\mathrm{vec}(X_i, Y_i, X_j, Y_j) \sim N_{2n}\left(0, \begin{pmatrix} \Sigma_2(\rho) & 0 \\ 0 & \Sigma_2(\rho) \end{pmatrix}\right)$ *for all $i \neq j \in [n]$.*
(iii) *The random component vector pairs $\mathrm{vec}(X_i, Y_i), \ldots, \mathrm{vec}(X_n, Y_n)$ are mutually independent, and $\mathrm{vec}(X_i, Y_i) \sim N_2(0, \Sigma_2(\rho))$ for all $i \in [n]$.*

Proof We just sketch the main ideas and leave the elaboration of the proof to the readers. Suppose that (i) holds. Since

$$\mathrm{vec}(X_i, Y_i, X_j, Y_j) = A(i, j)\mathrm{vec}(\mathbf{X}, \mathbf{Y}),$$

where $\mathbb{M}_{4,2n}(\mathbb{R}) \ni A(i,j) := \begin{pmatrix} e_i^\top & 0^\top \\ 0^\top & e_i^\top \\ e_j^\top & 0^\top \\ 0^\top & e_j^\top \end{pmatrix}$, it follows that

$$A(i,j)\Sigma_{2n}(\rho)A(i,j)^\top = \begin{pmatrix} \Sigma_2(\rho) & 0 \\ 0 & \Sigma_2(\rho) \end{pmatrix}$$

is the correlation matrix of the Gaussian random vector $\text{vec}(X_i, Y_i, X_j, Y_j)$ (cf. (2.1.1)), which proves (ii). If (ii) is given, the structure of the correlation matrix $\begin{pmatrix} \Sigma_2(\rho) & 0 \\ 0 & \Sigma_2(\rho) \end{pmatrix}$ implies that for $i \neq j$ the characteristic function of $\text{vec}(X_i, Y_i, X_j, Y_j)$ can be written as the product of the two characteristic functions of the pairs $\text{vec}(X_i, Y_i)$ and $\text{vec}(X_j, Y_j)$, respectively. Finally, if (iii) is given, it follows that for all $a, b \in \mathbb{R}^n$

$$(\text{vec}(a,b))^\top \text{vec}(\mathbf{X}, \mathbf{Y}) = \sum_{i=1}^n V_i,$$

where $V_i := a_i X_i + b_i Y_i \sim N_1(0, a_i^2 + 2\rho a_i b_i + b_i^2)$ for all $i \in [n]$ and $V_1, \ldots, V_n$ are mutually independent. Consequently, it follows that $(\text{vec}(a,b))^\top \text{vec}(\mathbf{X}, \mathbf{Y}) \sim N_1(0, \|a\|^2 + 2\rho a^\top b + \|b\|^2)$, and (i) is achieved. □

Regarding the underlying structure of the Haagerup equality (and its generalisation —see Theorem 7.1) an analysis of the structure of partitioned complex $2n$-dimensional Gaussian random vectors whose probability law is induced by the correlation matrix $\Sigma_{2n}(\zeta)$ leads to another important

Lemma 2.2 *Let $n \in \mathbb{N}$, $\zeta = x + iy \in \overline{\mathbb{D}}$ and $\text{vec}(\mathbf{Z}, \mathbf{W}) \sim \mathbb{C}N_{2n}(0, \Sigma_{2n}(\zeta))$, where the complex random vectors $\mathbf{Z}$ and $\mathbf{W}$ both map into $\mathbb{C}^n$. Then the following statements hold:*

(i) *For any $\alpha, \beta \in \mathbb{T}$, $\text{vec}(\alpha \mathbf{Z}, \beta \mathbf{W}) \sim \mathbb{C}N_{2n}(0, \Sigma_{2n}(\alpha \overline{\beta} \zeta))$.*
(ii) *$\text{vec}(\mathbf{W}, \mathbf{Z}) \sim \mathbb{C}N_{2n}(0, \Sigma_{2n}(\overline{\zeta}))$.*
(iii) *$\sqrt{2}\,\text{vec}(\text{Re}(\mathbf{Z}), \text{Re}(\mathbf{W})) \sim N_{2n}(0, \Sigma_{2n}(\text{Re}(\zeta)))$ and $\sqrt{2}\,\text{vec}(\text{Im}(\mathbf{Z}), \text{Im}(\mathbf{W})) \sim N_{2n}(0, \Sigma_{2n}(\text{Re}(\zeta)))$.*
(iv) *If $\zeta \in [-1, 1]$, then $\sqrt{2}\,\text{vec}(\text{Re}(\mathbf{Z}), \text{Im}(\mathbf{Z}), \text{Re}(\mathbf{W}), \text{Im}(\mathbf{W})) = \sqrt{2}\,\text{vec}(J_2(\mathbf{Z}), J_2(\mathbf{W})) \sim N_{4n}(0, \Sigma_{4n}(\zeta))$, and $\text{vec}(\text{Re}(\mathbf{Z}), \text{Re}(\mathbf{W}))$ and $\text{vec}(\text{Im}(\mathbf{Z}), \text{Im}(\mathbf{W}))$ are independent.*
(v) *If $\zeta \in \overline{\mathbb{D}} \setminus \{0\}$, then $\text{vec}(\text{sign}(\overline{\zeta})\mathbf{Z}, \mathbf{W}) \stackrel{d}{=} \text{vec}(\text{sign}(\zeta)\mathbf{Z}, \mathbf{W}) \sim \mathbb{C}N_{2n}(0, \Sigma_{2n}(|\zeta|))$. Moreover, $\sqrt{2}\,\text{vec}(J_2(\text{sign}(\overline{\zeta})\mathbf{Z}), J_2(\mathbf{W})) \stackrel{d}{=} \sqrt{2}\,\text{vec}(J_2(\text{sign}(\zeta)\mathbf{Z}), J_2(\mathbf{W})) \sim N_{4n}(0, \Sigma_{4n}(|\zeta|))$.*

Proof

(i) Let $\alpha, \beta \in \mathbb{T}$. Then

$$\begin{pmatrix} \alpha \mathbf{Z} \\ \beta \mathbf{W} \end{pmatrix} = \begin{pmatrix} \alpha I_n & 0 \\ 0 & \beta I_n \end{pmatrix} \begin{pmatrix} \mathbf{Z} \\ \mathbf{W} \end{pmatrix}$$

and

$$\begin{pmatrix} \alpha I_n & 0 \\ 0 & \beta I_n \end{pmatrix} \begin{pmatrix} I_n & z I_n \\ \overline{z} I_n & I_n \end{pmatrix} \begin{pmatrix} \overline{\alpha} I_n & 0 \\ 0 & \overline{\beta} I_n \end{pmatrix} = \begin{pmatrix} I_n & \alpha \overline{\beta} z I_n \\ \overline{\alpha} \beta \overline{z} I_n & I_n \end{pmatrix}.$$

The claim now follows from [5, Theorem 2.8].

(ii) Again, by analogy with the above approach, the claim from [5, Theorem 2.8].

(iii) Since $\sqrt{2} J_2(\text{vec}(\mathbf{Z}, \mathbf{W})) = \sqrt{2} \text{vec}(\text{Re}(\mathbf{Z}), \text{Re}(\mathbf{W}), \text{Im}(\mathbf{Z}), \text{Im}(\mathbf{W})) \sim N_{4n}(0, R_2(\Sigma_{2n}(\zeta)))$ and

$$R_2(\Sigma_{2n}(\zeta)) = \begin{pmatrix} \text{Re}(\Sigma_{2n}(\zeta)) & -\text{Im}(\Sigma_{2n}(\zeta)) \\ \text{Im}(\Sigma_{2n}(\zeta)) & \text{Re}(\Sigma_{2n}(\zeta)) \end{pmatrix} = \begin{pmatrix} \Sigma_{2n}(\text{Re}(\zeta)) & -\Sigma_{2n}(\text{Im}(\zeta)) \\ \Sigma_{2n}(\text{Im}(\zeta)) & \Sigma_{2n}(\text{Re}(\zeta)) \end{pmatrix},$$

claim (iii) follows.

(iv) Firstly, we fix an arbitrary $\zeta = x + iy \in \overline{\mathbb{D}}$. (1.2.5) implies that

$$\text{vec}(\text{Re}(\mathbf{Z}), \text{Im}(\mathbf{Z}), \text{Re}(\mathbf{W}), \text{Im}(\mathbf{W})) = G \, \text{vec}(\text{Re}(\mathbf{Z}), \text{Re}(\mathbf{W}), \text{Im}(\mathbf{Z}), \text{Im}(\mathbf{W})),$$

where $G \in O(4n)$ is the matrix, introduced in (1.2.6). Since $\sqrt{2} \text{vec}(\text{Re}(\mathbf{Z}), \text{Re}(\mathbf{W}), \text{Im}(\mathbf{Z}), \text{Im}(\mathbf{W})) = \sqrt{2} J_2(\text{vec}(\mathbf{Z}, \mathbf{W})) \sim N_{4n}(0, R_2(\Sigma_{2n}(\zeta)))$, an application of (2.1.1) therefore implies that

$$\sqrt{2} \, \text{vec}(\text{Re}(\mathbf{Z}), \text{Im}(\mathbf{Z}), \text{Re}(\mathbf{W}), \text{Im}(\mathbf{W})) \sim N_{4n}(0, G R_2(\Sigma_{2n}(\zeta)) G).$$

A straightforward block matrix multiplication shows that

$$G R_2(\Sigma_{2n}(\zeta)) G = \begin{pmatrix} I_n & 0 & x I_n & -y I_n \\ 0 & I_n & y I_n & x I_n \\ x I_n & y I_n & I_n & 0 \\ -y I_n & x I_n & 0 & I_n \end{pmatrix}$$

$$= \Sigma_{4n}(x) + \begin{pmatrix} 0 & 0 & 0 & -y I_n \\ 0 & 0 & y I_n & 0 \\ 0 & y I_n & 0 & 0 \\ -y I_n & 0 & 0 & 0 \end{pmatrix}.$$

Since by assumption $\zeta \in [-1,1] = \overline{\mathbb{D}} \cap \mathbb{R}$, $y = \text{Im}(\zeta) = 0$, and the first part of claim (iv) follows. To verify the independence assertion, put $\mathbf{X} := \sqrt{2}\,\text{vec}(\text{Re}(\mathbf{Z}), \text{Re}(\mathbf{W})) = \sqrt{2}\,\text{Re}(\text{vec}(\mathbf{Z}_1, \mathbf{Z}_2))$ and $\mathbf{Y} := \sqrt{2}\,\text{vec}(\text{Im}(\mathbf{Z}), \text{Im}(\mathbf{W})) = \sqrt{2}\,\text{Im}(\text{vec}(\mathbf{Z}_1, \mathbf{Z}_2))$. Since

$$\text{vec}(\mathbf{X}, \mathbf{Y}) = \sqrt{2} J_2(\text{vec}(\mathbf{Z}, \mathbf{W})) \sim N_{4n}(0, R_2(\Sigma_{2n}(\zeta))),$$

it follows that for all $(a, b) \in \mathbb{R}^{2n} \times \mathbb{R}^{2n}$

$$\phi_{\text{vec}(\mathbf{X},\mathbf{Y})}(\text{vec}(a, b)) = \exp(-\frac{1}{2}(a^\top \Sigma_{2n}(\zeta) a + b^\top \Sigma_{2n}(\zeta) b)) = \phi_{\mathbf{X}}(a)\phi_{\mathbf{Y}}(b).$$

Thus, $\mathbf{X}$ and $\mathbf{Y}$ are independent, and (iv) follows.
(v) follows from (i) and (iv). □

Next, we will recognise that for any $\rho \in [-1,1]$, the real probability law $N_{4n}(0, \Sigma_{4n}(\rho))$ actually originates from the complex probability law $\mathbb{C}N_{2n}(0, \Sigma_{2n}(\rho))$! Since:

Corollary 2.2 Let $\rho \in [-1,1]$. Let $\mathbf{X}_1, \mathbf{Y}_1, \mathbf{X}_2$ and $\mathbf{Y}_2$ be four $\mathbb{R}^n$-valued random vectors. Then the following statements are equivalent:

(i) $\text{vec}(\mathbf{X}_1, \mathbf{X}_2) \sim N_{2n}(0, \Sigma_{2n}(\rho))$, $\text{vec}(\mathbf{Y}_1, \mathbf{Y}_2) \sim N_{2n}(0, \Sigma_{2n}(\rho))$, and $\text{vec}(\mathbf{X}_1, \mathbf{X}_2)$ and $\text{vec}(\mathbf{Y}_1, \mathbf{Y}_2)$ are independent.
(ii) $\text{vec}(\mathbf{X}_1, \mathbf{X}_2, \mathbf{Y}_1, \mathbf{Y}_2) \sim N_{4n}(0, R_2(\Sigma_{2n}(\rho)))$.
(iii) $\text{vec}(\mathbf{X}_1, \mathbf{Y}_1, \mathbf{X}_2, \mathbf{Y}_2) \sim N_{4n}(0, \Sigma_{4n}(\rho))$.
(iv) $\text{vec}(\mathbf{X}_1, \mathbf{Y}_1, \mathbf{X}_2, \mathbf{Y}_2) = \sqrt{2}\text{vec}(\text{Re}(\mathbf{Z}_1), \text{Im}(\mathbf{Z}_1), \text{Re}(\mathbf{Z}_2), \text{Im}(\mathbf{Z}_2)) = \text{vec}(J_2(\mathbf{Z}_1), J_2(\mathbf{Z}_2))$, where $\text{vec}(\mathbf{Z}_1, \mathbf{Z}_2) \sim \mathbb{C}N_{2n}(0, \Sigma_{2n}(\rho))$.

Proof Let (i) be given. Since $\text{vec}(\mathbf{X}_1, \mathbf{X}_2, \mathbf{Y}_1, \mathbf{Y}_2) = \text{vec}(\mathbf{X}, \mathbf{Y})$, where $\mathbf{X} := \text{vec}(\mathbf{X}_1, \mathbf{X}_2)$ and $\mathbf{Y} := \text{vec}(\mathbf{Y}_1, \mathbf{Y}_2)$, the assumed independence of the random vectors $\mathbf{X}$ and $\mathbf{Y}$ implies that for all $a, b \in \mathbb{R}^{2n}$

$$\phi_{\text{vec}(\mathbf{X},\mathbf{Y})}(\text{vec}(a, b)) = \phi_{\mathbf{X}}(a)\phi_{\mathbf{Y}}(b)$$
$$= \exp(-\frac{1}{2}(a^\top \Sigma_{2n}(\rho) a + b^\top \Sigma_{2n}(\rho) b))$$
$$= \exp(-\frac{1}{2}\text{vec}(a, b)^\top R_2(\Sigma_{2n}(\rho))\,\text{vec}(a, b)).$$

Proposition 2.1 therefore concludes the proof of (ii).
Now, assume that (ii) holds. Then (2.1.1) implies that

$$\text{vec}(\mathbf{X}_1, \mathbf{Y}_1, \mathbf{X}_2, \mathbf{Y}_2) = G\,\text{vec}(\mathbf{X}_1, \mathbf{X}_2, \mathbf{Y}_1, \mathbf{Y}_2) \sim N_{4n}(0, G R_2(\Sigma_{2n}(\rho)) G^\top),$$

2.2 Partitioned Complex Gaussian Random Vectors in $\mathbb{C}^{2n}$ and the Probability... 37

where $G \in O(4n)$ is the matrix, introduced in (1.2.6). Since $GR_2(\Sigma_{2n}(\rho))G^\top = \Sigma_{4n}(\rho)$, (iii) follows.

Suppose that (iii) is given. Put $\mathbf{Z}_1 := \frac{1}{\sqrt{2}}(\mathbf{X}_1 + i\mathbf{Y}_1)$ and $\mathbf{Z}_2 := \frac{1}{\sqrt{2}}(\mathbf{X}_2 + i\mathbf{Y}_2)$. Then $\text{vec}(\mathbf{Z}_1, \mathbf{Z}_2) = \text{vec}(\frac{\mathbf{X}_1}{\sqrt{2}}, \frac{\mathbf{X}_2}{\sqrt{2}}) + i\,\text{vec}(\frac{\mathbf{Y}_1}{\sqrt{2}}, \frac{\mathbf{Y}_2}{\sqrt{2}})$. Hence,

$$\sqrt{2} J_2(\text{vec}(\mathbf{Z}_1, \mathbf{Z}_2)) = \text{vec}(\mathbf{X}_1, \mathbf{X}_2, \mathbf{Y}_1, \mathbf{Y}_2).$$

This, since $\text{vec}(\mathbf{X}_1, \mathbf{X}_2, \mathbf{Y}_1, \mathbf{Y}_2) = G\,\text{vec}(\mathbf{X}_1, \mathbf{Y}_1, \mathbf{X}_2, \mathbf{Y}_2)$ and $G\Sigma_{4n}(\rho)G^\top = R_2(\Sigma_{2n}(\rho))$, (iv) follows from Proposition 2.3.

Finally, if (iv) holds, then we may apply Lemma 2.2 and assertion (i) follows. □

Similarly, under the inclusion of Corollary 4.2, respectively [33, Proposition 8.7] (including a minor adjustment of the factor c_1 of the complex Gaussian probability measure in their proof, required due to the shape of the complex density function induced by the law $\mathbb{C}N_1(0,1) = \mathbb{R}N_2(0, \frac{1}{2}I_2))$, we obtain

Lemma 2.3 *Let* $n \in \mathbb{N}$, $z \in \overline{\mathbb{D}} \cap \mathbb{F}$, $\mathbf{X} = \text{vec}(X_1, \ldots, X_n) \sim \mathbb{F}N_n(0, I_n)$, $\text{vec}(\mathbf{Y}, \mathbf{Z}) \sim \mathbb{F}N_{2n}(0, \Sigma_{2n}(z))$ *and* $u, v \in S_{\mathbb{F}^n}$. *Then*

$$u^\top \mathbf{X} \stackrel{d}{=} u^* \mathbf{X} \sim \mathbb{F}N_1(0,1) \quad \text{and} \quad \begin{pmatrix} u^* \mathbf{Y} \\ v^* \mathbf{Z} \end{pmatrix} \sim \mathbb{F}N_2(0, \Sigma_2((u^*v)z)).$$

In particular, $\begin{pmatrix} u^* \mathbf{X} \\ v^* \mathbf{X} \end{pmatrix} \sim \mathbb{F}N_2(0, \Sigma_2(u^*v))$ *and*

$$\mathbb{E}\big[\big|\sum_{k=1}^n a_k X_k\big|^p\big] = \mathbb{E}\big[|a^\top \mathbf{X}|^p\big] = \|a\|_{\mathbb{F}_2^n}^p \mathbb{E}[|X_1|^p] = \|a\|_{\mathbb{F}_2^n}^p C_p^{\mathbb{F}}$$

for all $a \equiv (a_1, \ldots, a_n)^\top \in \mathbb{F}^n$ *and* $p \in (-1, \infty)$, *where* $C_p^{\mathbb{R}} := \frac{(\sqrt{2})^p}{\sqrt{\pi}} \Gamma(\frac{p+1}{2})$ *and* $C_p^{\mathbb{C}} := \Gamma(1 + \frac{p}{2})$.

Chapter 3
A Quantum Correlation Matrix Version of the Grothendieck Inequality

3.1 Gram Matrices, Quantum Correlation, and Beyond

In this section, we aim at another equivalent reformulation of the Grothendieck inequality (occasionally abbreviated by "GT") for both fields, built on the inclusion of correlation matrices; i.e., positive semidefinite matrices with entries in $\mathbb{F} \in \{\mathbb{R}, \mathbb{C}\}$ whose diagonal is occupied with 1's only. Since we need equivalent descriptions of a correlation matrix including its fundamental representation as a Gram matrix (cf. [70, Theorem 2.7.10]), we proceed with a fundamental acronym, to indicate a comprehensive class of matrices with entries in $\mathbb{F}$ which properly contains the class of all Gram matrices (cf. [70, page 441]) and reveals a deep connection to the foundations and philosophy of quantum mechanics.

Definition 3.1 Let $\mathbb{F} \in \{\mathbb{R}, \mathbb{C}\}$, $m, n \in \mathbb{N}$ and H be an $\mathbb{F}$-inner product space with inner product $\langle \cdot, \cdot \rangle \equiv \langle \cdot, \cdot \rangle_H$. Let $u \equiv (u_1, u_2, \ldots, u_m) \in H^m$ and $v \equiv (v_1, v_2, \ldots, v_n) \in H^n$. We put

$$\Gamma_H(u, v) := (\langle v_j, u_i \rangle)_{(i,j) \in [m] \times [n]} = \begin{pmatrix} \langle v_1, u_1 \rangle & \langle v_2, u_1 \rangle & \ldots & \langle v_n, u_1 \rangle \\ \langle v_1, u_2 \rangle & \langle v_2, u_2 \rangle & \ldots & \langle v_n, u_2 \rangle \\ \vdots & \vdots & \ddots & \vdots \\ \langle v_1, u_m \rangle & \langle v_2, u_m \rangle & \ldots & \langle v_n, u_m \rangle \end{pmatrix} \in \mathbb{M}_{m,n}(\mathbb{F}).$$

Observe that $\Gamma_H(u, v)^* = \Gamma_H(v, u)$ for all $(u, v) \in H^m \times H^n$. $\Gamma_H(u, v)$ should be viewed as an element of the image of the matrix-valued sesquilinear mapping

$$\Gamma_H : H^m \times H^n \longrightarrow \mathbb{M}_{m,n}(\mathbb{F}).$$

In particular, if Γ_H were restricted to the product $S_H^m \times S_H^n$, we would obtain a matrix-valued sesquilinear mapping, which is defined on the infinite-dimensional

C^∞-manifold $S_H^m \times S_H^n$; an interesting fact, which actually underlies our chosen notation (cf. [136, Example 1.34 and Definition 1.36]. If $m = n$ and $u = v \in H^m$, we get again the Gram matrix of the vectors $u_1, \ldots, u_m \in H$:

$$\Gamma_H(u, u) = \begin{pmatrix} \|u_1\|^2 & \langle u_2, u_1 \rangle & \cdots & \langle u_m, u_1 \rangle \\ \langle u_1, u_2 \rangle & \|u_2\|^2 & \cdots & \langle u_m, u_2 \rangle \\ \vdots & \vdots & \ddots & \vdots \\ \langle u_1, u_m \rangle & \langle u_2, u_m \rangle & \cdots & \|u_m\|^2 \end{pmatrix} \in M_m(\mathbb{F})^+.$$

If $(r, s) \in \mathbb{F}^m \times \mathbb{F}^n$, then

$$\Gamma_\mathbb{F}(r, s) \equiv \Gamma_{\mathbb{F}_2^1}(r, s) = \overline{r} s^\top = \overline{r} \otimes s^\top = \begin{pmatrix} \overline{r_1} s_1 & \overline{r_1} s_2 & \cdots & \overline{r_1} s_n \\ \overline{r_2} s_1 & \overline{r_2} s_2 & \cdots & \overline{r_2} s_n \\ \vdots & \vdots & \vdots & \vdots \\ \overline{r_m} s_1 & \overline{r_1} s_2 & \cdots & \overline{r_m} s_n \end{pmatrix}.$$

More generally, if $H = \mathbb{F}_2^d$ for some $d \in \mathbb{N}$, it follows that $\langle x, y \rangle_H = y^* x$ for all $x, y \in H$. Consequently, we obtain an important factorisation:

$$\Gamma_{\mathbb{F}_2^d}(u, v) = U^* V \text{ for all } (u, v) \in (\mathbb{F}_2^d)^m \times (\mathbb{F}_2^d)^n, \tag{3.1.1}$$

where $U := (u_1 | u_2 | \cdots | u_m) \in \mathbb{M}_{d,m}(\mathbb{F})$ and $V := (v_1 | v_2 | \cdots | v_n) \in \mathbb{M}_{d,n}(\mathbb{F})$. Thus,

$$\text{tr}(A^* \Gamma_{\mathbb{F}_2^d}(u, v)) = \overline{\text{tr}(\Gamma_{\mathbb{F}_2^d}(u, v)^* A)} = \overline{\text{tr}(V^*(UA))} = \overline{\langle UA, V \rangle_F} \tag{3.1.2}$$

for all $A \in \mathbb{M}_{m,n}(\mathbb{F})$ (cf. also Proposition 3.3). A straightforward proof shows that any Gram matrix is positive semidefinite (cf. [70, Theorem 2.7.10]). Moreover, if $(a_{ij}) \equiv A = (A^{1/2})^2 = A^{1/2}(A^{1/2})^\top \in M_n(\mathbb{F})^+$ is positive semidefinite and $\mathbf{X} \sim \mathbb{F}N_n(0, I_n)$, then $\mathbf{Z} := A^{1/2}\mathbf{X} \sim \mathbb{F}N_n(0, A)$, implying that $A = A^{1/2}\mathbb{E}[\mathbf{X}\mathbf{X}^*]A^{1/2} = \mathbb{E}[\mathbf{Z}\mathbf{Z}^*]$ (cf., e.g., [121, Lemma 12.10.]) and $a_{ij} = \langle A^{1/2} e_j, A^{1/2} e_i \rangle_{\mathbb{F}_2^n}$ for all $i, j \in [n]$. Let us also recall the following characterisation of the set $C(n; \mathbb{F})$ of all $n \times n$ correlation matrices (with entries in $\mathbb{F}$):

Lemma 3.1 *Let $\mathbb{F} \in \{\mathbb{R}, \mathbb{C}\}$, $n \in \mathbb{N}$ and $\Sigma = (\sigma_{ij}) \in \mathbb{M}_n(\mathbb{F})$. Then the following statements are equivalent:*

(i) $\Sigma \in C(n; \mathbb{F})$.
(ii) $\Sigma \subset \mathbb{M}_n(\mathbb{F})^+$ and $\sigma_{ll} = 1$ for all $i \subset [n]$.
(iii) *There exist vectors $x_1, \ldots, x_n \in S_{\mathbb{F}_2^n}$ such that*

$$\sigma_{ij} = \langle x_j, x_i \rangle_{\mathbb{F}_2^n} = x_i^* x_j \text{ for all } i, j \in [n].$$

3.1 Gram Matrices, Quantum Correlation, and Beyond

(iv) *There exist an $\mathbb{F}$-Hilbert space L and $x = (x_1, x_2, \ldots, x_n) \in L^n$ such that $\|x_i\|_L = 1$ for all $i \in [n]$ and*

$$\Sigma = \Gamma_L(x, x).$$

(v) $\Sigma = \mathbb{E}[\mathbf{Z}\mathbf{Z}^*]$ *for some n-dimensional Gaussian random vector* $\mathbf{Z} \sim \mathbb{F}N_n(0, \Sigma)$, *and* $\sigma_{ii} = 1$ *for all* $i \in [n]$.

In particular, the set $C(n; \mathbb{F})$ is convex, and $|\sigma_{ij}| \leq 1$ for all $i, j \in [n]$.

Remark 3.1 (The Elliptope $\mathcal{E}_n \equiv C(n; \mathbb{R})$) In the real case, the set of all $n \times n$-correlation matrices, which is very rich in geometrical and combinatorial structure, is also known as the so-called *elliptope* (standing for *ellip*soid and poly*tope*) $\mathcal{E}_n$, studied in detail by M. Laurent and S. Poljak (cf. [19, Example 5.44.], [32, Chapter 5.9.1] and [35, Chapter 31.5]). From these sources, we learn, among many other deep facts, that for any $n \in \mathbb{N}$ the set $\mathcal{E}_n \equiv C(n; \mathbb{R})$ is a convex polytope, which in general is not a polyhedron, so that it cannot be described as a finite intersection of weak half spaces (cf. [4, Chapter 5.10]). Since both sets, $C(n; \mathbb{R})$ and $C(n; \mathbb{C})$ are compact (with respect to the topology of pointwise convergence), and since norms on finite-dimensional $\mathbb{F}$-vector spaces are equivalent, it follows that any linear functional on the finite-dimensional Hilbert space $(\mathbb{M}_n(\mathbb{F}), \|\cdot\|_F) \cong \mathbb{F}_2^{n^2}$ attains its maximum (and minimum) on the compact set $C(n; \mathbb{F}) \subseteq \mathbb{M}_n(\mathbb{F})$, including the linear functional $\text{tr}(A^* \cdot) : \mathbb{M}_n(\mathbb{F}) \longrightarrow \mathbb{F}$, where $A \in \mathbb{M}_n(\mathbb{F})$ is given. From the point of view of real (convex) semidefinite optimisation, both, the primal SDP

$$\sup_{\Sigma \in C(m+n; \mathbb{R})} \text{tr}(A^\top \Sigma) = \sup \big\{ \text{tr}(A^\top \Sigma) : \text{tr}(e_\nu e_\nu^\top \Sigma)$$

$$= 1 \text{ for all } \nu \in [m+n], \Sigma \in \mathbb{M}_\nu(\mathbb{R})^+ \big\}$$

and its dual SDP have non-empty, compact sets of optimal solutions and hence attain their respective optima (cf. [19, Theorem 2.15, Theorem 2.29 and Exercise 2.41] and [21, Chapter 5]). We do not know whether this "strong duality" is also valid in the complex case (cf. Lemma 3.2).

Of particular relevance is the set

$$C_1(n; \mathbb{F}) := \{\Theta : \Theta \in C(n; \mathbb{F}) \text{ and } \text{rk}(\Theta) = 1\}$$

of all $n \times n$ correlation matrices of rank 1. The Gram matrix structure implies a neat characterisation of $C_1(n; \mathbb{F})$. At the same time, we recognise again the structure of all pure states on the Hilbert space $\mathbb{F}_2^n$. To this end, recall that $S_\mathbb{R} = \{-1, 1\}$ and $S_\mathbb{C} = \mathbb{T} = \{z \in \mathbb{C} : |z| = 1\}$.

Proposition 3.1 *Let $m, n \in \mathbb{N}$. Then the following statements hold:*

(i)
$$\{A : A \in \mathbb{M}_{m,n}(\mathbb{F}) \text{ and } rk(A) = 1\} = \{\overline{p}q^\top : (p, q) \in \mathbb{F}^m \setminus \{0\} \times \mathbb{F}^n \setminus \{0\}\}$$
$$= \{\Gamma_\mathbb{F}(p, q) : (p, q) \in \mathbb{F}^m \setminus \{0\} \times \mathbb{F}^n \setminus \{0\}\}.$$

(ii)
$$\{A : A \in \mathbb{M}_n(\mathbb{F})^+ \text{ and } rk(A) = 1\} = \{xx^* : x \in \mathbb{F}^n \setminus \{0\}\}$$
$$= \{\Gamma_\mathbb{F}(z, z) : z \in \mathbb{F}^n \setminus \{0\}\}.$$

In particular,
$$\{A : A \in \mathbb{M}_n(\mathbb{F})^+ \text{ and } rk(A) = 1 \text{ and } tr(A) = 1\} = \{xx^* : x \in S_{\mathbb{F}^n}\}$$
$$= \{\Gamma_\mathbb{F}(z, z) : z \in S_{\mathbb{F}^n}\}.$$

Moreover,
$$\mathbb{M}_n(S_\mathbb{F})^+ = \{xx^* : x \in S_\mathbb{F}^n\} = \{\Gamma_\mathbb{F}(z, z) : z \in S_\mathbb{F}^n\} = C_1(n; \mathbb{F}).$$

Proof

(i): Let $\mathrm{rank}(A) = 1$. Then $\{Ax : x \in \mathbb{F}^n\}$ is contained in the linear hull $[v]$ of some $v \in \mathbb{F}^m \setminus \{0\}$. Thus, for any $j \in [n]$, $Ae_j = \lambda_j v$ for some $\lambda_j \in \mathbb{F}$. Put $q := (\lambda_1, \ldots, \lambda_n)^\top \in \mathbb{F}^n$ and $p := \overline{v} \in \mathbb{F}^m \setminus \{0\}$. Then $a_{ij} = e_i^\top A e_j = \overline{p_i} q_j$ for all $(i, j) \in [m] \times [n]$. Hence, $A = \overline{p} q^\top$. Since $A \neq 0$, it also follows that $q \neq 0$. The remaining part of (i) is trivial.

(ii): Let $A \equiv (a_{ij}) \in \mathbb{M}_n(\mathbb{F})^+$. Then $A = B^2$ for some $B \in \mathbb{M}_n(\mathbb{F})^+$, where $rk(B) = 1$ (cf. [69, Theorem 7.2.6]). Let $b \in \mathbb{F}^n \setminus \{0\}$ such that $\{Bx : x \in \mathbb{F}^n\} \subseteq [b]$. For any $i \in [n]$, $Be_i = \lambda_i b$ for some $\lambda_i \in \mathbb{F}$. Put $y := (y_1, \ldots, y_n)^\top$, where $y_i := \|b\|_2 \lambda_i$. Then
$$(yy^*)_{ij} = y_i \overline{y_j} = \|b\|_2^2 \lambda_i \overline{\lambda_j} = \langle Be_i, Be_j \rangle_{\mathbb{F}_2^n} = e_j^\top B^2 e_i = a_{ji} = \overline{a_{ij}}$$
for all $i, j \in [n]$. Thus, $A = xx^*$, where $x := \overline{y} \neq 0$. In particular, if $|a_{ij}| = 1$ for all $i, j \in [n]$, then $|x_i|^2 = 1$ for all $i \in [n]$.

To conclude (ii), fix $\Theta \in \mathbb{M}_n(S_\mathbb{F})^+ = \mathbb{M}_n(\mathbb{F})^+ \cap \mathbb{M}_n(S_\mathbb{F})$. Nothing is to show for $n = 1$, of course. So, let $n \geq 2$. Since $\Theta^* = \Theta \equiv (\vartheta_{ij}) \in \mathbb{M}_n(\mathbb{F})^+$ is positive semidefinite and satisfies $|\vartheta_{ij}| = 1$ for all $i, j \in [n]$, it follows that $\vartheta_{ii} = e_i^\top \Theta e_i =$

3.1 Gram Matrices, Quantum Correlation, and Beyond

$e_i^* \Theta e_i \geq 0$, implying that $\vartheta_{ii} = |\vartheta_{ii}| = 1$ for all $i \in \mathbb{N}$ and hence $\Theta \in C(n; \mathbb{F})$. Thus,

$$\Theta = \begin{pmatrix} \Sigma & b \\ b^* & 1 \end{pmatrix}$$

for some correlation matrix $\Sigma \equiv (\sigma_{ij}) \in C(n-1; \mathbb{F})$ and $b \in S_{\mathbb{F}}^{n-1}$. Let $x \in \mathbb{F}^{n-1}$ be arbitrary and put $\mathbb{F}^n \ni \widetilde{y} := \text{vec}(x, -b^*x)$. Since

$$x^*(\Sigma - bb^*)x = \widetilde{y}^* \Theta \widetilde{y} \geq 0$$

(by construction), the matrix $(c_{ij}) \equiv C := \Sigma - bb^* \in \mathbb{M}_{n-1}(\mathbb{F})^+$ is positive semidefinite as well. Consequently, C can be represented as a Gram matrix: $C = \Gamma_H(w, w)$, $w = (w_1, \ldots, w_{n-1}) \in H^{n-1}$. The Cauchy-Schwarz inequality therefore implies that $|c_{ij}| = |\langle w_j, w_i \rangle_H| \leq \sqrt{|c_{ii}|}\sqrt{|c_{jj}|}$ for all $i, j \in [n-1]$. However, since $\sigma_{ii} = \vartheta_{ii} = 1 = |b_i|^2$ for all $i \in [n-1]$, it follows that $c_{ij} = 0$ for all $i, j \in [n-1]$. Thus, $\Sigma = bb^*$ and

$$\Theta = \begin{pmatrix} \Sigma & b \\ b^* & 1 \end{pmatrix} = \widetilde{b}\widetilde{b}^*,$$

where $\widetilde{b} := \text{vec}(b, 1) \in S_{\mathbb{F}}^n$. $\square$

Note that for all $m, n, \mu, \nu \in \mathbb{N}$, for all Hilbert spaces H, for all $(u, v) \in H^{m+n} \equiv H^m \times H^n$ and for all $(w, z) \in H^{\mu+\nu} \equiv H^\mu \times H^\nu$, the following block matrix representation in $\mathbb{M}_{m+n, \mu+\nu}(\mathbb{F})$ always is satisfied:

$$\begin{pmatrix} \Gamma_H(u, w) & \Gamma_H(u, z) \\ \Gamma_H(v, w) & \Gamma_H(v, z) \end{pmatrix} = \Gamma_H((u, v), (w, z)).$$

In particular,

$$\Gamma_{l_2^d}((u,v),(u,v)) = \begin{pmatrix} U^*U & U^*V \\ V^*U & V^*V \end{pmatrix} = \begin{pmatrix} U^* & 0 \\ V^* & 0 \end{pmatrix}\begin{pmatrix} U & V \\ 0 & 0 \end{pmatrix} = \begin{pmatrix} U & V \\ 0 & 0 \end{pmatrix}^*\begin{pmatrix} U & V \\ 0 & 0 \end{pmatrix}$$
(3.1.3)

for any $d \in \mathbb{N}$, where $U := (u_1|u_2|\cdots|u_m) \in \mathbb{M}_{d,m}(\mathbb{F})$ and $V := (v_1|v_2|\cdots|v_n) \in \mathbb{M}_{d,n}(\mathbb{F})$. (due to (3.1.1)). Regarding the topic of our work, the block structure of the elements of $C(m+n; \mathbb{F})$ is of particular interest. To this end, put

$$\mathcal{Q}_{m,n}(\mathbb{F}) := \{S : S = \Gamma_H(u, v) \text{ for some } \mathbb{F}\text{-Hilbert space } H \text{ and } (u, v) \in S_H^m \times S_H^n\}.$$

Any Hilbert space $(H, \langle \cdot, \cdot \rangle_H)$ can be isometrically embedded into the Hilbert space

$$(\widetilde{H}, \langle \cdot, \cdot \rangle_{\widetilde{H}}) := (H \oplus \mathbb{F}_2^2, \langle \cdot, \cdot \rangle_H + \langle \cdot, \cdot \rangle_{\mathbb{F}_2^2})$$

(external direct sum of H and $\mathbb{F}_2^2$). Thus, for any $(h, k) \in B_H \times B_H$, it follows by construction that

$$\widetilde{h} := (h, 0, \sqrt{1 - \|h\|_H}) \in S_{\widetilde{H}},$$
$$\widetilde{k} := (k, \sqrt{1 - \|k\|_H}, 0) \in S_{\widetilde{H}} \text{ and } \langle \widetilde{k}, \widetilde{h} \rangle_{\widetilde{H}} = \langle k, l \rangle_H. \tag{3.1.4}$$

Note that $\dim(\widetilde{H}) \geq 3$. Consequently,

$$\mathcal{Q}_{m,n}(\mathbb{F}) = \{S : S = \Gamma_H(x, y) \text{ for some } \mathbb{F}\text{-Hilbert space } H$$
$$\text{and } (x, y) \in B_H^m \times B_H^n\}. \tag{3.1.5}$$

Observe also that for $C(\nu; \mathbb{F}) \subseteq \mathcal{Q}_{\nu,\nu}(\mathbb{F})$ for all $\nu \in \mathbb{N}$. By "inflating" the set $\mathcal{Q}_{m,n}(\mathbb{F})$ to a $(m+n) \times (m+n)$-correlation matrix, we obtain the non-trivial fact that $\mathcal{Q}_{m,n}(\mathbb{F})$ is absolutely convex:

Corollary 3.1 *Let $m, n \in \mathbb{N}$ and $\Sigma \in \mathbb{M}_{m+n}(\mathbb{F})$. Then the following statements are equivalent:*

(i) $\Sigma \in C(m+n; \mathbb{F})$.

(ii) $\Sigma = \begin{pmatrix} \Gamma_H(u, u) & \Gamma_H(u, v) \\ \Gamma_H(u, v)^* & \Gamma_H(v, v) \end{pmatrix}$, *for some Hilbert space H over $\mathbb{F}$ and some* $(u, v) \in S_H^m \times S_H^n$.

(iii) $\Sigma = \begin{pmatrix} \Gamma_H(u, u) & \Gamma_H(u, v) \\ \Gamma_H(u, v)^* & \Gamma_H(v, v) \end{pmatrix}$, *for some Hilbert space H over $\mathbb{F}$ and some* $(u, v) \in B_H^m \times B_H^n$.

(iv) *There exist $d \in \mathbb{N}$, $U \in \mathbb{M}_{d,m}(\mathbb{F})$ and $V \in \mathbb{M}_{d,n}(\mathbb{F})$, such that $u_i := Ue_i \in S_{\mathbb{F}_2^d}$ for all $i \in [m]$, $v_j := Ve_j \in S_{\mathbb{F}_2^d}$ for all $j \in [n]$ and*

$$\Sigma = \begin{pmatrix} U^*U & U^*V \\ V^*U & V^*V \end{pmatrix} = \begin{pmatrix} U^* & 0 \\ V^* & 0 \end{pmatrix} \begin{pmatrix} U & V \\ 0 & 0 \end{pmatrix}.$$

In particular, $S \in \mathcal{Q}_{m,n}(\mathbb{F})$ if and only if there exist correlation matrices $A \in C(m; \mathbb{F})$ and $B \in C(n; \mathbb{F})$, such that $\begin{pmatrix} A & S \\ S^* & B \end{pmatrix} \in C(m+n; \mathbb{F})$ is a correlation matrix. The set $\mathcal{Q}_{m,n}(\mathbb{F})$ is absolutely convex.

Proof (i) $\Rightarrow$ (ii): If (i) holds, then $\Sigma = \Gamma_H(w, w)$ for some $\mathbb{F}$-Hilbert space H and some $w = (w_1, \ldots, w_m, w_{m+1}, \ldots, w_{m+n}) \in S_H^{m+n}$ (due to Lemma 3.1). Thus, if we put $u_i := w_i$ for $i \in [m]$ and $v_j := w_{m+j}$ for $j \in [n]$, then (ii) follows.

(iii) $\Rightarrow$ (iv): Since $H = [u_1, \ldots u_m, v_1, \ldots, v_n] \oplus [u_1, \ldots u_m, v_1, \ldots, v_n]^\perp$, the orthogonal projection from H onto the finite-dimensional Hilbert space $[u_1, \ldots u_m, v_1, \ldots, v_n]$ does not alter any of the inner product entries of $\Gamma_H(u, v)$, so that we may assume without loss of generality that the Hilbert space $H \equiv \mathbb{F}_2^k$ is finite-dimensional (for some $k \in [m+n]$). (iv) now follows from (a potential application of) (3.1.4) and (3.1.3), where $d \in \{k, k+2\}$.

(iv) $\Rightarrow$ (i): (iv), respectively (3.1.3) implies that $\Sigma \in \mathbb{M}_{m+n}(\mathbb{F})^+$. Since $(U^*U)_{ii} = (e_i^\top U^*)(Ue_i) = \overline{(Ue_i)}^\top (Ue_i) = \|u_i\|^2 = 1$ for all $i \in [m]$ and $(V^*V)_{jj} = 1$ for all $j \in [n]$, it also follows that $\Sigma_{vv} = 1$ for all $v \in [m+n]$.

To verify the absolute convexity statement, we have to show that $\mathcal{Q}_{m,n}(\mathbb{F})$ is convex and satisfies $(\mathbb{F} \cap \overline{\mathbb{D}}) \cdot \mathcal{Q}_{m,n}(\mathbb{F}) = \mathcal{Q}_{m,n}(\mathbb{F})$ (cf., e.g., [80, Proposition 6.1.1]). So, let $S_1, S_2 \in \mathcal{Q}_{m,n}(\mathbb{F})$. Choose $A_1, A_2 \in C(m; \mathbb{F})$ and $B_1, B_2 \in C(n; \mathbb{F})$, such that

$$\begin{pmatrix} A_1 & S_1 \\ S_1^* & B_1 \end{pmatrix} \in C(m+n; \mathbb{F}) \text{ and } \begin{pmatrix} A_2 & S_2 \\ S_2^* & B_2 \end{pmatrix} \in C(m+n; \mathbb{F}).$$

Let $\lambda \in [0, 1]$. Since $C(v; \mathbb{F})$ is convex for all $v \in \mathbb{N}$, it follows that $\lambda A_1 + (1-\lambda)A_2 \in C(m; \mathbb{F})$, $\lambda B_1 + (1-\lambda)B_2 \in C(n; \mathbb{F})$ and

$$\begin{pmatrix} \lambda A_1 + (1-\lambda)A_2 & \lambda S_1 + (1-\lambda)S_2 \\ (\lambda S_1 + (1-\lambda)S_2)^* & \lambda B_1 + (1-\lambda)B_2 \end{pmatrix}$$

$$= \lambda \begin{pmatrix} A_1 & S_1 \\ S_1^* & B_1 \end{pmatrix} + (1-\lambda) \begin{pmatrix} A_2 & S_2 \\ S_2^* & B_2 \end{pmatrix} \in C(m+n; \mathbb{F}).$$

Consequently, $\lambda S_1 + (1-\lambda)S_2 \in \mathcal{Q}_{m,n}(\mathbb{F})$. $(\mathbb{F} \cap \overline{\mathbb{D}}) \cdot \mathcal{Q}_{m,n}(\mathbb{F}) = \mathcal{Q}_{m,n}(\mathbb{F})$ follows from (3.1.5). $\square$

Remark 3.2 (Tsirel'son's Characterisation of Quantum Correlation Matrices) In the real case, i.e., if $\mathbb{F} = \mathbb{R}$, $\mathcal{Q}_{m,n} \equiv \mathcal{Q}_{m,n}(\mathbb{R})$ coincides with the class of so-called *quantum correlation matrices*. These matrices are particularly essential in the foundations and philosophy of quantum mechanics (cf. [9, Chapter 11], [95, 96], [147, Theorem 1] and [148, Section 4]). To this end, recall that in quantum mechanics a *matrix* $\rho \in \mathbb{M}_{n,n}(\mathbb{C})$ is called a *(quantum) state* if $\rho \in \mathbb{M}_{n,n}(\mathbb{C})^+$ and $\mathrm{tr}(\rho) = 1$ (cf., e.g., [9, Chapter 0.10]). In fact, we have (cf. [9, Chapter 11]):

Theorem (Tsirel'son [147]) *Let $m, n \in \mathbb{N}$ and $S \equiv (s_{ij}) \in \mathbb{M}_{m,n}(\mathbb{R})$. Then the following statements are equivalent:*

(i) $S \in \mathcal{Q}_{m,n}(\mathbb{R})$.
(ii) *There exists a unital C^*-algebra $\mathfrak{A}$, self-adjoint elements $A_1, \ldots, A_m$, $B_1, \ldots, B_n$ and a state τ on $\mathfrak{A}$, such that $A_i B_j = B_j A_i$, $\max\{\|A_i\|, \|B_j\|\} \leq 1$ and $s_{ij} = \tau(A_i B_j)$ for all $(i, j) \in [m] \times [n]$.*

(iii) *There is a state $\rho \in \mathbb{M}_{d_1 \cdot d_2}(\mathbb{C})$ (for some $d_1, d_2 \in \mathbb{N}$), Hermitian matrix families $(W_1, \ldots, W_m) \in \mathbb{M}_{d_1}(\overline{\mathbb{D}})^m$ and $(Z_1, \ldots, Z_n) \in \mathbb{M}_{d_2}(\overline{\mathbb{D}})^n$, such that*

$$s_{ij} = tr((W_i \otimes Z_j)\rho) \text{ for all } (i,j) \in [m] \times [n].$$

Here, according to the construction, the Hermitian matrix $W_i \otimes Z_j \in \mathbb{M}_{d_1 \cdot d_2}(\overline{\mathbb{D}})$ is given by the Kronecker product of $W_i \in \mathbb{M}_{d_1}(\overline{\mathbb{D}})$ and $Z_j \in \mathbb{M}_{d_2}(\overline{\mathbb{D}})$.

In particular, if $k \in \mathbb{N}_3$, then *every* real standard $(k \times k)$-correlation matrix—used in everyday statistical calculations—actually contains a *quantum* correlation matrix block part $\in \mathcal{Q}_{m,k-m}$ ($m \in [k-1]$) and its transpose ! Although, the Grothendieck inequality actually "compares" the set $\mathcal{Q}_{m,n}(\mathbb{F})$ of all real (respectively complex !) quantum correlation matrices with their extreme counterparts of rank 1 (cf. Proposition 3.1, (3.2.1), (3.2.2), Theorem 3.1, Corollary 3.3 and Theorem 3.5), Tsirel'son's groundbreaking result, *per se*, won't be discussed in detail in this book, though. Regarding a detailed introduction to this fascinating subject including full and detailed proofs of Tsirel'son's results, we particularly refer to [9, Ch. 11.2] and [54, 95, 96], and the references therein.

3.2 The Grothendieck Inequality, Correlation Matrices and the Matrix Norm $\|\cdot\|_{\infty,1}^{\mathbb{F}}$

Fix $\mathbb{F} \in \{\mathbb{R}, \mathbb{C}\}$ and $m, n \in \mathbb{N}$. Let $A \in \mathbb{M}_{m,n}(\mathbb{F})$ and H be an arbitrary $\mathbb{F}$-Hilbert space. If $(x, y) \in H^m \times H^n$, then

$$\overline{\sum_{i=1}^{m}\sum_{j=1}^{n} a_{ij}\langle x_i, y_j\rangle_H} = \sum_{i=1}^{m}\sum_{j=1}^{n} \overline{a_{ij}} \langle y_j, x_i\rangle_H = tr(A^* \Gamma_H(x,y)) = \overline{tr(A\, \Gamma_H(y,x))}.$$

Moreover, observe that (cf. [49, Lemma 2.2.], [79, Remark 10.1] and Remark 3.3 regarding the existence of the respective maxima)

$$\|A\|_H^G := \max_{\|u_i\|=1, \|v_j\|=1} \Big|\sum_{i=1}^{m}\sum_{j=1}^{n} a_{ij}\langle u_i, v_j\rangle_H\Big| = \max_{(u,v)\in S_H^m \times S_H^n} |tr(A^*\Gamma_H(u,v))|$$

$$= \max_{\|u_i\|\leq 1, \|v_j\|\leq 1} \Big|\sum_{i=1}^{m}\sum_{j=1}^{n} a_{ij}\langle u_i, v_j\rangle_H\Big|. \tag{3.2.1}$$

In particular (if $H = \mathbb{F}$, where $\langle z, w \rangle_H = \overline{w}z$ for all $z, w \in \mathbb{F}$), we have:

$$\|A\|_{\mathbb{F}}^G = \max_{(p,q) \in S_{\mathbb{F}}^m \times S_{\mathbb{F}}^n} |tr(A^* \Gamma_{\mathbb{F}}(p, q))| = \max_{|p_i| \leq 1, |q_j| \leq 1} \left| \sum_{i=1}^m \sum_{j=1}^n a_{ij} p_i \overline{q_j} \right|, \quad (3.2.2)$$

Consequently, if the matrix $A \in \mathbb{M}_{m,n}(\mathbb{F})$ is viewed as a bounded linear operator from l_∞^n into l_1^m, then (3.2.2) and the fact that l_∞^m is isometrically isomorphic to the dual space of l_1^m (via the linear map $\chi : l_\infty^m \longrightarrow (l_1^m)'$, defined as $l_\infty^m \ni z \mapsto \langle z, \chi(q) \rangle := q^\top z = \sum_{i=1}^m q_i z_i$), implies that

$$\begin{aligned}
\|A\|_{\mathbb{F}}^G &= \max_{(p,q) \in S_{\mathbb{F}}^m \times S_{\mathbb{F}}^n} |tr(A^* \Gamma_{\mathbb{F}}(p,q))| \stackrel{(3.2.2)}{=} \max_{(p,q) \in B_\infty^m \times B_\infty^n} |tr(A^* \Gamma_{\mathbb{F}}(p,q))| \\
&= \max_{(p,q) \in B_\infty^m \times B_\infty^n} |tr(A \overline{q} p^\top)| = \max_{(p,q) \in B_\infty^m \times B_\infty^n} |tr(A q p^\top)| \\
&= \max_{(p,q) \in B_\infty^m \times B_\infty^n} |\langle Aq, \chi(p) \rangle| \\
&= \|A\|_{\mathcal{L}(l_\infty^n, l_1^m)} =: \|A\|_{\infty, 1}^{\mathbb{F}},
\end{aligned} \quad (3.2.3)$$

where $B_\infty^\nu := B_{l_\infty^\nu}$, $\nu \in \mathbb{N}$. Consequently, Theorem 1.1 is equivalent to

Theorem 3.1 *Let $\mathbb{F} \in \{\mathbb{R}, \mathbb{C}\}$. There is an absolute constant $K > 0$ such that for any $m, n \in \mathbb{N}$, for any $A \in \mathbb{M}_{m,n}(\mathbb{F})$ and any $\mathbb{F}$-Hilbert space H, the following inequality is satisfied:*

$$\max_{(u,v) \in S_H^m \times S_H^n} |tr(A^* \Gamma_H(u, v))| = \|A\|_H^G \leq K \|A\|_{\infty, 1}^{\mathbb{F}}.$$

$K_G^{\mathbb{F}} > 1$ is the smallest possible value of the corresponding absolute constant K.

The operator norm $\|\cdot\|_{\infty,1}^{\mathbb{F}}$ on the right side of the Grothendieck inequality is a particular example of a *mixed subordinate matrix norm* (cf., e.g., [21, A.1.5] and [129, 143]). Here, we have to recall two key results, particularly regarding the computational complexity of $\|A\|_{\infty,1}^{\mathbb{R}}$ (cf. [67, 127, 129, 143]):

Theorem 3.2 (Rohn [129]) *Computing $\|A\|_{\infty,1}^{\mathbb{R}}$ is NP-hard in the class of Maximum Cut Matrices.*

Even an approximation of $\|A\|_{\infty,1}^{\mathbb{R}}$ is NP-hard ([129, Theorem 6]):

Theorem 3.3 (Hendrickx and Olshevsky [67]) *Unless $P = NP$, there is no polynomial time algorithm which, given a real matrix A with entries in $\{-1, 0, 1\}$, approximates $\|A\|_{\infty,1}^{\mathbb{R}}$ to some fixed error with polynomial running time in the dimensions of the matrix.*

These observations immediately result in another important well-known fact which will be used later in this book to show that *for both fields* the calculation of $K_G^{\mathbb{F}}$ can also be elaborated by means of semidefinite programming, which is a *convex optimisation problem* (cf. Corollary 3.3, Proposition 3.2 and [21, Chapter 4.6.2]). Namely,

Lemma 3.2 *Let* $\mathbb{F} \in \{\mathbb{R}, \mathbb{C}\}$, $m, n \in \mathbb{N}$ *and* $A \in \mathbb{M}_{m,n}(\mathbb{F})$. *Let* $HIL^{\mathbb{F}}$ *denote the class of all $\mathbb{F}$-Hilbert spaces. Put*

$$\Delta(A) \equiv \Delta^{\mathbb{F}}(A) := \tfrac{1}{2}\begin{pmatrix} 0 & A \\ A^* & 0 \end{pmatrix}. \tag{3.2.4}$$

Then $\Delta(A)$ *is Hermitian and*

$$vec(x, y)^* \Delta(A) vec(x, y) = \operatorname{Re}(x^* A y) \text{ for all } (x, y) \in \mathbb{F}^m \times \mathbb{F}^n. \tag{3.2.5}$$

In particular, $\Delta(A) \in \mathbb{M}_{m+n}(\mathbb{F})^+$ *if and only if* $A = 0$. *Moreover,*

$$tr\left(\Delta(A)\begin{pmatrix} C & S \\ R^* & D \end{pmatrix}\right) = \tfrac{1}{2}(\overline{tr(A^*R)} + tr(A^*S)) \tag{3.2.6}$$

for all $(C, D) \in \mathbb{M}_m(\mathbb{F}) \times \mathbb{M}_n(\mathbb{F})$ *and* $S, R \in \mathbb{M}_{m,n}(\mathbb{F})$. $0 \leq \max_{\Sigma \in C(m+n;\mathbb{F})} tr(\Delta(A)\Sigma) < \infty$, *and*

$$\sup_{H \in HIL^{\mathbb{F}}} \|A\|_H^G = \max_{S \in \mathcal{Q}_{m,n}(\mathbb{F})} |tr(A^*S)| = \max_{S \in \mathcal{Q}_{m,n}(\mathbb{F})} \operatorname{Re}(tr(A^*S))$$

$$= \max_{\Sigma \in C(m+n;\mathbb{F})} tr(\Delta(A)\Sigma) \leq K_G^{\mathbb{F}}(m,n)\|A\|_{\infty,1} \tag{3.2.7}$$

Proof Equation (3.2.5) and its impact on the positive semidefiniteness of $\Delta(A)$ follows immediately. Since

$$\Delta(A)\begin{pmatrix} C & S \\ R^* & D \end{pmatrix} = \tfrac{1}{2}\begin{pmatrix} AR^* & AD \\ A^*C & A^*S \end{pmatrix},$$

it follows that

$$tr\left(\Delta(A)\begin{pmatrix} C & S \\ R^* & D \end{pmatrix}\right) = \tfrac{1}{2}(tr(AR^*) + tr(A^*S)) = \tfrac{1}{2}(\overline{tr(A^*R)} + tr(A^*S)).$$

Observe, that (3.2.6) does not at all depend on the choice of the matrices C and D! Regarding the proof of (3.2.7), we firstly verify the second equality. To this end, fix $S \in \mathcal{Q}_{m,n}(\mathbb{F})$. Since any $z \in \mathbb{F}$ satisfies $|z| = \operatorname{Re}(\zeta z)$, where $\zeta := 1$ if $z = 0$ and $\zeta := \frac{\bar{z}}{|z|}$ if $z \neq 0$, it follows in particular that $|tr(A^*S)| = \operatorname{Re}(\alpha\, tr(A^*S)) = \operatorname{Re}(tr(A^*\alpha S)) \geq 0$ for some $\alpha \in S_{\mathbb{F}}$. Since $\alpha S \in \mathcal{Q}_{m,n}(\mathbb{F})$, the completion of the proof of (3.2.7) now follows by applying (3.2.6) twice, including the fact that $|\text{tr}(\Delta(A)\Sigma)| \leq \max\limits_{V \in \mathcal{Q}_{m+n,m+n}(\mathbb{F})} |\text{tr}(\Delta(A)V)|$ for any $\Sigma \in C(m+n;\mathbb{F}) \subseteq \mathcal{Q}_{m+n,m+n}(\mathbb{F})$. □

Let $A \in \mathbb{M}_n(\mathbb{F})$. Put $d_1(A) \equiv d_1^{\mathbb{F}}(A) := \sup\limits_{\Theta \in C_1(n;\mathbb{F})} |\text{tr}(A^*\Theta)|$. Because of Proposition 3.1 it follows that

$$d_1(A) = \max_{x \in S_{\mathbb{F}}^n} |x^*Ax| = \max_{x \in S_{\mathbb{F}}^n} |\text{tr}(A^*xx^*)| \leq \|A\|_{\infty,1}^{\mathbb{F}}.$$

Observe that $d_1 \neq \|\cdot\|_{\infty,1}^{\mathbb{F}}$ (due to [48, Corollary 2.11]). If B is symmetric, respectively Hermitian, then $d_1(B)$ coincides with the seminorm $\|B\|_{\gamma,1}$ of S. Friedland and L.-H. Lim (cf. [48, Proposition 2.5]). Moreover, in the positive semidefinite case [48, Proposition 2.8] implies that

$$d_1(M) = \max_{x \in S_{\mathbb{F}}^n} |x^*Mx| = \max_{x \in [-1,1]^n} |x^*Mx| \text{ for all } M \in \mathbb{M}_n(\mathbb{F})^+. \quad (3.2.8)$$

Recall from Lemma 3.2 the Hermitian matrix $\Delta(A) \equiv \Delta^{\mathbb{F}}(A) := \frac{1}{2}\begin{pmatrix} 0 & A \\ A^* & 0 \end{pmatrix}$, where $m, n \in \mathbb{N}$ and $A \in M(m \times n;\mathbb{F})$. Observe that in the following inequalities, which are an immediate application of [48, Corollary 2.6., (29) and Proposition 2.8., (31)], seemingly no Hilbert space presence is required (cf. also (3.2.7) and Corollary 6.10).

Proposition 3.2 *Let* $\mathbb{F} \in \{\mathbb{R}, \mathbb{C}\}$ *and* $m, n \in \mathbb{N}$. *Then* $K_G^{\mathbb{F}}(m, n)$ *is the smallest constant* $c > 0$, *satisfying*

$$|tr(\Delta(A)\,\Sigma)| \leq c\,\|A\|_{\infty,1} \text{ for all } \Sigma \in C(m+n;\mathbb{F}) \text{ and } A \in \mathbb{M}_{m,n}(\mathbb{F}). \quad (3.2.9)$$

The little Grothendieck constant $k_G^{\mathbb{F}}$ *is the smallest constant* $\gamma > 0$, *such that*

$$|tr(B\,\Sigma)| \leq \gamma\,\|B\|_{\infty,1} \text{ for all } \Sigma \in C(n;\mathbb{F}) \text{ and } B \in \mathbb{M}_n(\mathbb{F})^+. \quad (3.2.10)$$

In fact, if we allow the implementation of a possibly strictly larger absolute constant than $K_G^{\mathbb{F}}$, our approach leads to a further, more general inequality, which encompasses the real and the complex Grothendieck inequality as a special case (cf. Theorem 6.8 (real case), respectively Theorem 7.6 (complex case)). Moreover, it extends the symmetric Grothendieck equality of Friedland and Lim in [48] from symmetric $\mathbb{F}$-matrices to arbitrary $\mathbb{F}$-matrices (cf. (3.3.13) and Remark 3.8):

Theorem 3.4 *Let* $\mathbb{F} \in \{\mathbb{R}, \mathbb{C}\}$. *Then there exists an absolute constant* $K_*^{\mathbb{F}} > 1$ *such that*

$$|tr(B^*\,\Sigma)| \leq K_*^{\mathbb{F}} d_1(B).$$

for any $k \in \mathbb{N}$, any $\Sigma \in C(k; \mathbb{F})$ and any $B \in \mathbb{M}_k(\mathbb{F})$. Moreover,

$$C(k; \mathbb{F}) \subseteq K_*^{\mathbb{F}} acx(C_1(k; \mathbb{F})) \text{ for all } k \in \mathbb{N},$$

$K_*^{\mathbb{R}} \in [K_G^{\mathbb{R}}, \sinh(\frac{\pi}{2})]$ and $K_*^{\mathbb{C}} \in [K_G^{\mathbb{C}}, \frac{8}{\pi} - 1]$.

3.3 Characterisation of $K_G^{\mathbb{F}}$ Through Operator Ideals and Violation of Bell Inequalities (A Brief Digression)

Readers who are familiar with operator ideals in the sense of A. Pietsch (cf. [33, 37, 123]) should take notice of Remark 3.4 below regarding the Grothendieck norm (3.2.1) on the left side of the Grothendieck inequality. To round out the picture, we list a rather elementary result (Proposition 3.3), which however unveals the link between matrices in $\mathcal{Q}_{m,n}(\mathbb{F})$ and a well-known functional analytic formulation of the Grothendieck inequality; namely as an inequality between Banach ideal matrix norms, when matrices are viewed as bounded linear operators from l_∞^n into l_1^m (respectively from l_1^n into l_∞^m), being elements of certain 1-Banach ideals (in the sense of A. Pietsch), or equivalently as an inequality between certain tensor product norms on tensor products of Banach spaces. The latter approach was developed by Grothendieck (cf. Remark 3.5). Regarding the underlying functional analytic details, we refer the readers to [33, 123]. Primarily, we need an intuitive understanding of bounded linear operators between (finite-dimensional) Banach spaces, *factoring* through a Hilbert space and the basics of nuclear operators (cf. [123, Chapter 6.3]).

So, let E and F be $\mathbb{F}$-Banach spaces and $S \in \mathfrak{L}(E, F)$. By definition, $S \in \mathfrak{L}_2(E, F)$ if and only if there exist an $\mathbb{F}$-Hilbert space H and bounded linear operators $R \in \mathfrak{L}(H, F)$, $T \in \mathfrak{L}(E, H)$, such that $S = RT$. $S \in \mathfrak{L}_2(E, F)$ is said to be *2-factorable* (cf., e.g., [33, Corollary 18.6.2]).

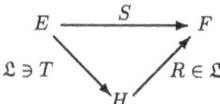

It is well-known that $(\mathfrak{L}_2, \|\cdot\|_{\mathfrak{L}_2})$ is a 1-Banach ideal. The norm is defined as

$$\|S : E \longrightarrow F\|_{\mathfrak{L}_2} \equiv \|S\|_{\mathfrak{L}_2(E,F)} := \inf \|R\| \|T\|,$$

where the infimum is taken over all factorisations $S = RT$ through any Hilbert space.

3.3 Characterisation of $K_G^{\mathbb{F}}$ Through Operator Ideals and Violation of Bell... 51

The unit ball of the Banach space $\mathfrak{L}_2(l_1^n, l_\infty^m)$ completely characterises the convex set $\mathcal{Q}_{m,n}(\mathbb{F})$, since:

Proposition 3.3 *Let $m, n \in \mathbb{N}$ and $S \in \mathbb{M}_{m,n}(\mathbb{F})$. Then the following statements are equivalent:*

(i) $S \in B_{\mathfrak{L}_2(l_1^n, l_\infty^m)}$.
(ii) *There exist $d \in \mathbb{N}$ and $(u, v) \in (S_{\mathbb{F}_2^d})^m \times (S_{\mathbb{F}_2^d})^n$ such that $S = \Gamma_{\mathbb{F}_2^d}(u, v)$.*
(iii) *There exist an $\mathbb{F}$-Hilbert space H and $(u, v) \in B_H^m \times B_H^n$ such that $S = \Gamma_H(u, v)$.*
(iv) *There exist $d \in \mathbb{N}$, $U \in \mathbb{M}_{d,m}(\mathbb{F})$ and $V \in \mathbb{M}_{d,n}(\mathbb{F})$ such that $\|U : l_1^m \longrightarrow l_2^d\| = 1$, $\|V : l_1^n \longrightarrow l_2^d\| = 1$ and $S = U^*V$.*

Proof (i) $\Rightarrow$ (ii): Let

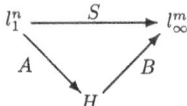

for some Hilbert space H and $\|B\| \|A\| = \|S\|_{\mathfrak{L}_2} \leq 1$ (cf. [33, Corollary 18.6.2]). Obviously, we may assume without loss of generality that $H = l_2^k$, for some $k \in \mathbb{N}$, $\|B\| \leq 1$ and $\|A\| \leq 1$. As usual, we identify H' and H (by the Riesz representation theorem). Let $\Lambda \in \mathfrak{L}(l_1^m, (l_\infty^m)')$ be the well-known linear mapping (which even is an isometric isomorphism, since $\dim(l_\infty^m) = m < \infty$), whose duality bracket is given by

$$\langle z, \Lambda x \rangle := x^\top z = \sum_{i=1}^m z_i x_i \qquad ((z, x) \in l_\infty^m \times l_1^m).$$

Obviously, $\|\Lambda\| \leq 1$. Let $B' \in \mathfrak{L}((l_\infty^m)', (l_2^k)')$ the dual operator of B. Then $\|B'\| = \|B\| \leq 1$. By construction, it follows that the matrix of S (with respect to the standard bases) is given entrywise as

$$S_{ij} = (BA)_{ij} = e_i^\top BAe_j = \langle BAe_j, \Lambda e_i \rangle$$
$$= \langle Ae_j, (B'\Lambda)e_i \rangle = \langle y_j, x_i \rangle_H \text{ for all } (i, j) \in [m] \times [n],$$

where $y_j := Ae_j \in B_H$ and $x_i := (B'\Lambda)e_i \in B_{H'} \equiv B_H$. Consequently, $S = \Gamma_{\mathbb{F}_2^d}(u, v)$ for some $(u, v) \in (S_{\mathbb{F}_2^d})^m \times (S_{\mathbb{F}_2^d})^n$, where $d \in \{k, k+2\}$ (due to (3.1.4)).
(iii) $\Rightarrow$ (iv): Since $H = [u_1, \ldots u_m, v_1, \ldots, v_n] \oplus [u_1, \ldots u_m, v_1, \ldots, v_n]^\perp$, the orthogonal projection from H onto the finite-dimensional Hilbert space $[u_1, \ldots u_m, v_1, \ldots, v_n]$ does not alter any of the inner product entries of $\Gamma_H(u, v)$, so that we may assume without loss of generality that the Hilbert space $H \equiv l_2^d$ is finite-dimensional. Due to (3.1.4), we can assume (after a possible transition from l_2^d to l_2^{d+2}) that $(u, v) \in S_H^m \times S_H^n$. Hence, we may apply the factorisation (3.1.1),

implying that $S = \Gamma_H(u, v) = U^*V$, where $U := (u_1 | u_2 | \cdots | u_m) \in \mathbb{M}_{d,m}(\mathbb{F})$ and $V := (v_1 | v_2 | \cdots | v_n) \in \mathbb{M}_{d,n}(\mathbb{F})$. If we view U as $U \in \mathfrak{L}(l_1^m, l_2^d)$ and V as $V \in \mathfrak{L}(l_1^n, l_2^d)$, then a straightforward calculation shows that for any $y = (y_1, \ldots, y_n)^\top = \sum_{j=1}^n y_j e_j \in l_1^n$,

$$\|Vy\|_2 = \Big\|\sum_{j=1}^n y_j(Ve_j)\Big\|_2 = \Big\|\sum_{j=1}^n y_j v_j\Big\|_2 \leq \|y\|_{l_1^n} (\max_{j \in [n]} \|v_j\|_2) = \|y\|_{l_1^n},$$

whence $\|V\| \leq 1 = \|v_1\|_2 = \|Ve_1\|_2 \leq \|V\|$. Thus, $\|V\| = 1$. Similarly, we obtain that $\|U\| = 1$, and (iv) follows.

(iv) $\Rightarrow$ (i): Again, we identify l_2^d with its dual. It is also well-known that $\Psi : (l_1^m)' \xrightarrow{\cong} l_\infty^m$ is an isometric isomorphism, defined as

$$\Psi a := \sum_{i=1}^m \langle e_i, a \rangle e_i \quad (a \in (l_1^m)').$$

Put $W := \Psi \overline{U}' : l_2^d \equiv (l_2^d)' \longrightarrow l_\infty^m$. ($l_2^d \ni \overline{U}x := \overline{Ux}$ for all $x \in l_1^m$, of course.) The mapping rule of Ψ implies that for any $j \in [n]$,

$$(WV)e_j = \sum_{i=1}^m \langle e_i, \overline{U}' V e_j \rangle e_i = \sum_{i=1}^m \langle \overline{U}e_i, Ve_j \rangle_{l_2^d} e_i$$

$$= \sum_{i=1}^m \langle e_i, (U^*V)e_j \rangle_{l_2^d} e_i = \sum_{i=1}^m (e_i^\top (U^*V)e_j)e_i = Se_j.$$

Consequently, $S = WV \in \mathfrak{L}_2(l_1^n, l_\infty^m)$, and $\|S\|_{\mathfrak{L}_2} \leq \|W\|\|V\| \leq \|\overline{U}'\|\|V\| = \|U\|\|V\| \leq 1$. □

In other words, if $m, n \in \mathbb{N}$ and $\mathbb{F} \in \{\mathbb{R}, \mathbb{C}\}$, then

$$B_{\mathfrak{L}_2(l_1^n, l_\infty^m)} = \mathcal{Q}_{m,n}(\mathbb{F}) = \bigcup_{d=1}^\infty \{\Gamma_{\mathbb{F}_2^d}(u, v) : (u, v) \in (S_{\mathbb{F}_2^d})^m \times (S_{\mathbb{F}_2^d})^n\}. \quad (3.3.1)$$

Corollary 3.2 Let $d, m, n \in \mathbb{N}$, $(z, w) \in (\mathbb{C}_2^d)^m \times (\mathbb{C}_2^d)^n$ and $(a, b) \in (\mathbb{R}_2^{2d})^m \times (\mathbb{R}_2^{2d})^n$. Then

$$\Gamma_{\mathbb{C}_2^d}(z, w) = \Gamma_{\mathbb{R}_2^{2d}}(x, y) + i\,\Gamma_{\mathbb{R}_2^{2d}}(x, y'),$$

and

$$\Gamma_{\mathbb{R}_2^{2d}}(a,b) = \text{Re}(\Gamma_{\mathbb{C}_2^d}(\zeta,\xi)),$$

where $(\zeta,\xi) \in (\mathbb{C}_2^d)^m \times (\mathbb{C}_2^d)^n$, $(x,y) \in (\mathbb{R}_2^{2d})^m \times (\mathbb{R}_2^{2d})^n$ and $y' \in (\mathbb{R}_2^{2d})^n$ are given as $x_i := J_2(z_i)$, $y_j := J_2(w_j)$, $y'_j := J_2(-i\,w_j) = R_2(-i\,Id_d)y_j$, $\zeta_i := J_2^{-1}(a)$ and $\xi_j := J_2^{-1}(b)$ $((i,j) \in [m] \times [n])$. *In particular,* $\{\text{Re}(S) : S \in \mathcal{Q}_{m,n}(\mathbb{C})\} \subseteq \mathcal{Q}_{m,n}(\mathbb{R})$, $\{\text{Im}(S) : S \in \mathcal{Q}_{m,n}(\mathbb{C})\} \subseteq \mathcal{Q}_{m,n}(\mathbb{R})$ *and* $\mathcal{Q}_{m,n}(\mathbb{C}) \subseteq \mathcal{Q}_{m,n}(\mathbb{R}) + i\,\mathcal{Q}_{m,n}(\mathbb{R})$. *Moreover,* $\{\text{Re}(\Sigma) : \Sigma \in C(n;\mathbb{C})\} \subseteq C(n;\mathbb{R})$.

Proof We just have to apply (3.3.1), (1.2.8) and (1.2.10), taking into account the algebraic properties of the mappings J_2 and R_2. □

Let $(u,v) \in S_H^m \times S_H^n$. Since $\Gamma_H(u,v)^* = \Gamma_H(v,u)$, it follows that also $\Gamma_H(u,v)^* \in \mathcal{Q}_{n,m}(\mathbb{F})$. Consequently, if we recall Theorem 3.1 and Lemma 3.2, we arrive at the following crucial implication of Proposition 3.3 (cf. [54]):

Corollary 3.3 *Let* $\mathbb{F} \in \{\mathbb{R},\mathbb{C}\}$ *and* $m,n \in \mathbb{N}$. *Let* $A \in \mathbb{M}_{m,n}(\mathbb{F})$ *such that* $\|A\|_{\infty,1}^{\mathbb{F}} \leq 1$. *Then the following statements are equivalent to each other and to* (1.1.3):

(i)

$$|tr(A^* \Gamma_H(u,v))| \leq K_G^{\mathbb{F}}(m,n) \text{ for all Hilbert spaces } H \text{ and for all}$$
$$(u,v) \in S_H^m \times S_H^n.$$

(ii)

$$\max_{S \in \mathcal{Q}_{m,n}(\mathbb{F})} |tr(A^* S)| = \max_{R \in B_{\mathfrak{L}_2(l_1^m, l_\infty^n)}} |tr(A\,R)| \leq K_G^{\mathbb{F}}(m,n).$$

(iii)

$$\max_{\Sigma \in C(m+n;\mathbb{F})} tr(\Delta(A)\Sigma) \leq K_G^{\mathbb{F}}(m,n).$$

Recall that for any $\nu \in \mathbb{N}$, the canonical isometric isomorphisms $(l_1^\nu)' \cong l_\infty^\nu$, and $(l_\infty^\nu)' \cong l_1^\nu$ (since l_∞^ν is finite-dimensional) explicitly characterise the respective dual spaces. Put

$$X := \mathfrak{L}_2(l_1^n, l_\infty^m),\ Y := \mathfrak{L}(l_\infty^m, l_1^n),\ Z := \mathfrak{N}(l_1^n, l_\infty^m) \text{ and } W := \mathfrak{L}(l_\infty^n, l_1^m), \tag{3.3.2}$$

where $(\mathfrak{N}, \|\cdot\|_{\mathfrak{N}})$ denotes the 1-Banach ideal of all nuclear operators, which is the smallest 1-Banach ideal (originally created by A. Grothendieck in his famous thesis

[56]). Let us quickly recall that a linear operator $T : E \longrightarrow F$ between two $\mathbb{F}$-Banach spaces E and F is said to be *nuclear* if there exist sequences $(a_n)_{n \in \mathbb{N}} \subseteq B_{E'}$, $(y_n)_{n \in \mathbb{N}} \subseteq B_F$ and $(\lambda_n)_{n \in \mathbb{N}} \in l_1$ such that

$$T = \sum_{n=1}^{\infty} \lambda_n \langle \cdot, a_n \rangle y_n,$$

and the series converges in $\mathfrak{L}(E, F)$ (cf. Remark 3.6, [56] and [123, Chapter 6.3 and Theorem 9.2.1.]). It is a well-known fact that the Banach space Z is isometrically isomorphic to the dual Banach space Y'. The isometric isomorphism $\tau : Z \longrightarrow Y'$ is given by canonical trace duality

$$Z \ni S \mapsto \tau_S \equiv \tau(S), \text{ where } \langle B, \tau_S \rangle := \mathrm{tr}(BS) \text{ for all } B \in Y \qquad (3.3.3)$$

(see [123, Theorem 9.2.1]). Readers, who are familiar with tensor norms and Banach ideals could verify the above trace duality very quickly, (since $Y' \cong (l_1^m \otimes_\varepsilon l_1^n)' \cong (l_1^n \otimes_\varepsilon l_1^m)' \cong \mathfrak{I}(l_1^n, l_\infty^m) = Z$ (cf. [33, Corollary 5.7.1 and Proposition 16.7])). Observe that Proposition 3.3 implies that $B_Z \subseteq B_X = \mathcal{Q}_{m,n}(\mathbb{F})$. However, since X consists of elementary operators only (i.e., linear operators between finite-dimensional $\mathbb{F}$-vector spaces), it follows that we may identify the equivalently normed finite-dimensional Banach spaces $(Z, \|\cdot\|_Z)$ and $(X, \|\cdot\|_X)$ topologically (yet not isometrically !), implying that $\mathcal{Q}_{m,n}(\mathbb{F}) = B_X \subseteq X \subseteq Z$. Fix $B \in Y$. Then $A := B^* \in W$, and $\|B\|_Y = \|A^*\|_Y = \|A\|_W = \|A\|_{\infty,1}$. (3.2.7) implies that

$$|\mathrm{tr}(BS)| = |\mathrm{tr}(A^*S)| \leq K_G^\mathbb{F}(m, n) \|A\|_{\infty,1}$$

$$= K_G^\mathbb{F}(m, n) \|B\|_Y \text{ for all } (B, S) \in Y \times \mathcal{Q}_{m,n}(\mathbb{F}).$$

By making use of polarisation with respect to the dual pairing (Y, Z) (cf., e.g., [80, Chapter 8.2]), the latter inequality is equivalent to

$$\mathcal{Q}_{m,n}(\mathbb{F}) \subseteq K_G^\mathbb{F}(m, n) B_Y^\circ = K_G^\mathbb{F}(m, n) B_Z.$$

So, we get again a well-known norm inequality variant of the Grothendieck inequality (cf. also [33, Corollary 14.3] and [54, Section 1.2]); namely:

$$B_{\mathfrak{L}_2(l_1^n, l_\infty^m)} \stackrel{(3.3.1)}{=} \mathcal{Q}_{m,n}(\mathbb{F}) \subseteq K_G^\mathbb{F}(m, n) B_{\mathfrak{N}(l_1^n, l_\infty^m)} \subseteq K_G^\mathbb{F} B_{\mathfrak{N}(l_1^n, l_\infty^m)} \qquad (3.3.4)$$

and $\|S\|_{\mathfrak{N}(l_1^n, l_\infty^m)} \leq K_G^\mathbb{F}(m, n) < K_G^\mathbb{F}$ for all $m, n \in \mathbb{N}$ and $S \in \mathcal{Q}_{m,n}(\mathbb{F})$.

3.3 Characterisation of $K_G^{\mathbb{F}}$ Through Operator Ideals and Violation of Bell...

Thus, we recognise that the isometric isomorphism $\tau : Z \longrightarrow Y'$ can be *extended* to the well-defined bounded linear operator $\tilde{\tau} : X \longrightarrow Y'$, where the latter is given as $\tilde{\tau}(R) := K_G^{\mathbb{F}}(m, n) \, \tau\left(\frac{1}{K_G^{\mathbb{F}}(m,n)} R\right)$ for all $R \in X$; i.e.,

$$Y \times X \ni (B, R) \mapsto \langle B, \tilde{\tau}(R) \rangle := K_G^{\mathbb{F}}(m, n) \left\langle B, \tau\left(\frac{1}{K_G^{\mathbb{F}}(m, n)} R\right)\right\rangle = \operatorname{tr}(BR). \tag{3.3.5}$$

Note that $\|\tilde{\tau}(R)\| = K_G^{\mathbb{F}}(m, n) \, \|\frac{1}{K_G^{\mathbb{F}}(m,n)} R\|_Z \leq K_G^{\mathbb{F}}(m, n) \, \|R\|_X \leq K_G^{\mathbb{F}} \, \|R\|_X$ for all $R \in X$ (due to (3.3.4)), whence $\|\tilde{\tau}\| \leq K_G^{\mathbb{F}}(m, n) \leq K_G^{\mathbb{F}}$.

Consequently, since $K_G^{\mathbb{F}}(m, n)$ is the smallest constant which satisfies inequality (1.1.3) (or equivalently (3.2.7)), it even follows that

$$K_G^{\mathbb{F}}(m, n) = \sup_{S \in \mathcal{Q}_{m,n}(\mathbb{F})} \|S\|_Z \tag{3.3.6}$$

(since $|\operatorname{tr}(A^*S)| \stackrel{(3.3.5)}{=} |\langle A^*, \tilde{\tau}(S) \rangle| \leq K_G^{\mathbb{F}}(m, n) \, \|\tau\left(\frac{1}{K_G^{\mathbb{F}}(m,n)} S\right)\| \, \|A^*\|_Y \stackrel{(3.3.3)}{\leq} K_G^{\mathbb{F}}(m, n) \, \|\frac{1}{K_G^{\mathbb{F}}(m,n)} S\|_Z$ for all $S \in \mathcal{Q}_{m,n}(\mathbb{F})$ and $A \in B_W$).

Remark 3.3 (Attainability of Maximum in GT) For any $A \in W$ the linear functional $f_A : X \longrightarrow \mathbb{R}$, $R \mapsto \langle A^*, \tilde{\tau}(R) \rangle = \operatorname{tr}(A^*R)$ satisfies $|f_A(R)| \leq \|\tilde{\tau}(R)\| \|A\|_W \leq K_G^{\mathbb{F}} \|R\|_X \|A\|_{\infty,1}$ for all $R \in X$. Hence, $f_A : X \longrightarrow \mathbb{R}$ is continuous and attains it maximum on the *compact* unit ball $B_X = \mathcal{Q}_{m,n}(\mathbb{F})$ (since X is finite-dimensional). Thus, we may indeed replace the supremum by the maximum in Theorem 3.1. Similarly, the maximum is attained in (3.3.6), whence

$$K_G^{\mathbb{F}}(m, n) = \|S_0\|_Z > 1$$

for some $S_0 \in B_X \setminus B_Z = \mathcal{Q}_{m,n}(\mathbb{F})$. Recall that $\tau : Z \xrightarrow{\cong} Y'$ is an isometric isomorphism (cf. (3.3.3)) and observe that

$$\langle S, (\tau' j_Y)(B) \rangle = \langle B, \tau(S) \rangle = \operatorname{tr}(BS) \text{ for all } (S, B) \in Z \times Y.$$

Thus, $\tau' j_Y : Y \xrightarrow{\cong} Z'$ again is an isometric isomorphism (since Y is finite-dimensional). It therefore follows the existence of some $B_0 \in S_Y$, such that

$$K_G^{\mathbb{F}}(m, n) = \|S_0\|_Z = |\operatorname{tr}(A_0^* S_0)|,$$

where $A_0 := B_0^* \in S_W$. Moreover, the polar of B_Z satisfies

$$B_Z^\circ = B_{Z'} = \tau' j_Y(B_Y). \tag{3.3.7}$$

Consequently, for *any* matrix $A \in \mathbb{M}_{m,n}(\mathbb{F})$, the following equivalence holds: $|\text{tr}(A^*R)| \leq 1$ for all $R \in B_Z$ *if and only if* $\|A\|_W = \|A\|_{\infty,1} \leq 1$. In summary, we have:

$$K_G^\mathbb{F}(m,n) = |\text{tr}(A_0^* S_0)| > 1 \text{ and } |\text{tr}(A_0^* R)| \leq 1 \text{ for all } R \in B_Z; \qquad (3.3.8)$$

a fact which plays a key role in quantum mechanics (cf. Remark 3.7). Moreover, due to (3.3.1), $S_0 = \Gamma_{\mathbb{F}_2^{d_0}}(u_0, v_0)$, for some $d_0 \equiv d_0(m,n) \in \mathbb{N}$ and $(u_0, v_0) \in (S_{\mathbb{F}_2^{d_0}})^m \times (S_{\mathbb{F}_2^{d_0}})^n$. Hence, $K_G^\mathbb{F}(m,n) \leq \sup\limits_{(u,v) \in (S_{\mathbb{F}_2^{d_0}})^m \times (S_{\mathbb{F}_2^{d_0}})^n} |\text{tr}(A_0^* \Gamma_{\mathbb{F}_2^{d_0}}(u,v))| \leq K_G^\mathbb{F}(d_0)$.

Remark 3.4 (Adjoining GT) Readers who are familiar with adjoint normed operator ideals and trace duality in general (cf. [123, Chapter 9.1]) immediately recognise that Corollary 3.3 implies that the Grothendieck inequality actually is equivalent to an inequality between two matrix norms, induced by two (adjoint) Banach ideals; namely:

$$\sup_{H \in \text{HIL}^\mathbb{F}} \|A\|_H^G \stackrel{(3.2.7)}{=} \|A\|_{\mathfrak{D}_2} \leq K_G^\mathbb{F}(m,n) \|A\|_{\infty,1} \leq K_G^\mathbb{F} \|A\|_{\infty,1}$$

for all matrices $A \in \mathbb{M}_{m,n}(\mathbb{F})$, $m,n \in \mathbb{N}$, or, equivalently,

$$B_{\mathfrak{L}(l_\infty^n, l_1^m)} \subseteq K_G^\mathbb{F}(m,n) \, B_{\mathfrak{D}_2(l_\infty^n, l_1^m)} \subseteq K_G^\mathbb{F} \, B_{\mathfrak{D}_2(l_\infty^n, l_1^m)} \qquad (3.3.9)$$

for all $m,n \in \mathbb{N}$, where $(\mathfrak{D}_2, \|\cdot\|_{\mathfrak{D}_2}) = (\mathfrak{L}_2^*, \|\cdot\|_{\mathfrak{L}_2^*}) = (\mathfrak{P}_2^d \circ \mathfrak{P}_2, \|\cdot\|_{\mathfrak{P}_2^d \circ \mathfrak{P}_2})$ characterises the Banach ideal of 2-dominated operators (cf. [33, Table 17.12, Theorem 17.14 and Chapter 19] and Remark 4.2). The latter inclusion also follows directly from "adjoining" (3.3.4) above. In fact, if K and L are arbitrary compact sets, the following deep result of Grothendieck holds:

$$\mathfrak{L}(C(K), C(L)') \subseteq \mathfrak{D}_2(C(K), C(L)') \subseteq \mathfrak{P}_2(C(K), C(L)') \subseteq \mathfrak{L}_2(C(K), C(L)'),$$

and

$$\|T\|_{\mathfrak{L}_2} \leq \|T\|_{\mathfrak{P}_2} \leq \|T\|_{\mathfrak{D}_2} \leq K_G^\mathbb{F} \|T\|$$

for all $T \in \mathfrak{L}(C(K), C(L)')$. To recognise this highly noteworthy statement, we just have to note that [126, Theorem 2.1] implies that any $T \in \mathfrak{L}(C(K), C(L)')$ can be represented as $T = (J_{\mathbb{P}_L})' U J_{\mathbb{P}_K}$, where for $\Delta \in \{K, L\}$, $\mathbb{P}_\Delta$ is a well-defined *probability* measure on Δ, $J_{\mathbb{P}_\Delta} : C(\Delta) \hookrightarrow L^2(\mathbb{P}_\Delta)$ denotes the canonical (norm 1) inclusion and $\|U\| \leq K_G^\mathbb{F} \|T\|$. Each of the two operators $J_{\mathbb{P}_\Delta}$ is absolutely 2-summing (such as their biduals—cf. [33, Corollary 17.8.4]) and satisfies $\|J_{\mathbb{P}_\Delta}\|_{\mathfrak{P}_2} = \mathbb{P}_\Delta(\Delta)^{1/2} = 1$ (cf. [33, Subsection 11.2])). In particular, we reobtain [126, Corollary

3.3 Characterisation of $K_G^{\mathbb{F}}$ Through Operator Ideals and Violation of Bell...

2.2]. It is quite instructive to compare this result with [33, Corollary 14.5.2 and Theorem 17.14], [72, Section 5] and [81, Theorem G].

Remark 3.5 (Tensor Norm Representation of GT) Let $m, n \in \mathbb{N}$. Readers who are familiar with both, Banach operator ideals and tensor norms, very likely re-recognise the following *equivalent* tensor norm representations of the inequality (3.3.4) (respectively (3.3.9)) at once (cf. [33, Theorem 14.4 and Theorem 20.17]):

(i) $\pi(\cdot; l_\infty^n, l_\infty^m) \leq K_G^{\mathbb{F}}(m,n) \, w_2(\cdot; l_\infty^n, l_\infty^m) \leq K_G^{\mathbb{F}} \, w_2(\cdot; l_\infty^n, l_\infty^m)$.
(ii) $w_2^*(\cdot; l_1^n, l_1^m) \leq K_G^{\mathbb{F}}(m,n) \, \varepsilon(\cdot; l_1^n, l_1^m) \leq K_G^{\mathbb{F}} \, \varepsilon(\cdot; l_1^n, l_1^m)$.

We just have to apply the representation theorem for minimal Banach operator ideals to the minimal kernels of the maximal Banach operator ideals $\mathfrak{J} \sim \pi$ and $\mathfrak{L}_2 \sim w_2$ (cf. [33, Corollary 22.2.1]). (i) then follows from the isometric equalities

$$l_\infty^n \otimes_\pi l_\infty^m \cong \mathfrak{J}^{min}(l_1^n, l_\infty^m) \stackrel{1}{=} \mathfrak{N}(l_1^n, l_\infty^m)$$

and

$$l_\infty^n \otimes_{w_2} l_\infty^m \cong \mathfrak{L}_2^{min}(l_1^n, l_\infty^m) \stackrel{1}{=} \mathfrak{L}_2(l_1^n, l_\infty^m).$$

Since $(M \otimes_\alpha N)' \cong M' \otimes_{\alpha'} N'$ for all finitely generated tensor norms α and all finite-dimensional Banach spaces M and N (prove it!), it follows from trace duality that (i) and (ii) in fact are equivalent. Here, it should be noted that quite often the tensor norm w_2 is also denoted as γ_2, and the maximal Banach ideal $\mathfrak{L}_2$ is also known as Γ_2.

Given that trace duality view, the role of the "free parameters" $m, n, d \in \mathbb{N}$, where the pair $(m, n) \in \mathbb{N}^2$ describes the size of the matrices and d is the dimension of the underlying finite-dimensional Hibert space $\mathbb{F}_2^d$ is explicitly described in

Proposition 3.4 *Let* $\mathbb{F} \in \{\mathbb{R}, \mathbb{C}\}$ *and* $m, n, d \in \mathbb{N}$. *Put*

$$K_G^{\mathbb{F}}(m, n; d) := \sup \left\{ \| \Gamma_{\mathbb{F}_2^d}(u, v) : l_1^n \longrightarrow l_\infty^m \|_{\mathfrak{N}} : (u, v) \in (S_{\mathbb{F}_2^d})^m \times (S_{\mathbb{F}_2^d})^n \right\}.$$

Then

$$K_G^{\mathbb{F}}(d) = \sup_{(m,n) \in \mathbb{N}^2} K_G^{\mathbb{F}}(m, n; d). \qquad (3.3.10)$$

and

$$K_G^{\mathbb{F}}(m, n) = \sup_{S \in \mathcal{Q}_{m,n}(\mathbb{F})} \| S \|_{\mathfrak{N}} = \sup_{d \in \mathbb{N}} K_G^{\mathbb{F}}(m, n; d). \qquad (3.3.11)$$

The sequence $(K_G^{\mathbb{F}}(d))_{d \in \mathbb{N}}$ is non-decreasing, and

$$K_G^{\mathbb{F}} = \sup \{\|S\|_{\mathfrak{N}} : m, n \in \mathbb{N}, S \in \mathcal{Q}_{m,n}(\mathbb{F})\} = \sup_{(m,n) \in \mathbb{N}^2} K_G^{\mathbb{F}}(m, n)$$

$$= \sup_{d \in \mathbb{N}} K_G^{\mathbb{F}}(d) = \lim_{d \to \infty} K_G^{\mathbb{F}}(d). \tag{3.3.12}$$

In particular, $K_G^{\mathbb{F}}(1, 1) = 1$ and $K_G^{\mathbb{F}}(m, n; 1) = K_G^{\mathbb{F}}(1) = 1$ for all $m, n \in \mathbb{N}$.

Proof First of all, it follows from (3.3.1) and (3.3.6) that

$$K_G^{\mathbb{F}}(m, n) = \sup_{S \in \mathcal{Q}_{m,n}(\mathbb{F})} \|S\|_{\mathfrak{N}}$$

$$= \sup_{d \in \mathbb{N}} K_G^{\mathbb{F}}(m, n; d) \leq \sup\{\|S\|_{\mathfrak{N}} : m, n \in \mathbb{N}, S \in \mathcal{Q}_{m,n}(\mathbb{F})\}.$$

Since

$$\sup_{A \in B_W} |\mathrm{tr}(A^* S)| \stackrel{(3.3.3)}{=} \sup_{A \in B_W} |\langle A^*, \tau_S \rangle| = \|\tau_S\| = \|S\|_{\mathfrak{N}}$$

for any $S \in \mathcal{Q}_{m,n}(\mathbb{F})$, Theorem 3.1 and (1.1.3) imply that

$$\sup\{\|S\|_{\mathfrak{N}} : m, n \in \mathbb{N}, S \in \mathcal{Q}_{m,n}(\mathbb{F})\} = K_G^{\mathbb{F}} \leq \sup_{(m,n) \in \mathbb{N}^2} K_G^{\mathbb{F}}(m, n).$$

Hence,

$$\sup_{d \in \mathbb{N}} \left(\sup_{(m,n) \in \mathbb{N}^2} K_G^{\mathbb{F}}(m, n; d) \right) = \sup_{(m,n) \in \mathbb{N}^2} \left(\sup_{d \in \mathbb{N}} K_G^{\mathbb{F}}(m, n; d) \right)$$

$$= \sup_{(m,n) \in \mathbb{N}^2} K_G^{\mathbb{F}}(m, n)$$

$$= \sup\{\|S\|_{\mathfrak{N}} : m, n \in \mathbb{N}, S \in \mathcal{Q}_{m,n}(\mathbb{F})\} = K_G^{\mathbb{F}}.$$

Moreover, if $m, n, d \in \mathbb{N}$ are given, recall that

$$K_G^{\mathbb{F}}(m, n; d) = \sup\{\|\Gamma_{\mathbb{F}_2^d}(u, v) : l_1^n \longrightarrow l_\infty^m\|_{\mathfrak{N}} : (u, v) \in (S_{\mathbb{F}_2^d})^m \times (S_{\mathbb{F}_2^d})^n\}$$

$$= \sup\{|\mathrm{tr}(A^* \Gamma_{\mathbb{F}_2^d}(u, v))| : A \in B_W, (u, v) \in (S_{\mathbb{F}_2^d})^m \times (S_{\mathbb{F}_2^d})^n\},$$

whence $\sup_{(m,n) \in \mathbb{N}^2} K_G^{\mathbb{F}}(m, n; d) \leq K_G^{\mathbb{F}}(d)$ (due to (1.1.2) and (3.2.1)). By definition, $K_G^{\mathbb{F}}(d)$ is the smallest constant $K(d)$ which satisfies inequality (1.1.2). Thus, $\sup_{(m,n) \in \mathbb{N}^2} K_G^{\mathbb{F}}(m, n; d) = K_G^{\mathbb{F}}(d)$. Finally, since $\langle v, u \rangle_{\mathbb{F}_2^d} = \langle Jv, Ju \rangle_{\mathbb{F}_2^{d+1}}$ for all $d \in$

3.3 Characterisation of $K_G^\mathbb{F}$ Through Operator Ideals and Violation of Bell... 59

$\mathbb{N}$ and $u, v \in \mathbb{F}_2^d$, where $J : \mathbb{F}_2^d \longrightarrow \mathbb{F}_2^{d+1}$, $(x_1, \ldots, x_d) \mapsto (x_1, \ldots, x_d, 0)$ denotes the canonical isometric injection from $\mathbb{F}_2^d$ into $\mathbb{F}_2^{d+1}$, it follows that $K_G^\mathbb{F}(d) \leq K_G^\mathbb{F}(d+1)$ for all $d \in \mathbb{N}$. □

Remark 3.6 Even in the matrix case, the Banach space $(\mathfrak{N}(l_1^n, l_\infty^m), \|\cdot\|_\mathfrak{N})$ should not be confused with the Banach space $(\mathfrak{N}(\mathbb{F}_2^n, \mathbb{F}_2^m), \|\cdot\|_\mathfrak{N}) = (\mathfrak{S}_1(\mathbb{F}_2^n, \mathbb{F}_2^m), \sigma_1)$! The latter space namely consists of matrices—viewed as operators—between (finite-dimensional) *Hilbert spaces*, contained in the so-called *Schatten-von Neumann class of index 1*, also known as *trace-class operators* (cf., e.g., [69] and [80, Chapter 20.2]). In particular, if we view a given matrix $M \in \mathbb{M}_{m,n}(\mathbb{F})$ as linear operator *from* l_1^n *to* l_∞^m, the norm $\|M\|_\mathfrak{N}$ in general does *not* coincide with the so-called *trace norm* of M (also known as *nuclear norm*). The latter is given by $\|M\|_* := \mathrm{tr}(|M|)$, where $|M| := (M^*M)^{1/2}$. Since the trace norm of M coincides with the Schatten 1-norm $\sigma_1(M)$, it equals the sum of the singular values of the matrix M (cf., e.g., [15, Chapter IV.2], [30, Exercises IX.2.19, IX.2.20 and IX.2.21] and [70, Chapter 5.6 and Chapter 7.4.7]).

Let us recall the isometric isomorphism $\tau : Z \xrightarrow{\cong} Y'$, induced by the Banach spaces $Y := \mathfrak{L}(l_\infty^m, l_1^n)$ and $Z := \mathfrak{N}(l_1^n, l_\infty^m) \cong Y'$ (cf. (3.3.3)). Put $\delta := \tau' j_Y : Y \xrightarrow{\cong} Z'$. As we have seen, also δ is an isometric isomorphism (cf. Remark 3.3). Its inverse is given by $\delta^{-1} = j_Y^{-1}(\tau^{-1})'$ (since Y is finite-dimensional). An application of the bipolar theorem to the dual pairing $\langle Z, Z' \rangle = \langle Z, \delta(Y) \rangle$, induced by the bilinear form $Z \times Z' \longrightarrow \mathbb{F}$, $(R, z') \mapsto \langle R, z' \rangle := \mathrm{tr}(R\,\delta^{-1}(z'))$ (cf. [30, V.1.8] and [80, Chapter 8.2]) implies the following explicit representation result for the so-called "local correlation polytope". To the best of our knowledge, the outcome for the complex case (i.e., if $\mathbb{F} = \mathbb{C}$) is new. A (different) part of our proof for the real case can be found in the proof of [9, Proposition 11.7]. In particular, we are going to shed some light on the geometry of the unit ball of $\mathfrak{N}(l_1^n, l_\infty^m)$. To this end, recall that for any subset S of an $\mathbb{F}$-vector space

$$\mathrm{cx}(S) := \bigcup_{n \in \mathbb{N}} \Big\{ \sum_{i=1}^n \alpha_i x_i : x_1, \ldots, x_n \in S, \alpha_1, \ldots, \alpha_n \geq 0 \text{ and } \sum_{i=1}^n \alpha_i = 1 \Big\}$$

denotes the convex hull of S and that

$$\mathrm{acx}(S) := \bigcup_{n \in \mathbb{N}} \Big\{ \sum_{i=1}^n \alpha_i x_i : x_1, \ldots, x_n \in S, \alpha_1, \ldots, \alpha_n \in \mathbb{F} \text{ and } \sum_{i=1}^n |\alpha_i| \leq 1 \Big\}$$

(3.3.13)

marks the absolute convex hull of S. Here we adopt the notation, introduced right below [80, Proposition 6.1.3]. Recall also that $\mathrm{acx}(S) = \mathrm{cx}(\check{S})$, where $\check{S} := (\mathbb{F} \cap \overline{\mathbb{D}})S$ denotes the circled hull of S (cf. [80, Proposition 6.1.4]).

Theorem 3.5 *Let* $\mathbb{F} \in \{\mathbb{R}, \mathbb{C}\}$ *and* $m, n \in \mathbb{N}$. *Put*

$$\mathcal{G}_{m,n}(\mathbb{F}) := \{\overline{p}q^\top : (p,q) \in S_\mathbb{F}^m \times S_\mathbb{F}^n\} = \{\Gamma_\mathbb{F}(p,q) : (p,q) \in S_\mathbb{F}^m \times S_\mathbb{F}^n\}$$

and

$$\mathcal{H}_{m,n}(\mathbb{F}) := \{\overline{x}y^\top : (x,y) \in (\mathbb{F} \cap \overline{\mathbb{D}})^m \times (\mathbb{F} \cap \overline{\mathbb{D}})^n\}$$
$$= \{\Gamma_\mathbb{F}(x,y) : (x,y) \in (\mathbb{F} \cap \overline{\mathbb{D}})^m \times (\mathbb{F} \cap \overline{\mathbb{D}})^n\}.$$

Then

$$B_{\mathfrak{N}(l_1^n, l_\infty^m)} = acx(\mathcal{G}_{m,n}(\mathbb{F})) = cx(\widetilde{\mathcal{G}_{m,n}(\mathbb{F})}) = cx(\mathcal{H}_{m,n}(\mathbb{F})). \tag{3.3.14}$$

(i) *If* $\mathbb{F} = \mathbb{R}$, *then the set* $\mathcal{G}_{m,n}(\mathbb{R})$ *even coincides with the set of all extreme points of* $B_{\mathfrak{N}(l_1^n, l_\infty^m)}$ *and*

$$B_{\mathfrak{N}(l_1^n, l_\infty^m)} = cx(\mathcal{G}_{m,n}(\mathbb{F}))$$
$$= \Big\{ \sum_{i=1}^{mn+1} \lambda_i x_i : x_i \in \mathcal{G}_{m,n}(\mathbb{R}), 0 \le \lambda_i \le 1, \sum_{i=1}^{mn+1} \lambda_i = 1 \Big\}$$
$$= \big\{ \mathbb{E}[\mathbf{X}\mathbf{Y}^\top] : \max\{|X_i|, |Y_j|\} \le 1 \text{ a.s., for all } (i,j) \in [m] \times [n] \big\}.$$

(ii) *If* $\mathbb{F} = \mathbb{C}$, *then*

$$B_{\mathfrak{N}(l_1^n, l_\infty^m)} = \Big\{ \sum_{i=1}^{2mn+1} \lambda_i x_i : x_i \in \widetilde{\mathcal{G}_{m,n}(\mathbb{C})}, 0 \le \lambda_i \le 1, \sum_{i=1}^{2mn+1} \lambda_i = 1 \Big\}.$$

Proof Put $Y := \mathfrak{L}(l_\infty^m, l_1^n)$ and $Z := \mathfrak{N}(l_1^n, l_\infty^m)$. Fix $(p,q) \in S_\mathbb{F}^m \times S_\mathbb{F}^n$. Recall the well-known isometric isomorphism $\chi : l_\infty^n \xrightarrow{\cong} (l_1^n)'$, defined as

$$\langle z, \chi(q) \rangle := z^\top q = \sum_{j=1}^n q_j z_j \quad ((z,q) \in l_1^n \times l_\infty^n).$$

Since $\overline{p}_i q_j = e_i^\top \big(\langle \cdot, \chi(q) \rangle \overline{p} \big) e_j$ for all $(i,j) \in [m] \times [n]$, it follows that $\overline{p}q^\top$ is the matrix representation of the linear operator $\langle \cdot, \chi(q) \rangle \overline{p} : \mathbb{F}^n \longrightarrow \mathbb{F}^m$ with respect

3.3 Characterisation of $K_G^\mathbb{F}$ Through Operator Ideals and Violation of Bell... 61

to the standard bases. By definition of the nuclear norm it therefore follows that $\|\overline{p}q^\top\|_\mathfrak{N} = \|\langle \cdot, \chi(q)\rangle \overline{p}\|_\mathfrak{N} = \|q\|_\infty \|p\|_\infty = 1$ (cf. [123, Chapter 6.3]). Hence,

$$\mathrm{acx}(\mathcal{G}_{m,n}(\mathbb{F})) \subseteq B_Z.$$

Moreover, $\mathcal{G}_{m,n}(\mathbb{F}) = f(S_\mathbb{F}^m \times S_\mathbb{F}^n)$, where $f : l_\infty^m \times l_\infty^n \longrightarrow Z$ is defined as $f(p,q) := \overline{p}q^\top$. A standard application of the triangle inequality implies that f is continuous. Hence, $\mathcal{G}_{m,n}(\mathbb{F})$ is a compact subset of the Banach space Z. Similarly it follows that $\mathcal{H}_{m,n}(\mathbb{F}) \subseteq Z$ is compact.

Let $z' \in (\mathcal{G}_{m,n}(\mathbb{F}))^\circ \subseteq Z'$. Consider the isometric isomorphism $\delta : Y \longrightarrow Z'$. Then in particular $|\mathrm{tr}(\overline{p}q^\top B_0)| = |\langle \overline{p}q^\top, z'\rangle| \leq 1$ for all $(p, q) \in S_\mathbb{F}^m \times S_\mathbb{F}^n$, where $B_0 := \delta^{-1}(z') \in Y$. Consequently, $B_0 \in B_Y$ (due to (3.2.3)), whence

$$(\mathrm{acx}(\mathcal{G}_{m,n}(\mathbb{F})))^\circ \subseteq (\mathcal{G}_{m,n}(\mathbb{F}))^\circ \subseteq \delta(B_Y) = B_{Z'} = B_Z^\circ.$$

Consequently, polarisation again (now applied to subsets of Z'—cf., e.g., [30, Definition V.1.6]) implies that

$$B_Z \subseteq B_Z^{\circ\circ} \subseteq (\mathrm{acx}(\mathcal{G}_{m,n}(\mathbb{F})))^{\circ\circ} = \overline{\mathrm{acx}(\mathcal{G}_{m,n}(\mathbb{F}))}^{\|\cdot\|_Z}. \quad (3.3.15)$$

The last equality follows from the bipolar theorem (cf., e.g., [30, Theorem V.1.8] or [80, Theorem 8.2.2]). However, since actually Z is a finite-dimensional space, it follows that also $\mathrm{acx}(\mathcal{G}_{m,n}(\mathbb{F}))$ is a compact subset of Z (see [80, Proposition 6.7.4]). In particular, it is closed, whence $B_Z = \mathrm{acx}(\mathcal{G}_{m,n}(\mathbb{F}))$.

The inclusion $\mathrm{acx}(\mathcal{G}_{m,n}(\mathbb{F})) \subseteq \mathrm{cx}(\mathcal{H}_{m,n}(\mathbb{F}))$ is trivial. Let $(\lambda_1, \ldots, \lambda_k) \in [0,1]^k$, such that $\sum_{\nu=1}^k \lambda_\nu = 1$. Since

$$\|\sum_{\nu=1}^k \lambda_\nu \overline{x} y^\top\|_\mathfrak{N} \leq \sum_{\nu=1}^k \lambda_\nu \|x\|_\infty \|y\|_\infty \leq \sum_{\nu=1}^k \lambda_\nu = 1,$$

for all $(x, y) \in (\mathbb{F} \cap \mathbb{D})^m \times (\mathbb{F} \cap \mathbb{D})^n$, $l \in \mathbb{N}$, it follows that $\mathrm{cx}(\mathcal{H}_{m,n}(\mathbb{F})) \subseteq B_Z$. Finally, [80, Proposition 6.2.4] implies that $\mathrm{acx}(\mathcal{G}_{m,n}(\mathbb{F})) = \mathrm{cx}(\widetilde{\mathcal{G}_{m,n}(\mathbb{F})})$, which concludes the proof of (3.3.14).

(i) The non-trivial statement that—in the real case—$\mathcal{G}_{m,n}(\mathbb{R})$ precisely describes the set of all extreme points of the convex compact unit ball B_Z (whose proof requires tensor product methods and trace duality) follows from [33, Proposition 16.7], together with [133, Theorem 1.3]. Equipped with this deep fact, we may apply the finite-dimensional version of the Krein-Milman Theorem to the non-empty convex compact set B_Z (cf. [4, Theorem 7.68]), whence $B_Z = \mathrm{cx}(\mathcal{G}_{m,n}(\mathbb{F}))$. Now, we may apply Carathéodory's Convexity Theorem (cf. [4, Theorem 5.32]), which gives us the second equality. The last equality follows from [9, Definition 11.5 and Proposition 11.7].

(ii) To proceed in a similar way as in the real case (i), we view $Z \cong \mathbb{M}_{m,n}(\mathbb{C}) \cong \mathbb{C}^{mn}$ as a real finite-dimensional vector space, implying that $\dim_{\mathbb{R}}(Z) = 2mn$. Consequently, we may apply Carathéodory's Convexity Theorem again (cf. [4, Theorem 5.32] and the proof of [80, Proposition 6.7.4]); namely to $\mathrm{cx}(\widetilde{\mathcal{G}_{m,n}(\mathbb{F})}) = B_Z \subseteq Z$, from which (ii) follows (due to (3.3.14)). □

Based on [9, Definition 11.5 and remark right below], we have shown that at least in the real case (i.e., if $\mathbb{F} = \mathbb{R}$), B_Z precisely coincides with the set of all *classical (or local) "correlation" matrices*.

Consequently, (3.3.4) implies that for all $m, n \in \mathbb{N}$, for all $\mathbb{F}$-Hilbert spaces $H^{\mathbb{F}}$ and $(u, v) \in S_{H^{\mathbb{F}}}^m \times S_{H^{\mathbb{F}}}^n$ there are $x_1^{\mathbb{F}}, \ldots, x_{k^{\mathbb{F}}}^{\mathbb{F}} \in (\mathbb{F} \cap \overline{\mathbb{D}})^m$, $y_1^{\mathbb{F}}, \ldots, y_{k^{\mathbb{F}}}^{\mathbb{F}} \in (\mathbb{F} \cap \overline{\mathbb{D}})^n$ and $(\lambda_1^{\mathbb{F}}, \ldots, \lambda_{k^{\mathbb{F}}}^{\mathbb{F}}) \in [0, 1]^{k^{\mathbb{F}}}$, such that $\sum_{\nu=1}^{k^{\mathbb{F}}} \lambda_\nu^{\mathbb{F}} = 1$ and

$$\Gamma_{H^{\mathbb{F}}}(u, v) = K_G^{\mathbb{F}} \sum_{\nu=1}^{k^{\mathbb{F}}} \lambda_\nu^{\mathbb{F}} \overline{x_\nu^{\mathbb{F}}} (y_\nu^{\mathbb{F}})^{\top}, \qquad (3.3.16)$$

where $k_{\mathbb{R}} := mn + 1$ and $k_{\mathbb{C}} := 2mn + 1$. If $\mathbb{F} = \mathbb{R}$, we may assume that $|x_\nu^{\mathbb{R}}| = 1$ and $|y_\nu^{\mathbb{R}}| = 1$ for all $\nu \in [k^{\mathbb{R}}]$. In particular (if $m = n = 1$),

$$\langle u, v \rangle_{H^{\mathbb{R}}} = K_G^{\mathbb{R}} (\lambda_1^{\mathbb{R}} x_1^{\mathbb{R}} y_1^{\mathbb{R}} + (1 - \lambda_1^{\mathbb{R}}) x_2^{\mathbb{R}} y_2^{\mathbb{R}})$$

and

$$\langle b, a \rangle_{H^{\mathbb{C}}} = K_G^{\mathbb{C}} \sum_{\nu=1}^{3} \lambda_\nu^{\mathbb{C}} \overline{x_\nu^{\mathbb{C}}} y_\nu^{\mathbb{C}}$$

for all $u, v \in S_{H^{\mathbb{R}}}$ and $a, b \in S_{H^{\mathbb{C}}}$. Since $K_G^{\mathbb{R}} > 1$, it follows that $\mathrm{sign}(x_1^{\mathbb{R}} y_1^{\mathbb{R}}) \neq \mathrm{sign}(x_2^{\mathbb{R}} y_2^{\mathbb{R}})$. Thus, if $\lambda_1 \neq \frac{1}{2}$, then

$$\frac{\pi}{2} < K_G^{\mathbb{R}} \leq \frac{1}{|2\lambda_1 - 1|}, \text{ respectively } |\lambda_1| \leq \frac{1 + K_G^{\mathbb{R}}}{2 K_G^{\mathbb{R}}} < \frac{\pi + 2}{2\pi} \approx 0.818.$$

Remark 3.7 (Quantum Violation of a Bell Inequality) Fix an *arbitrarily given* $S \in \mathcal{Q}_{m,n}(\mathbb{F}) \setminus B_Z$. Recall again the construction of the isometric isomorphism $\tau : Z \xrightarrow{\cong} Y'$ via trace duality (cf. (3.3.2) and (3.3.3)). Since B_Z is a non-empty closed and absolutely convex subset of the Banach space Z, we may apply [80, Corollary 7.3.6]. The latter is an implication of hyperplane separation (which is a geometric version of the Hahn-Banach theorem and thus an implication of Zorn's lemma). Hence, due to (3.3.7), it follows the existence of a matrix $B_0 \in B_Y$, such that

$$\mathrm{tr}(A_0^* S) = \mathrm{tr}(B_0 S) > 1 \text{ and } |\mathrm{tr}(A_0^* R)| \leq 1 \text{ for all } R \in B_Z,$$

3.3 Characterisation of $K_G^{\mathbb{F}}$ Through Operator Ideals and Violation of Bell...

where $A_0 := B_0^* \in B_W$ (compare also with (3.3.8)). In particular, we have been provided with a matrix $A_0 \in \mathbb{M}_{m,n}(\mathbb{F})$, such that $\|A_0\|_{\infty,1} \leq 1$ and

$$1 < \sup_{H \in \text{HIL}^{\mathbb{F}}} \|A_0\|_H^G = \max_{S \in \mathcal{Q}_{m,n}(\mathbb{F})} |\text{tr}(A_0^* S)| = \max_{S \in \mathcal{Q}_{m,n}(\mathbb{F})} \text{Re}(\text{tr}(A_0^* S))$$

$$= \max_{\Sigma \in C(m+n;\mathbb{F})} \text{tr}(\Delta(A_0)\Sigma)$$

(due to (3.2.7)). In quantum mechanics (if $\mathbb{F} = \mathbb{R}$), the latter inequality is somewhat vaguely referred to as the "maximal quantum violation of a Bell (correlation) inequality". The "maximal violation of the related Bell inequality" coincides precisely with the inequality

$$\|f_{A_0}\| = \max_{S \in \mathcal{Q}_{m,n}(\mathbb{F})} |f_{A_0}(S)| = \max_{S \in \mathcal{Q}_{m,n}(\mathbb{F})} |\text{tr}(A_0^* S)| > 1,$$

where f_{A_0} is defined as in Remark 3.3 (for both fields). The term "Bell inequality" (now with respect to any given $A \in B_W$, of course) is therefore to be understood as the inequality

$$\max_{R \in B_Z} |f_A(R)| = \max_{R \in B_Z} |\text{tr}(A^* R)| \leq 1,$$

which holds for all $A \in B_W$ (due to (3.3.7)). That Bell inequality is "violated", if and only if $|f_A(S_0)| > 1$ for some quantum correlation matrix $S_0 \in \mathcal{Q}_{m,n}(\mathbb{F}) \setminus B_Z$. (cf. also Remark 3.3, respectively Theorem 3.5, together with [96, Lemma 2] if $\mathbb{F} = \mathbb{R}$). In this respect, the maximum value $\|f_A\|$ is then referred to as "maximal violation of the Bell inequality". Consequently, due to (3.3.1) and Remark 3.3, there are $S_* \in \mathcal{Q}_{m,n}(\mathbb{F}) \setminus B_Z$ and—hence—$d_* \in \mathbb{N}$, such that "the maximal violation of the Bell inequality" (i.e., $\|f_A\|$) is uniformly bounded above by $K_G^{\mathbb{F}}(d_*) \leq K_G^{\mathbb{F}}$. More precisely, if $S_* = \Gamma_{\mathbb{F}_2^{d_*}}(u_*, v_*)$, for some $d_* \in \mathbb{N}$ and $(u_*, v_*) \in (S_{\mathbb{F}_2^{d_*}})^m \times (S_{\mathbb{F}_2^{d_*}})^n$, then

$$1 < \|f_A\| = |\text{tr}(A^* S_*)| = \sup_{(u,v) \in (S_{\mathbb{F}_2^{d_*}})^m \times (S_{\mathbb{F}_2^{d_*}})^n} |\text{tr}(A^* \Gamma_{\mathbb{F}_2^{d_*}}(u,v))| \leq K_G^{\mathbb{F}}(d_*) \leq K_G^{\mathbb{F}}.$$

Remark 3.8 We don't know whether in (3.2.9) we may substitute the block matrix $\Delta(A) \in \mathbb{M}_{m+n}(\mathbb{F})$ through an *arbitrary* matrix $B \in \mathbb{M}_{m+n}(\mathbb{F})$. If this were the case, a further application of the bipolar theorem (cf. [80, Theorem 8.2.2]) shows that the latter would be equivalent to

$$C(k; \mathbb{F}) \subseteq K_G^{\mathbb{F}} \text{acx}(C_1(k; \mathbb{F})) \text{ for all } k \in \mathbb{N}_2.$$

Theorem 3.5 therefore would imply that

$$C(k; \mathbb{F}) \subseteq K_G^{\mathbb{F}} \operatorname{acx}(\{\overline{p}q^\top : (p,q) \in S_{\mathbb{F}}^k \times S_{\mathbb{F}}^k\}) \stackrel{(3.3.14)}{=} K_G^{\mathbb{F}} B_{\mathfrak{N}(l_1^k, l_\infty^k)} \text{ for all } k \in \mathbb{N}_2,$$

where $B_{\mathfrak{N}(l_1^k, l_\infty^k)}$ again denotes the unit ball of the Banach space of nuclear operators between l_1^k and l_∞^k, equipped with the nuclear norm.

3.4 $K_G^{\mathbb{R}}(2)$ and the Walsh-Hadamard Transform: Krivine's Approach Revisited

Regarding explicit constructions of elements of $\mathcal{Q}_{m,n}(\mathbb{F})$, a rigorous description of the *entries* of the Kronecker product of matrices proves to be a very useful tool (cf. Example 3.3). To this end, we consider the mapping:

$$\mathbb{Z} \times \mathbb{N} \ni (\nu, n) \mapsto r_n(\nu) := \begin{cases} n & \text{if } n \text{ is a divisor of } \nu \\ \operatorname{rem}_n(\nu) & \text{if } n \text{ is not a divisor of } \nu, \end{cases}$$

where $\operatorname{rem}_n(\nu) \in \{0, 1, \ldots, n-1\}$ denotes the uniquely determined remainder in Euclidean division of ν by n, implying that $r_n(\nu) \in [n]$ (by construction). Thus, if $p \in \mathbb{Z}$ and $\nu = pn + \operatorname{rem}_n(\nu)$, then $p + 1 \geq \frac{\nu+1}{n}$ and

$$f_n(\nu) := \frac{\nu - r_n(\nu)}{n} + 1 = \begin{cases} p + 1 < \frac{\nu}{n} + 1 & \text{if } n \text{ is not a divisor of } \nu \\ \frac{\nu}{n} & \text{if } n \text{ is a divisor of } \nu. \end{cases}$$

Consequently, if $l \in \mathbb{N}$ and $\nu \in [ln]$, then $f_n(\nu) \in [l]$. In particular, $r_n(\nu) = \nu$ if $\nu \in [n]$.

Especially with regard to Example 3.3 the "Boolean" case n=2 is of particular importance to us. Here, we obviously obtain:

$$f_2(\nu) = \left\lceil \frac{\nu}{2} \right\rceil = \begin{cases} \frac{\nu}{2} & \text{if } \nu \text{ is even} \\ \frac{\nu+1}{2} & \text{if } \nu \text{ is odd} \end{cases} \text{ and } b_1(\nu) := r_2(\nu) - 1 = \mathbb{1}_{2\mathbb{N}}(\nu)$$

$$= \begin{cases} 1 & \text{if } \nu \text{ is even} \\ 0 & \text{if } \nu \text{ is odd}. \end{cases} \quad (3.4.1)$$

Note that the structure of f_2 implies that $f_2([2^i]) = [2^{i-1}]$ for all $i \in \mathbb{N}$. In particular, for any $m \in \mathbb{N}_2$ and $i \in \{2, 3, \ldots, m\}$, the well-defined function $b_i := b_1 \circ \underbrace{f_2 \circ \cdots \circ f_2}_{(i-1)-\text{times}}$ is the ith component of the $\{0, 1\}^m$-valued function

3.4 $K_G^{\mathbb{R}}(2)$ and the Walsh-Hadamard Transform: Krivine's Approach Revisited

$\pi_m : \mathbb{Z} \longrightarrow \{0, 1\}^m$, defined as

$$\pi_m(\nu) := (b_1(\nu), b_2(\nu), \ldots, b_m(\nu))^\top \text{ for all } \nu \in \mathbb{Z}. \tag{3.4.2}$$

The actual role of the sequence of functions $(r_n)_{n \in \mathbb{N}}$ is encoded in

Lemma 3.3 *Let $n \in \mathbb{N}$. Then the mapping*

$$\Psi_n : \mathbb{Z} \times [n] \xrightarrow{\cong} \mathbb{Z}, (i, j) \mapsto (i-1)n + j$$

is bijective. Its inverse is given by $\Psi_n^{-1} = \Lambda_n$, where

$$\Lambda_n : \mathbb{Z} \longrightarrow \mathbb{Z} \times [n], \nu \mapsto (f_n(\nu), r_n(\nu)).$$

Moreover, $\Psi_n([l] \times [n]) = [ln]$ for all $l \in \mathbb{N}$.

Proof Fix $(i, j) \in \mathbb{Z} \times [n]$. If $\Psi(i, j) = (i-1)n + j = ln$ for some $l \in \mathbb{Z}$, it follows that $0 < j = (l - (i-1))n \leq n$, whence $0 < l - (i-1) \leq 1$. Thus, $l = i$, and hence $j = n$. Thus, n is a divisor of $(i-1)n + j$ if and only if $j = n$, implying that $r_n((i-1)n + j) = n$ (by construction of the mapping r). Hence, if n is not a divisor of $(i-1)n + j$, then $j < n$, and it follows that $r_n((i-1)n + j) = j$ is the uniquely determined remainder of $(i-1)n + j$. Now, we only have to use a bit of elementary algebra, to show that $\Lambda_n \circ \Psi_n = \mathrm{id}_{\mathbb{Z} \times [n]}$ and $\Psi_n \circ \Lambda_n = \mathrm{id}_{\mathbb{Z}}$, where $\Lambda(\nu) := \left(\frac{\nu - r_n(\nu)}{n} + 1, r_n(\nu)\right)$ for all $\nu \in \mathbb{Z}$. Consequently, $\Lambda_n = \Psi_n^{-1}$. Finally, if $\Psi_n(i, j) = (i-1)n + j \leq ln$, it follows that $n(l - (i-1)) \geq j > 0$, whence $i - 1 < l$. Thus, $i \in [l]$, and it follows that $[ln] = [ln] \cap \Psi_n(\mathbb{Z} \times [n]) \subseteq \Psi_n([l] \times [n])$ (since Ψ_n is onto). Conversely, if $i \in [l]$ and $j \in [n]$, then $\Psi_n(i, j) = (i-1)n + j \leq in \leq ln$. $\square$

Equipped with the remainder mapping r and the both bijections $\Psi_n : \mathbb{Z} \times [n] \longrightarrow \mathbb{Z}$ and $\Psi_m : \mathbb{Z} \times [m] \longrightarrow \mathbb{Z}$, we are now able to describe both, the bijective linear operator $\mathrm{vec} : \mathbb{M}_{m,n}(\mathbb{F}) \longrightarrow \mathbb{F}^{mn}$ and the Kronecker product explicitly *entrywise*. So, fix $m, n, p, q \in \mathbb{N}$. If $A = (a_{ij}) \in \mathbb{M}_{m,n}(\mathbb{F})$, $B = (b_{kl}) \in \mathbb{M}_{p,q}(\mathbb{F})$, $\alpha \in [mp]$, $\beta \in [nq]$ and $\gamma \in [mn]$, put

$$(A \otimes B)_{\alpha, \beta} := a_{f_p(\alpha), f_q(\beta)} \cdot b_{r_p(\alpha), r_q(\beta)} = ((A^\top \otimes B^\top)^\top)_{\alpha, \beta} \tag{3.4.3}$$

and

$$\mathrm{vec}(A)_\gamma \equiv \mathrm{vec}_m(A)_\gamma := a_{r_m(\gamma), f_m(\gamma)} = (A^\top)_{\Psi_m^{-1}(\gamma)} = \left(A_{\tau \circ \Psi_m^{-1}}\right)_\gamma, \tag{3.4.4}$$

where $(i, j) \mapsto \tau(i, j) := (j, i)$ denotes transposition. In other words, $\mathrm{vec} \equiv \mathrm{vec}_m = C_{\tau \circ \Psi_m^{-1}}$. In particular,

$$\mathrm{vec}\left(e_i^{(m)}(e_j^{(n)})^\top\right) = e_{(j-1)m+i}^{(nm)} \text{ for all } (i, j) \in [m] \times [n].$$

Consequently, Lemma 3.3 implies that

$$e_v^{(nm)} = \text{vec}\left(e_{r_m(v)}^{(m)} e_{f_m(v)}^{(n)\top}\right) \text{ for all } v \in [nm]. \tag{3.4.5}$$

Moreover, if $n = 1$, it follows that $\text{vec}(A)_i = a_{i,1}$ for all $i \in [m]$. Observe that the vectorisation of the matrix A involves its transpose $A^\top = A_\tau \in \mathbb{M}_{n,m}(\mathbb{F})$. Not too surprisingly, it follows that

$$\text{vec}_n \circ C_\tau(A) = \text{vec}_n(A^\top) = \text{vec}_n(A_\tau) = A_{\Psi_n^{-1}} = C_{\Psi_n^{-1}}(A)$$

(since $\tau \circ \tau = \text{id}$). The construction implies at once the non-trivial fact that $\text{vec}_n(A^\top) = C_{\Psi_m \circ \tau \circ \Psi_n^{-1}}(\text{vec}_m(A))$, where $C_{\Psi_m \circ \tau \circ \Psi_n^{-1}} : \mathbb{F}^{mn} \longrightarrow \mathbb{F}^{nm} = \mathbb{F}^{mn}$ is the related linear composition operator. Consequently,

$$\text{vec}_n(A^\top) = K_{m,n} \text{vec}_m(A),$$

where the matrix $K_{n,m}^\top = K_{m,n} \in O(mn)$ satisfies

$$(K_{m,n})_{v\mu} = \delta_{(\Psi_m \circ \tau \circ \Psi_n^{-1})(v),\,\mu} = \delta_{(r_n(v)-1)m+f_n(v),\,\mu}$$

$$\stackrel{(!)}{=} \delta_{f_m(\mu),r_n(v)} \cdot \delta_{f_n(v),r_m(\mu)}$$

$$\stackrel{(3.4.3)}{=} \left(\sum_{i=1}^{m} \sum_{j=1}^{n} e_i e_j^\top \otimes e_j e_i^\top\right)_{v\mu}$$

for all $v, \mu \in [mn]$. The second equality follows from Lemma 3.3: $(r_n(v) - 1)m + f_n(v) = \mu$ if and only if $\Psi_m(r_n(v), f_n(v)) = \mu = \Psi_m(f_m(\mu), r_m(\mu))$. Since $f_n(v) \in [m]$, it follows that $r_n(v) = f_m(\mu)$ and $f_n(v) = r_m(\mu)$. Therefore, Lemma 3.3 (respectively Euclidean division with remainder) allows to extend the results in [1, Chapter 11, including Exercise 11.8] by an explicit entrywise (and hence implementable) description of the commutation matrix $K_{m,n} = \sum_{i=1}^{m} \sum_{j=1}^{n} e_i e_j^\top \otimes e_j e_i^\top$; namely in form of a product of two Kronecker delta symbols.

Moreover, $\text{vec}^{-1} = \text{mat}$, where the "matrixation operator" $\text{mat} : \mathbb{F}^{mn} \longrightarrow \mathbb{M}_{m,n}(\mathbb{F})$ is given by $\text{mat} \equiv \text{mat}_m := C_{\Psi_m \circ \tau}$; i.e.,

$$\text{mat}(x) := x_{\Psi_m \circ \tau} = \left(x_{\Psi_m(\tau(i,j))}\right)_{(i,j)} = \left(x_{(j-1)m+i}\right)_{(i,j)}$$

$$= \begin{pmatrix} x_1 & x_{m+1} & \cdots & x_{(n-1)m+1} \\ x_2 & x_{m+2} & \cdots & x_{(n-1)m+2} \\ \vdots & \vdots & \cdots & \vdots \\ x_m & x_{2m} & \cdots & x_{mn} \end{pmatrix}$$

3.4 $K_G^{\mathbb{R}}(2)$ and the Walsh-Hadamard Transform: Krivine's Approach Revisited

for all $x \in \mathbb{F}^{mn}$. Again, if $n = 1$, we recognise that $\mathrm{mat}(x)_{i1} = x_i$ for all $i \in [m]$. Consequently, we may identify $\mathrm{mat}(\mathbb{F}^m) \equiv \mathrm{vec}(\mathbb{M}_{m,1}(\mathbb{F})) \equiv \mathbb{F}^m$ for all $m \in \mathbb{N}$, so that we may assume without loss of generality that $(m, n) \in \mathbb{N}_2 \times \mathbb{N}_2$. That assumption also avoids necessarily the review, whether vec maps $\mathbb{F}^{mn}$ into $\mathbb{M}_{m,n}(\mathbb{F})$, or into $\mathbb{M}_{mn,1}(\mathbb{F}) \equiv \mathbb{F}^{mn}$, or into $\mathbb{M}_{1,mn}(\mathbb{F}) \equiv \{x^\top : x \in \mathbb{F}^{mn}\}$!

Example 3.1 Consider $A := \begin{pmatrix} a_{11} & a_{12} & a_{13} \\ a_{21} & a_{22} & a_{23} \end{pmatrix} \in \mathbb{M}_{2,3}(\mathbb{F})$. Then $4 \in [2 \cdot 3] = [6]$ and $\Psi_3^{-1}(4) = (\frac{4-r_3(4)}{3} + 1, r_3(4)) = (2, 1)$. Thus, $\mathrm{vec}(A^\top)_4 = a_{21}$.

Therefore, we reobtain the following well-known, easily computable statements

$$xy^\top \equiv x \otimes y^\top \text{ for all } (x, y) \in \mathbb{F}^m \times \mathbb{F}^n.$$

and

$$\mathrm{vec}(xy^\top) \equiv y \otimes x \text{ for all } (x, y) \in \mathbb{F}^m \times \mathbb{F}^n.$$

In particular,

$$e_\nu^{(nm)} \stackrel{(3.4.5)}{=} e_{f_m(\nu)}^{(n)} \otimes e_{r_m(\nu)}^{(m)}$$

$$\| \qquad\qquad \|$$

$$e_{(j-1)m+i}^{(nm)} \qquad e_j^{(n)} \otimes e_i^{(m)}$$

(3.4.6)

for all $\nu \in [nm] = \Psi_m([n] \times [m]) = \{(j-1)m+i : (j, i) \in [n] \times [m]\}$. Equivalently (now translated into Dirac's bra-ket language):

$$|\alpha \beta\rangle \equiv |\alpha\rangle |\beta\rangle = |\alpha m + \beta\rangle \text{ for all } (\alpha, \beta) \in \{0, 1, \ldots, n-1\} \times \{0, 1, \ldots, m-1\}.$$

(3.4.7)

Consequently, if $C = \sum_{j=1}^{n} \sum_{k=1}^{p} c_{jk} e_j e_k^\top \in \mathbb{M}_{n,p}(\mathbb{F})$ is a third given matrix, then $ACB = \sum_{j=1}^{n} \sum_{k=1}^{p} c_{jk} (A e_j)(B^\top e_k)^\top$ and $\mathrm{vec}(C) = \sum_{j=1}^{n} \sum_{k=1}^{p} c_{jk} e_k \otimes e_j$, leading to another important, well-known matrix equality:

$$\mathrm{vec}(ACB) = (B^\top \otimes A)\mathrm{vec}(C).$$

Example 3.2 (Werner State) Let $p \in [0, 1]$. Put

$$\mathbb{R}^4 \ni \psi^- := \frac{1}{\sqrt{2}}(0, 1, -1, 0)^\top \stackrel{(3.4.7)}{=} \frac{1}{\sqrt{2}}(|0\,1\rangle - |1\,0\rangle).$$

Consider the matrix

$$\mathbb{M}_4(\mathbb{R}) \ni \rho_p^W := p\,\psi^-(\psi^-)^\top + \frac{1-p}{4} I_4 = \frac{1}{4}\begin{pmatrix} 1-p & 0 & 0 & 0 \\ 0 & 1+p & -2p & 0 \\ 0 & -2p & 1+p & 0 \\ 0 & 0 & 0 & 1-p \end{pmatrix}.$$

Since the rank one matrix $p\,\psi^-(\psi^-)^\top$ is positive semidefinite and $\frac{1-p}{4} I_4$ is positive definite if $p < 1$, it immediately follows that ρ_p^W is positive definite if $p < 1$ and $\rho_1^W = \psi^-(\psi^-)^\top$ is positive semidefinite (yet not invertible). Moreover, by construction, it trivially follows that in general $\mathrm{tr}(\rho_p^W) = 1$. Note that ρ_p^W can also be written as

$$\rho_p^W = \frac{3 - 2\lambda(p)}{6} I_4 - \frac{3 - 4\lambda(p)}{6} G_1 = \frac{3 - 2\lambda(p)}{6}\left(I_4 - \frac{3 - 4\lambda(p)}{3 - 2\lambda(p)} G_1\right),$$

where $\lambda(p) := \frac{3}{4}(1 - p) \in [0, 1]$ and $G_1 \in O(4)$ is the "flip operator" (cf. (1.2.6)). In quantum physics, the matrix $\rho_p^W \in \mathbb{M}_4(\mathbb{R})^+$ is known as the so-called "two-qubit Werner state". A very detailed discussion of the origin and the meaning of Werner states in the foundations and philosophy of quantum mechanics, particularly in relation to the topic of entanglement and "local hidden-variable theories" can be found, for example, in [9, 25, 45] and in the relevant references therein.

Proposition 3.5 Let $m, n, p, q \in \mathbb{N}$, $S \in \mathbb{M}_{m,n}(\mathbb{F})$ and $R \in \mathbb{M}_{p,q}(\mathbb{F})$. If $S \in \mathcal{Q}_{m,n}(\mathbb{F})$ and $R \in \mathcal{Q}_{p,q}(\mathbb{F})$, then $S \otimes R \in \mathcal{Q}_{mp,nq}(\mathbb{F})$.

Proof We only have to apply Proposition 3.3 and the entrywise description (3.4.3) of the Kronecker product $S \otimes R$. So, choose $\mathbb{F}$-Hilbert spaces H_1, H_2, $(u^{(1)}, v^{(1)}) \in S_{H_1}^m \times S_{H_1}^n$ and $(u^{(2)}, v^{(2)}) \in S_{H_2}^p \times S_{H_2}^q$, such that $S = \Gamma_{H_1}(u^{(1)}, v^{(1)})$ and $R = \Gamma_{H_2}(u^{(2)}, v^{(2)})$. Fix $(\alpha, \beta) \in [mp] \times [nq]$. (3.4.3) consequently implies that

$$(S \otimes R)_{\alpha,\beta} = \left\langle v^{(1)}_{\frac{\beta - r_q(\beta)}{q}+1}, u^{(1)}_{\frac{\alpha - r_p(\alpha)}{p}+1}\right\rangle_{H_1} \cdot \left\langle v^{(2)}_{r_q(\beta)}, u^{(2)}_{r_p(\alpha)}\right\rangle_{H_2} = \langle v_\beta, u_\alpha\rangle_H,$$

where the Hilbert space $H := H_1 \otimes H_2$ denotes the standard tensor product of the Hilbert spaces H_1 and H_2, $v_\beta := v^{(1)}_{\frac{\beta - r_q(\beta)}{q}+1} \otimes v^{(2)}_{r_q(\beta)}$ and $u_\alpha := u^{(1)}_{\frac{\alpha - r_p(\alpha)}{p}+1} \otimes u^{(2)}_{r_p(\alpha)}$. Since $u_\alpha \in S_H$ and $v_\beta \in S_H$ (by construction), it follows that $S \otimes R = \Gamma_H(u, v)$, where $(u, v) \in S_H^{mp} \times S_H^{nq}$. Thus, $S \otimes R \in \mathcal{Q}_{mp,nq}(\mathbb{F})$ (again, a consequence of Proposition 3.3). □

3.4 $K_G^{\mathbb{R}}(2)$ and the Walsh-Hadamard Transform: Krivine's Approach Revisited

An important example of a matrix $A \in \mathbb{M}_{2^m}(\{-1,1\})$, which satisfies $\|A\|_{\infty,1}^{\mathbb{R}} \leq 1$ and delivers $\sqrt{2}$ as a lower bound of $K_G^{\mathbb{R}}$ (cf. Proposition 3.7), and also plays a key role in the foundations of quantum mechanics and quantum information is the so-called *Walsh-Hadamard transform* (also known as *quantum gate*—cf. [25]). In the following enlightening example, we extend the Walsh-Hadamard transform $H_m \in \mathbb{M}_{2^m}(\{-1,1\})$ to a complex Walsh-Hadamard transform $H_m^{\mathbb{C}} \in \mathbb{M}_{2^m}(\mathbb{T})$ and disclose some surprising properties of that matrix. In particular, we will show that the (value of the) sign of any of the 4^m entries of the real Walsh-Hadamard transform can be specified precisely, in exactly m calculation steps—for any $m \in \mathbb{N}$! To this end, recall the construction of the function $\pi_m : \mathbb{Z} \longrightarrow \{0,1\}^m$ (cf. (3.4.2)) and put

$$N_m(\nu, \mu) := \langle \pi_m(\nu), \pi_m(\mu) \rangle_{\mathbb{F}_2^m} = \sum_{i=1}^{m} b_i(\nu) b_i(\mu)$$

$$= b_1(\nu) b_1(\mu) + N_{m-1}(f_2(\nu), f_2(\mu)),$$

where $(\nu, \mu) \in [2^m] \times [2^m]$ and $N_0 := 0$. $N_m(\nu, \mu)$ counts the number of all $i \in [m]$, such that $b_i(\nu) b_i(\mu) = 1$. In particular, $N_m(1, \mu) = 0$ for all $\mu \in [2^m]$ (since $b_1(1) = 0$).

Example 3.3 (Real and Complex Walsh-Hadamard Transform) Let

$$H_1 := \tfrac{1}{\sqrt{2}} \begin{pmatrix} 1 & 1 \\ 1 & -1 \end{pmatrix} \text{ and } H_1^{\text{op}} := R_2(i) H_1 = \tfrac{1}{\sqrt{2}} \begin{pmatrix} -1 & 1 \\ 1 & 1 \end{pmatrix}.$$

For $m \in \mathbb{N}$, put

$$H_{m+1} := H_m \otimes H_1 = H_1 \otimes H_m = \tfrac{1}{\sqrt{2}} \begin{pmatrix} H_m & H_m \\ H_m & -H_m \end{pmatrix}$$

and

$$H_{m+1}^{\text{op}} := H_m \otimes H_1^{\text{op}} = H_1 \otimes H_m^{\text{op}} = \tfrac{1}{\sqrt{2}} \begin{pmatrix} H_m^{\text{op}} & H_m^{\text{op}} \\ H_m^{\text{op}} & -H_m^{\text{op}} \end{pmatrix}.$$

Then the following properties are satisfied:

(i) $(H_1)_{\alpha\beta} = \tfrac{1}{\sqrt{2}} (-1)^{(\alpha-1)(\beta-1)}$ for all $(\alpha, \beta) \in [2] \times [2]$. $H_1^{\top} = H_1 \in O(2)$ and $(H_1^{\text{op}})^{\top} = H_1^{\text{op}} \in O(2)$. If $m \in \mathbb{N}_2$, then $(H_m)^{\top} = H_m \in SO(2^m)$ and $(H_m^{\text{op}})^{\top} = H_m^{\text{op}} \in SO(2^m)$.

(ii) Let $m \in \mathbb{N}_2$ and $(\nu, \mu) \in [2^m] \times [2^m]$. Then

$$(H_m)_{\nu\mu} = \tfrac{1}{\sqrt{2}} (H_{m-1})_{f_2(\nu) f_2(\mu)} \cdot (-1)^{b_1(\nu) b_1(\mu)} = \tfrac{1}{\sqrt{2^m}} (-1)^{N_m(\nu,\mu)}, \tag{3.4.8}$$

In particular,
$$(H_m)_{1\mu} = (H_m)_{\mu 1} = \frac{1}{\sqrt{2^m}}.$$

(iii) Let $m \in \mathbb{N}$. Then $H_m = \text{Re}(H_m^{\mathbb{C}}) \in \mathcal{Q}_{2^m, 2^m}(\mathbb{R})$ and $H_m^{\text{op}} = \text{Im}(H_m^{\mathbb{C}}) \in \mathcal{Q}_{2^m, 2^m}(\mathbb{R})$, where
$$H_m^{\mathbb{C}} := H_m + i H_m^{\text{op}} \in \mathcal{Q}_{2^m, 2^m}(\mathbb{C}).$$

In particular the matrix,
$$H_1^{\mathbb{C}} = \begin{pmatrix} 1 & -i \\ i & 1 \end{pmatrix} H_1 = \frac{1}{\sqrt{2}} \begin{pmatrix} 1-i & 1+i \\ 1+i & -1+i \end{pmatrix} = \overline{p} q^{\top} = \Gamma_{\mathbb{C}}(p, q)$$

is of rank 1 and satisfies $\|H_1^{\mathbb{C}}\|_{\infty, 1} = \|p\|_1 \|q\|_1 = 4$, where $p := (i, 1)^{\top} \in \mathbb{T}^2$ and $q := (\frac{1+i}{\sqrt{2}}, \frac{-1+i}{\sqrt{2}})^{\top} \in \mathbb{T}^2$ and $H_{m+1}^{\mathbb{C}} = H_1 \otimes H_m^{\mathbb{C}}$ for all $m \in \mathbb{N}$.

(iv) For any $m \in \mathbb{N}$, $\|H_m\|_{\infty, 1} = \|H_m^{\text{op}}\|_{\infty, 1}$. The sequence $(\|H_m\|_{\infty, 1})_{m \in \mathbb{N}}$ is non-decreasing. Moreover,

$$\sqrt{2} \|H_m\|_{\infty, 1} \leq \|H_{m+1}\|_{\infty, 1} \leq 2\sqrt{2} \|H_m\|_{\infty, 1} \qquad (3.4.9)$$

and

$$(\sqrt{2})^m \leq \|H_m\|_{\infty, 1} \leq (\sqrt{2})^{3m-2} \qquad (3.4.10)$$

for all $m \in \mathbb{N}$. In particular, $\|A_m^{\text{Had}}\|_{\infty, 1} \leq 1$ for all $m \in \mathbb{N}$, where

$$A_m^{\text{Had}} := \frac{1}{(\sqrt{2})^{3m-2}} H_m = \frac{1}{2^{2m-1}}((\sqrt{2})^m H_m)$$

and

$$\frac{1}{\sqrt{2}} \|H_1\|_{\infty, 1} = \|A_1^{\text{Had}}\|_{\infty, 1} = 1. \qquad (3.4.11)$$

Proof First of all, it should be noted that our entire proof (also) involves induction on $m \in \mathbb{N}$.

(i) It is sufficient to verify (i) for H_m only. The case $m = 1$ is trivial. Given the recursive construction of H_m, we only have to apply the rule for the transpose of a Kronecker product of two matrices to recognise that $H_m = H_m^{\top}$ (cf. (3.4.3)

3.4 $K_G^{\mathbb{R}}(2)$ and the Walsh-Hadamard Transform: Krivine's Approach Revisited

and [69, 4.2.4]). The recursive construction of H_m also directly implies that $H_m^2 = H_m H_m = I_{2^m}$, whence $H_m \in O(2^m)$. However, since

$$\det(H_{m+1}) = \det(H_1 \otimes H_m) = (\det(H_1))^{2^m} (\det(H_m))^2 = (\det(H_m))^2,$$

(cf. [69, Problem 4.2.1]), the induction hypothesis implies that $\det(H_{m+1}) = 1$, whence $H_{m+1} \in SO(2^{m+1})$.

(ii) The first equality in (3.4.8) instantly follows from the construction of b_1 and f_2 (cf. (3.4.1)) and (3.4.3), where the latter is applied to $p = q = 2$. Because of the obvious fact that $\pi_m \circ f_2 = (b_2, \ldots, b_{m+1})^\top$, a remaining and very elementary proof by induction on m immediately leads to the explicit entrywise determination of the sign of H_m in (3.4.8).

(iii) Due to Proposition 3.5, it is sufficient to verify the base case $m = 1$ only. Since by construction, $H_{m+1}^{\mathbb{C}} = H_1 \otimes H_m + i(H_1 \otimes H_m^{op}) = H_1 \otimes (H_m + i H_m^{op})$ for all $m \in \mathbb{N}$, we just have to prove the claims for H_1 and H_1^{op}. The complex case follows then immediately. To this end, consider the (real) Hilbert space $\mathbb{R}_2^2$. Put $u := (u_1, u_2)$ and $v := (v_1, v_2)$, where

$$u_1 := \begin{pmatrix} 0 \\ 1 \end{pmatrix}, u_2 := \begin{pmatrix} 1 \\ 0 \end{pmatrix}, v_1 := \frac{1}{\sqrt{2}} \begin{pmatrix} 1 \\ 1 \end{pmatrix} \text{ and } v_2 := \frac{1}{\sqrt{2}} \begin{pmatrix} -1 \\ 1 \end{pmatrix}.$$

Then $(u, v) \in (\mathbb{S}^1)^2 \times (\mathbb{S}^1)^2$, and

$$H_1 = \frac{1}{\sqrt{2}} \begin{pmatrix} 1 & 1 \\ 1 & -1 \end{pmatrix} = \Gamma_{\mathbb{R}_2^2}(u, v).$$

Thus, $H_1 \in \mathcal{Q}_{2,2}(\mathbb{R})$ (due to (3.3.1)). Since obviously, $R_2(i) = \begin{pmatrix} 0 & -1 \\ 1 & 0 \end{pmatrix}$ is an isometry from l_∞^2 to l_∞^2, it follows that $\|H_1^{op}\|_{\mathscr{L}_2} = \|R_2(i)H_1\|_{\mathscr{L}_2} \leq \|H_1\|_{\mathscr{L}_2} \leq 1$ (due to (3.3.1)). Consequently, a further application of (3.3.1) implies that $H_1^{op} \in \mathcal{Q}_{2,2}(\mathbb{R})$. Finally, since for any $\xi \in \mathbb{C}^2$, $\|(\overline{p} q^\top)\xi\|_1 = |\overline{q}^* \xi| \|p\|_1 = 2|\overline{q}^* \xi|$, and since $(l_\infty^2)' \cong l_1^2$, it follows that $\|H_1^{\mathbb{C}}\|_{\infty,1} = \|p\|_1 \|q\|_1 = 4$.

(iv) Firstly, let $m = 1$ (the base case). Let $p = (x_1, x_2)^\top \in \{-1, 1\}^2$ and $q = (y_1, y_2)^\top \in \{-1, 1\}^2$ be arbitrarily given. Then

$$\left|\operatorname{tr}((A_1^{\text{Had}})^\top p q^\top)\right| = \left|\operatorname{tr}(A_1^{\text{Had}} p q^\top)\right| = \left|q^\top A_1^{\text{Had}} p\right|$$
$$= \tfrac{1}{2} |(x_2 y_1 - x_2 y_2) + (x_1 y_1 + x_1 y_2)|.$$

Since $\max\{|x_1|, |x_2|, |y_1|, |y_2|\} \leq 1$, the triangle inequality clearly implies that

$$|(x_2 y_1 - x_2 y_2) + (x_1 y_1 + x_1 y_2)| \leq |y_1 - y_2| + |y_1 - (-y_2)|.$$

However, since $-1 \leq y_2$, it further follows that $y_1(1 + y_2) \leq |y_1(1 + y_2)| \leq 1 + y_2$, whence $y_1 - y_2 \leq 1 - y_1 y_2 = 1 - y_2 y_1$. Consequently, we obtain that $|y_1 - y_2| \leq 1 - y_1 y_2$. Similarly, since also $-1 \leq -y_2$, it follows that $|y_1 - (-y_2)| \leq 1 - y_1(-y_2) = 1 + y_1 y_2$. Hence,

$$\left|\text{tr}((A_1^{\text{Had}})^\top pq^\top)\right| \leq \tfrac{1}{2}(1 - y_1 y_2 + 1 + y_1 y_2) = 1.$$

On the other hand, an easy calculation shows that

$$\text{tr}((A_1^{\text{Had}})^\top \widetilde{p}\widetilde{q}^\top) = \widetilde{q}^\top A_1^{\text{Had}} \widetilde{p} = 1 \leq \|A_1^{\text{Had}}\|_{\infty,1},$$

where $\widetilde{p} := (1, -1)^\top$ and $\widetilde{q} := (1, 1)^\top$, which finishes the proof of the base case $m = 1$ as well as of (3.4.11). So let us now assume that (3.4.10) is satisfied for a fixed $m \in \mathbb{N}$ (induction hypothesis). Firstly, we consider the upper bound. Let $p = \text{vec}(p_1, p_2) \in \{-1, 1\}^{2^{m+1}} \equiv \{-1, 1\}^{2^m} \times \{-1, 1\}^{2^m}$ and $q = \text{vec}(q_1, q_2) \in \{-1, 1\}^{2^{m+1}} \equiv \{-1, 1\}^{2^m} \times \{-1, 1\}^{2^m}$. Since

$$A_{m+1}^{\text{Had}} = \tfrac{1}{2}(A_1^{\text{Had}} \otimes A_m^{\text{Had}}) = \tfrac{1}{4}\begin{pmatrix} A_m^{\text{Had}} & A_m^{\text{Had}} \\ A_m^{\text{Had}} & -A_m^{\text{Had}} \end{pmatrix},$$

it follows that

$$|\text{tr}(A_{m+1}^{\text{Had}} pq^\top)| = \tfrac{1}{4}|\text{tr}(A_m^{\text{Had}} p_1 q_1^\top + A_m^{\text{Had}} p_2 q_1^\top + A_m^{\text{Had}} p_1 q_2^\top - A_m^{\text{Had}} p_2 q_2^\top)|$$
$$\leq \|A_m^{\text{Had}}\|_{\infty,1}.$$

The induction hypothesis therefore implies that $\|A_{m+1}^{\text{Had}}\|_{\infty,1} \leq \|A_m^{\text{Had}}\|_{\infty,1} \leq 1$, and hence $\|H_{m+1}\|_{\infty,1} \leq 2\sqrt{2}\|H_m\|_{\infty,1} \leq (\sqrt{2})^{3(m+1)-2}$. Now, let us turn to the lower bound. To this end, let $s, r \in \{-1, 1\}^{2^m}$ be arbitrarily chosen. Consider $\widetilde{s} := \text{vec}(s, s) \in \{-1, 1\}^{2^{m+1}}$ and $\widetilde{r} := \text{vec}(r, r) \in \{-1, 1\}^{2^{m+1}}$. Then

$$|\text{tr}(A_{m+1}^{\text{Had}} \widetilde{s}\widetilde{r}^\top)| = \frac{1}{4}|\text{tr}(A_m^{\text{Had}} sr^\top + 2A_m^{\text{Had}} sr^\top - A_m^{\text{Had}} sr^\top)| = \frac{1}{2}|\text{tr}(A_m^{\text{Had}} sr^\top)|.$$

Consequently, it follows that

$$\tfrac{1}{2}\|A_m^{\text{Had}}\|_{\infty,1} \leq \|A_{m+1}^{\text{Had}}\|_{\infty,1},$$

whence $\|H_m\|_{\infty,1} \leq \sqrt{2}\|H_m\|_{\infty,1} \leq \|H_{m+1}\|_{\infty,1}$. The induction hypothesis therefore implies that $(\sqrt{2})^{m+1} \leq \|H_{m+1}\|_{\infty,1}$, and it follows that the estimates (3.4.9) and (3.4.10) are valid.

□

3.4 $K_G^{\mathbb{R}}(2)$ and the Walsh-Hadamard Transform: Krivine's Approach Revisited

Remark 3.9 After some "skillful searching", a then simple calculation shows that also

$$\|A_2^{\text{Had}}\|_{\infty,1} = 1.$$

If we namely consider the vectors $\widetilde{p} := (1, 1, -1, 1)^\top \in \{-1, 1\}^4$ and $\widetilde{q} := (1, -1, 1, 1)^\top \in \{-1, 1\}^4$, it follows that

$$\text{tr}(A_2^{\text{Had}} \widetilde{p}\widetilde{q}^\top) = \langle A_2^{\text{Had}} \widetilde{p}, \widetilde{q}\rangle_{\mathbb{R}_2^4} = \frac{1}{8}\left\langle \begin{pmatrix} 2 \\ -2 \\ 2 \\ 2 \end{pmatrix}, \begin{pmatrix} 1 \\ -1 \\ 1 \\ 1 \end{pmatrix} \right\rangle = 1.$$

In particular,

$$\left|(A_m^{\text{Had}} \widetilde{p})_i\right| = \frac{1}{2^m} \text{ for all } i \in [2^m] \qquad (3.4.12)$$

(since $m = 2$). This naturally leads to the (open) question, whether $\|A_m^{\text{Had}}\|_{\infty,1} = 1$ for all $m \in \mathbb{N}$ and whether (3.4.12) holds for all $m \in \mathbb{N}$. It seems that we cannot make use of induction on $m \in \mathbb{N}$ here. In fact, if $\nu \neq 2$ and $\|p\|_{l_\infty^{2^\nu}} \leq 1$, then $\left|(A_\nu^{\text{Had}} p)_{i_0}\right| \neq \frac{1}{2^\nu}$ for some $i_0 \in [2^\nu]$! The case $\nu = 1$ follows from the fact that for any $a, b \in [-1, 1]$, $|a + b| = 1 = |a - b|$ if and only if $\begin{pmatrix} a \\ b \end{pmatrix} \in \left\{ \begin{pmatrix} 0 \\ 1 \end{pmatrix}, \begin{pmatrix} 0 \\ -1 \end{pmatrix}, \begin{pmatrix} 1 \\ 0 \end{pmatrix}, \begin{pmatrix} -1 \\ 0 \end{pmatrix} \right\}$. In order to verify the claim for $\nu > 2$, assume by contradiction that there exist $m \in \mathbb{N}_3$ and $\widetilde{p} \in B_{l_\infty^{2^m}}$, such that $\frac{1}{2^m} = \left|(A_m^{\text{Had}} \widetilde{p})_i\right| = \frac{1}{(\sqrt{2})^{3m-2}} |(H_m \widetilde{p})_i| = \frac{1}{2^{2m-1}} \left| \sum_{j=1}^{2^m} (-1)^{N_m(i,j)} \widetilde{p}_j \right|$ for all $i \in [2^m]$. Put $\widetilde{q} := 2^m A_m^{\text{Had}} \widetilde{p}$. Then $\widetilde{q} \in \{-1, 1\}^{2^m}$ (due to the assumption) and

$$1 = 2^m \|A_m^{\text{Had}} \widetilde{p}\|^2_{l_2^{2^m}} = \langle A_m^{\text{Had}} \widetilde{p}, \widetilde{q}\rangle_{\mathbb{R}_2^{2^m}} = 2^m \langle \widetilde{p}, (A_m^{\text{Had}})^2 \widetilde{p}\rangle_{\mathbb{R}_2^{2^m}}$$

$$= \frac{2^m}{2^{3m-2}} \|\widetilde{p}\|^2_{l_2^{2^m}} \leq \frac{1}{2^{3m-2}} 2^{2m} = \frac{1}{2^{m-2}}.$$

On the other hand, $\frac{1}{2^{m-2}} \leq \frac{1}{2} < 1$ (since $m \geq 3$ by assumption), which is absurd. Observe that in any case $\left|(A_m^{\text{Had}} x)_i\right| \leq \frac{2^m}{2^{2m-1}} = \frac{2}{2^m}$ for all $i \in [2^m]$ and $x \in B_{l_\infty^{2^m}}$. Consequently,

$$\left\|A_m^{\text{Had}} \frac{p}{2}\right\|_{l_\infty^{2^m}} < \frac{1}{2^m}$$

for all $m \in \mathbb{N}_3$ and $p \in B_{l_\infty^{2^m}}$. Our conjecture is that there exists $\tilde{m} \in \mathbb{N}_3$ such that $\|A_{\tilde{m}}^{\text{Had}}\|_{\infty,1} < 1$.

To be more explicit, note e.g. that

$$H_2 = \frac{1}{2}\begin{pmatrix} 1 & 1 & 1 & 1 \\ 1 & -1 & 1 & -1 \\ 1 & 1 & -1 & -1 \\ 1 & -1 & -1 & 1 \end{pmatrix}, \quad H_2^{\text{op}} = \frac{1}{2}\begin{pmatrix} -1 & 1 & -1 & 1 \\ 1 & 1 & 1 & 1 \\ -1 & 1 & 1 & -1 \\ 1 & 1 & -1 & -1 \end{pmatrix}$$

and

$$H_3 = \frac{1}{(\sqrt{2})^3}\begin{pmatrix} 1 & 1 & 1 & 1 & 1 & 1 & 1 & 1 \\ 1 & -1 & 1 & -1 & 1 & -1 & 1 & -1 \\ 1 & 1 & -1 & -1 & 1 & 1 & -1 & -1 \\ 1 & -1 & -1 & 1 & 1 & -1 & -1 & 1 \\ 1 & 1 & 1 & 1 & -1 & -1 & -1 & -1 \\ 1 & -1 & 1 & -1 & -1 & 1 & -1 & 1 \\ 1 & 1 & -1 & -1 & -1 & -1 & 1 & 1 \\ 1 & -1 & -1 & 1 & -1 & 1 & 1 & -1 \end{pmatrix}.$$

To see how smoothly and quickly (3.4.8) can be applied to H_3, let us perform a calculation for the two matrix entries $(H_3)_{6,4}$ and $(H_3)_{7,3}$ as an example. There are two ways to perform the calculation process. Either we go through each step of the recursion relation, or we count, step by step, how often the products $b_i(\nu)b_i(\mu)$ are equal to 1. So, either we proceed with

$$(H_3)_{6,4} = -\frac{1}{\sqrt{2}}(H_2)_{3,2} = -\frac{1}{\sqrt{2^2}}(H_1)_{2,1} = \frac{1}{\sqrt{2^3}}(-1)$$

and

$$(H_3)_{7,3} = \frac{1}{\sqrt{2}}(H_2)_{4,2} = -\frac{1}{\sqrt{2^2}}(H_1)_{2,1} = \frac{1}{\sqrt{2^3}}(-1),$$

or we count: $N_3(6, 4) = 1$ (since $b_1(6)b_1(4) = 1$, $b_2(6)b_2(4) = b_1(3)b_1(2) = 0$ and $b_3(6)b_3(4) = b_2(3)b_2(2) = b_1(2)b_1(1) = 0$). Similarly, we obtain that $N_3(7, 3) = 1$. However, since $\frac{1}{\sqrt{2^3}}(-1) = (H_3)_{7,3} \neq \frac{1}{\sqrt{2^3}} = \frac{1}{\sqrt{2^3}}(-1)^{(7-1)(3-1)}$, [108, Exercise 2.33, (2.55)] seems to be wrong.

Remark 3.10 (CHSH Inequalities) As we have seen (just by making use of elementary calculus on the real line), the following inequality holds

$$|x_1 y_1 + x_1 y_2 + x_2 y_1 - x_2 y_2| \leq 2 \text{ for all } (x_1, x_2, y_1, y_2) \in [-1, 1]^4, \quad (3.4.13)$$

3.4 $K_G^{\mathbb{R}}(2)$ and the Walsh-Hadamard Transform: Krivine's Approach Revisited

which is equivalent to $A_1^{\text{Had}} = \frac{1}{\sqrt{2}} H_1 \in B_{\mathcal{L}(l_\infty^2, l_1^2)}$. Even $A_1^{\text{Had}} \in S_{\mathcal{L}(l_\infty^2, l_1^2)}$ holds (cf. (3.4.11)). We also know that $\|A_1^{\text{Had}}\|_{\infty, 1} \le 1$ is equivalent to

$$|\text{tr}(A_1^{\text{Had}} B)| \le 1 \text{ for all } B \in B_{\mathfrak{N}(l_1^2, l_\infty^2)}$$

(cf. (3.3.3)). On the other hand,

$$\sqrt{2} = \frac{1}{\sqrt{2}} |\text{tr}(I_2)| = \frac{1}{\sqrt{2}} |\text{tr}(H_1^2)|$$

$$= |\text{tr}(A_1^{\text{Had}} \widetilde{S})| > 1 \text{ for some } \widetilde{S} \in \mathcal{Q}_{2,2} = B_{\mathcal{L}_2(l_1^2, l_\infty^2)}$$

(namely, $\widetilde{S} := H_1$), implying again that $B_{\mathfrak{N}(l_1^2, l_\infty^2)}$ is strictly contained in $B_{\mathcal{L}_2(l_1^2, l_\infty^2)}$. Particularly, physicists, who are working in the foundations and philosophy of quantum mechanics recognise that (3.4.13)—which are just inequalities between certain real numbers—*instantly imply* the famous CHSH inequalities. CHSH stands for John Clauser, Michael Horne, Abner Shimony, and Richard Holt, who introduced the inequalities (between expectation values) in [27] (cf. https://www.nobelprize.org/prizes/physics/2022/clauser/facts/) and used them as a means of proving Bell's theorem. In the 2-dimensional case (i.e., if $m = n = 2$) the CHSH inequalities coincide with the so-called "Bell inequalities", assigned to the matrix A_1^{Had}. Somewhat vaguely, it is said that the matrix $\widetilde{S} = H_1 \in \mathcal{Q}_{2,2}(\mathbb{R})$ "violates the Bell inequalities". That "violation" implies that certain consequences of spatial entanglement in quantum mechanics can not be reproduced by classical probability theory in the sense of A. Kolmogorov (i.e., it cannot be reduced to "local hidden-variable theories").

Remark 3.11 (An Application of H_m in Evolutionary Biology) The Walsh-Hadamard transform can even be found in evolutionary biology, specifically in relation to the challenge of reconstructing evolutionary trees from events several million years in the past. (cf. [68])! In order to recognise this, we consider the orthogonal matrix $I_1^{(-)} := \begin{pmatrix} 1 & 0 \\ 0 & -1 \end{pmatrix} \in O(2)$. A straightforward proof by induction shows that the matrix family $\{H^{(m)} : m \in \mathbb{N}\} = \{(1)\} \cup \{H^{(m)} : m \in \mathbb{N}_2\}$, consisting of invertible matrices $H^{(m)} \in \mathbb{M}_{2^{m-1}}(\mathbb{R})$, introduced in [68], in fact can be represented as

$$H^{(1)} := (1)$$

and

$$H^{(m)} := \begin{pmatrix} H^{(m-1)} & -H^{(m-1)} \\ H^{(m-1)} & H^{(m-1)} \end{pmatrix} = H^{(2)} \otimes H^{(m-1)} = \bigotimes_{i=1}^{m-1} H^{(2)} = \sqrt{2^{m-1}} \, H_{m-1} \, I_{m-1}^{(-)}$$

if $m \in \mathbb{N}_2$, where $I_l^{(-)} := I_1^{(-)} \otimes I_{l-1}^{(-)} = \bigotimes_{i=1}^{l} I_1^{(-)} \in O(2^l)$ for all $l \in \mathbb{N}_2$.

It is quite instructive to compare (3.3.16) to (6.4.12) (real case), (7.3.5) (complex case) and the following

Proposition 3.6 *Suppose there exist* $c \in (1, \infty)$, *a sequence* $(r_\nu)_{\nu \in \mathbb{N}} \in B_{l_1(\mathbb{F})}$, *a probability space* $(\Omega, \mathscr{F}, \mathbb{P})$ *and sequences* $(P_{1,\nu})_{\nu \in \mathbb{N}}, \ldots, (P_{m,\nu})_{\nu \in \mathbb{N}}$, $(Q_{1,\nu})_{\nu \in \mathbb{N}}, \ldots, (Q_{n,\nu})_{\nu \in \mathbb{N}}$ *of random variables which map into* $S_{\mathbb{F}}$ $\mathbb{P}$-*a.s., such that for all* $m, n \in \mathbb{N}$, *for all* $\mathbb{F}$-*Hilbert spaces* H, *for all* $(u, v) \in S_H^m \times S_H^n$, *and for all* $(i, j) \in [m] \times [n]$ *the following equality holds:*

$$\langle v_j, u_i \rangle_H = \Gamma_H(u, v)_{ij} = c \sum_{\nu=1}^\infty r_\nu \, \mathbb{E}_\mathbb{P}[\overline{P_{i,\nu}} Q_{j,\nu}].$$

Then

$$K_G^{\mathbb{F}} \leq c.$$

Proof

$$|\mathrm{tr}(A^* \Gamma_H(u, v))| = c \Big| \sum_{\nu=1}^\infty r_\nu \Big(\sum_{i=1}^m \sum_{j=1}^n \overline{a_{ij}} \, \mathbb{E}_\mathbb{P}[\overline{P_{i,\nu}} Q_{j,\nu}] \Big) \Big|$$

$$\leq c \sum_{\nu=1}^\infty |r_\nu| \Big| \sum_{i=1}^m \sum_{j=1}^n \overline{a_{ij}} \, \mathbb{E}_\mathbb{P}[\overline{P_{i,\nu}} Q_{j,\nu}] \Big|$$

$$\leq c \sum_{\nu=1}^\infty |r_\nu| \, \mathbb{E}_\mathbb{P}\Big[\Big| \sum_{i=1}^m \sum_{j=1}^n \overline{a_{ij}} \, \overline{P_{i,\nu}} Q_{j,\nu} \Big| \Big]$$

$$= c \sum_{\nu=1}^\infty |r_\nu| \, \mathbb{E}_\mathbb{P} |\mathrm{tr}(A^* \Gamma_\mathbb{F}(\mathbf{P}_\nu, \mathbf{Q}_\nu))|,$$

where the components of the random vectors $\mathbf{P}_\nu : \Omega \longrightarrow \mathbb{F} \cap \overline{\mathbb{D}}^m$ and $\mathbf{Q}_\nu : \Omega \longrightarrow \mathbb{F} \cap \overline{\mathbb{D}}^n$ are defined as $(\mathbf{P}_\nu)_i := P_{i,\nu}$ and $(\mathbf{Q}_\nu)_j := Q_{j,\nu}$. Since $|\mathrm{tr}(A^* \Gamma_\mathbb{F}(\mathbf{P}_\nu(\omega), \mathbf{Q}_\nu(\omega)))| \leq \|A\|_{\infty,1}$ for almost all $\omega \in \Omega$, it therefore follows that

$$|\mathrm{tr}(A^* \Gamma_H(u, v))| \leq c \sum_{\nu=1}^\infty |r_\nu| \|A\|_{\infty,1} \leq c \|A\|_{\infty,1}$$

(since $\sum_{\nu=1}^\infty |r_\nu| \leq 1$, by assumption). □

Of particular interest is the value of $K_G^\mathbb{R}(2)$. A *lower* bound is rather easy to detect: $\sqrt{2} \leq K_G^\mathbb{R}(2)$, implying the important fact that $K_G^\mathbb{R} > 1$. We just have to work with the Walsh-Hadamard transform $H_1 = \Gamma_{\mathbb{R}_2^2}(u_1, u_2, v_1, v_2) \in \mathcal{Q}_{2,2}(\mathbb{R}) \cap O(2)$

3.4 $K_G^{\mathbb{R}}(2)$ and the Walsh-Hadamard Transform: Krivine's Approach Revisited

(cf. Example 3.3). To this end, consider again the symmetric matrix $A_1^{\text{Had}} = \frac{1}{\sqrt{2}} H_1$. Recall that $\|A_1^{\text{Had}}\|_{\infty,1} = 1$ (due to (3.4.11)), and note that $\text{tr}(A_1^{\text{Had}} H_1) = \frac{1}{\sqrt{2}} \text{tr}(H_1^2) = \frac{1}{\sqrt{2}} \text{tr}(I_2) = \sqrt{2}$. Consequently, it follows that

$$\sqrt{2} = |\text{tr}((A_1^{\text{Had}})^\top H_1)| \leq K_G^{\mathbb{R}}(2, 2; 2) \leq \min\{K_G^{\mathbb{R}}(2, 2), K_G^{\mathbb{R}}(2)\} \leq K_G^{\mathbb{R}}.$$

Much less trivial is the proof of the reverse direction, performed by Krivine in [90, 92]. Within the scope of Proposition 3.4 he namely represented—in the real 2-dimensional case—any $\Gamma_{\mathbb{R}_2^2}(u, v) \in \mathcal{Q}_{2,2}(\mathbb{R})$ as matrix $\sum_{\nu=1}^{\infty} b_\nu T_\nu : l_1^n \longrightarrow l_\infty^m$ such that $\|\Gamma_{\mathbb{R}_2^2}(u,v)\|_{\mathfrak{N}} = \|\sum_{\nu=1}^{\infty} b_\nu T_\nu\|_{\mathfrak{N}} \leq \sqrt{2}$ (which he called "norme de la fonction $\cos(x - y)$ dans le produit tensoriel projectif $C[-\pi, \pi] \widehat{\otimes} C[-\pi, \pi]$" in [90]). Actually, the main building block in his proof is an intricate sophisticated representation of the function $\mathbb{R} \times \mathbb{R} \ni (x, y) \mapsto \cos(x - y)$ by convolution (cf. Theorem 3.6). However, that representation allows us to provide a short, straightforward proof of Corollary 3.4—without the use of any tensor product structure.

Theorem 3.6 (Krivine [90]) *Consider the probability space* $([-\pi, \pi], \mathcal{B}([-\pi, \pi]), \mu)$, *where* $\mu := \frac{1}{2\pi} \lambda_1|_{\mathcal{B}([-\pi,\pi])}$. *Then there exist two functions* $p, q : \mathbb{R} \longrightarrow [-1, 1]$ *and a sequence* $(r_n)_{n \in \mathbb{N}}$ *of real numbers such that* $\sum_{n=1}^{\infty} |r_n| = 1$ *and*

$$\cos(x - y) = \sqrt{2} \sum_{n=1}^{\infty} r_n \mathbb{E}_\mu[p(nx - S)q(ny - S)]$$

$$= \frac{\sqrt{2}}{2\pi} \sum_{n=1}^{\infty} r_n \int_{[-\pi,\pi]} p(nx - \omega)q(ny - \omega) \lambda_1(d\omega)$$

for all $x, y \in \mathbb{R}$, *where* $\mathbb{R} \ni t \mapsto q(t) := \text{sign}(\cos(t))$ *and* $[-\pi, \pi] \ni \omega \mapsto S(\omega) := \omega$.

Corollary 3.4 (Krivine [90])

$$K_G^{\mathbb{R}}(2) = \sqrt{2}.$$

Proof Consider the 2-dimensional standard Euclidean vector space $H := \mathbb{R}_2^2$. Let $u = \begin{pmatrix} u_1 \\ u_2 \end{pmatrix} \in S_H = \mathbb{S}^1$ and $v = \begin{pmatrix} v_1 \\ v_2 \end{pmatrix} \in S_H = \mathbb{S}^1$. Since

$$\langle u, v \rangle_H = \langle \begin{pmatrix} \cos(x) \\ \sin(x) \end{pmatrix}, \begin{pmatrix} \cos(y) \\ \sin(y) \end{pmatrix} \rangle_H = \cos(x - y),$$

for some $x, y \in \mathbb{R}$, Theorem 3.6 unveils as a special case of Proposition 3.6, and the claim follows. $\square$

Proposition 3.4, together with Corollary 3.4 help us to find a "fitting" relation between $K_G^{\mathbb{R}}(2d)$ and $K_G^{\mathbb{C}}(d)$. To this end, we firstly supplement and prove once again (for the sake of completeness) [48, Corollary 2.14., (38)].

Proposition 3.7 *Let $d, m, n \in \mathbb{N}$, $A \in \mathbb{M}_{m,n}(\mathbb{R})$ and $B \in \mathbb{M}_{m,n}(\mathbb{C})$. Then*

$$\|A\|_{\mathbb{R}_2^d}^G \leq \|A\|_{\mathbb{C}_2^d}^G = \|A\|_{\mathbb{R}_2^{2d}}^G \leq K_G^{\mathbb{R}}(m, n; 2d) \|A\|_{\infty, 1}^{\mathbb{R}} . \tag{3.4.14}$$

In particular,

$$\|A\|_{\infty,1}^{\mathbb{R}} \leq \|A\|_{\infty,1}^{\mathbb{C}} = \|A\|_{\mathbb{R}_2^2}^G \leq \sqrt{2} \|A\|_{\infty,1}^{\mathbb{R}} \tag{3.4.15}$$

and

$$\|\operatorname{Re}(B)\|_{\mathbb{C}_2^d}^G \leq \|B\|_{\mathbb{C}_2^d}^G \text{ and } \|\operatorname{Im}(B)\|_{\mathbb{C}_2^d}^G \leq \|B\|_{\mathbb{C}_2^d}^G . \tag{3.4.16}$$

Moreover,

$$\|A_1^{Had}\|_{\infty,1}^{\mathbb{R}} = 1 < \sqrt{2} = \|A_1^{Had}\|_{\infty,1}^{\mathbb{C}} .$$

Proof Since

$$xy = \left(\frac{1-i}{\sqrt{2}}\right) x \left(\frac{1+i}{\sqrt{2}}\right) y \text{ for all } x, y \in \mathbb{R},$$

it follows that for any $(u, v) \in S_{\mathbb{R}_2^d}^m \times S_{\mathbb{R}_2^d}^n$, $\Gamma_{\mathbb{R}_2^d}(u, v) = \Gamma_{\mathbb{C}_2^d}(z, w)$ for some $(z, w) \in S_{\mathbb{C}_2^d}^m \times S_{\mathbb{C}_2^d}^n$. Thus, $\|A\|_{\mathbb{R}_2^d}^G \leq \|A\|_{\mathbb{C}_2^d}^G$. In particular, if $d = 1$, it follows that $\|A\|_{\infty,1}^{\mathbb{R}} \leq \|A\|_{\infty,1}^{\mathbb{C}}$. Since any $z \in \mathbb{C}$ satisfies $|z| = \operatorname{Re}(\alpha z)$, for some $\alpha \in \mathbb{T}$, we may assume that $\|A\|_{\mathbb{C}_2^d}^G = \operatorname{Re}(\operatorname{tr}(A^\top \Gamma_{\mathbb{C}_2^d}(z, w)))$, for some $(z, w) \in S_{\mathbb{C}_2^d}^m \times S_{\mathbb{C}_2^d}^n$. Since A has real entries by assumption, it follows that $\|A\|_{\mathbb{C}_2^d}^G = \operatorname{tr}(A^\top \operatorname{Re}(\Gamma_{\mathbb{C}_2^d}(z, w)))$. Corollary 3.2 therefore implies that

$$\|A\|_{\mathbb{C}_2^d}^G = \operatorname{tr}(A^\top \Gamma_{\mathbb{R}_2^{2d}}(u, v)) \leq \|A\|_{\mathbb{R}_2^{2d}}^G \leq K_G^{\mathbb{R}}(m, n; 2d) \|A\|_{\infty,1}^{\mathbb{R}}$$

for some $(u, v) \in (\mathbb{S}^{2d-1})^m \times (\mathbb{S}^{2d-1})^n$. In particular,

$$\|A\|_{\infty,1}^{\mathbb{C}} \leq \|A\|_{\mathbb{R}_2^2}^G \leq K_G^{\mathbb{R}}(2) \|A\|_{\infty,1}^{\mathbb{R}} .$$

Thus, we may apply Krivine's remarkable result (cf. Corollary 3.4), and it follows that

$$\|A\|_{\infty,1}^{\mathbb{C}} \leq \|A\|_{\mathbb{R}_2^2}^G \leq \sqrt{2} \|A\|_{\infty,1}^{\mathbb{R}} .$$

3.4 $K_G^{\mathbb{R}}(2)$ and the Walsh-Hadamard Transform: Krivine's Approach Revisited

On the other hand, Corollary 3.2 clearly implies also that $\|A\|_{\mathbb{R}_2^{2d}}^G = \mathrm{Re}(\mathrm{tr}(A^\top \widetilde{S}))$ for some $\widetilde{S} = \Gamma_{\mathbb{C}_2^d}(z, w) \in \mathcal{Q}_{m,n}(\mathbb{C})$, whence $\|A\|_{\mathbb{R}_2^{2d}}^G \leq \|A\|_{\mathbb{C}_2^d}^G$. In order to verify the complex case (3.4.16), we must only note that $\|B\|_{\mathbb{C}_2^d}^G = \|\overline{B}\|_{\mathbb{C}_2^d}^G$, implying that

$$\|\mathrm{Re}(B)\|_{\mathbb{C}_2^d}^G \leq \frac{1}{2}(\|B\|_{\mathbb{C}_2^d}^G + \|\overline{B}\|_{\mathbb{C}_2^d}^G) = \|B\|_{\mathbb{C}_2^d}^G$$

(since $|\mathrm{tr}(B^* \Gamma_{\mathbb{C}_2^d}(z,w))| = |\mathrm{tr}((B^*)^\top \Gamma_{\mathbb{C}_2^d}(\overline{w}, \overline{z}))|$ for all $(z, w) \in (\mathbb{C}^d)^m \times (\mathbb{C}^d)^n$ — cf. also with (3.1.1)).

Finally, put $\zeta_0 := (z_0, w_0)^\top := (\frac{1}{\sqrt{2}}(1+i), \frac{1}{\sqrt{2}}(1-i))^\top$. Then $\zeta_0 \in \mathbb{T}^2 \subseteq S_{\mathbb{C}_\infty^2}$, and a further application of the Walsh-Hadamard transform A_1^{Had} therefore leads to

$$\|A_1^{\mathrm{Had}}\|_{\infty,1}^{\mathbb{R}} \stackrel{(3.4.11)}{=} 1 < \sqrt{2} = \frac{1}{2}(|z_0 + w_0| + |z_0 - w_0|) = \|A_1^{\mathrm{Had}} \zeta_0\|_{\mathbb{C}_1^2}$$

$$\leq \|A_1^{\mathrm{Had}}\|_{\infty,1}^{\mathbb{C}} \leq \sqrt{2}\, \|A_1^{\mathrm{Had}}\|_{\infty,1}^{\mathbb{R}} = \sqrt{2}.$$

$\square$

Since

$$\|A\|_{\mathbb{R}_2^{2d}}^G \stackrel{(3.4.14)}{=} \|A\|_{\mathbb{C}_2^d}^G \leq K_G^{\mathbb{C}}(d)\, \|A\|_{\infty,1}^{\mathbb{C}} \stackrel{(3.4.15)}{\leq} \sqrt{2}\, K_G^{\mathbb{C}}(d)\, \|A\|_{\infty,1}^{\mathbb{R}}$$

for all $m, n \in \mathbb{N}$ and $A \in \mathbb{M}_{m,n}(\mathbb{R})$ and

$$\|B\|_{\mathbb{C}_2^d}^G = \|\mathrm{Re}(B) + i\,\mathrm{Im}(B)\|_{\mathbb{C}_2^d}^G \leq \|\mathrm{Re}(B)\|_{\mathbb{C}_2^d}^G + \|\mathrm{Im}(B)\|_{\mathbb{C}_2^d}^G$$

$$\stackrel{(3.4.14)}{\leq} K_G^{\mathbb{R}}(2d)(\|\mathrm{Re}(B)\|_{\infty,1}^{\mathbb{R}} + \|\mathrm{Im}(B)\|_{\infty,1}^{\mathbb{R}}) \stackrel{(3.4.16)}{\leq} 2\, K_G^{\mathbb{R}}(2d)\, \|B\|_{\infty,1}^{\mathbb{C}}$$

for all $m, n \in \mathbb{N}$ and $B \in \mathbb{M}_{m,n}(\mathbb{C})$, we obtain

Corollary 3.5

$$\tfrac{1}{\sqrt{2}}\, K_G^{\mathbb{R}}(2d) \leq K_G^{\mathbb{C}}(d) \leq 2\, K_G^{\mathbb{R}}(2d)\ \text{for all}\ d \in \mathbb{N}.$$

In particular,

$$\tfrac{1}{\sqrt{2}} K_G^{\mathbb{R}} \leq K_G^{\mathbb{C}} \leq 2\, K_G^{\mathbb{R}}. \qquad (3.4.17)$$

Note that the implication (3.4.17) contains [79, Theorem 10.6]. We do not know whether the second estimation could be improved to $K_G^{\mathbb{C}}(d) \stackrel{?}{\leq} \sqrt{2}\, K_G^{\mathbb{R}}(2d)$ for all $d \in \mathbb{N}$. In particular, since $\mathrm{Re}(B)$ and $\mathrm{Im}(B)$ do not commute, we do not know whether $(\|\operatorname{Re}(B) + i\operatorname{Im}(B)\|_{\mathbb{C}_2^d}^G)^2 \overset{?}{\leq} (\|\operatorname{Re}(B)\|_{\mathbb{C}_2^d}^G)^2 + (\|\operatorname{Im}(B)\|_{\mathbb{C}_2^d}^G)^2$ holds for all $m, n \in \mathbb{N}$ and $B \in \mathbb{M}_{m,n}(\mathbb{C})$ (in analogy to $|x + iy|^2 = x^2 + y^2$ for all $x, y \in \mathbb{R}$).

3.5 The Gaussian Inner Product Splitting Property

Next, we are going to disclose a crucial *joint* multivariate Gaussian "splitting property" of inner products of vectors on the unit sphere of an arbitrary *separable* $\mathbb{F}$-Hilbert space. That result (which should be compared to the construction of Gaussian Hilbert spaces or Moore's theorem on the characterisation of kernel functions (cf. [121, Theorem 2.14])) runs like a thread throughout the whole book, including its implementation in Theorem 6.2 and Theorem 7.4. It holds for both fields, $\mathbb{F} = \mathbb{R}$ and $\mathbb{F} = \mathbb{C}$, and plays a significant role, when we are looking for a specific Gaussian random structure in quantum correlation matrices. So, let $k, m, n \in \mathbb{N}$. Fix $S = \Gamma_H(u, v) = (u_i^* v_j)_{ij} \in Q_{m,n}(\mathbb{F})$, where $(u, v) \in S_H^m \times S_H^n$ and $(i, j) \in [m] \times [n]$. Based on our analysis so far, if $\zeta_{ij} := u_i^* v_j = \langle v_j, u_i \rangle_H$ is given, then we only know about the existence of a joint Gaussian random vector $\operatorname{vec}(\mathbf{Z}_{ij}, \mathbf{W}_{ij}) \sim \mathbb{F} N_{2k}(0, \Sigma_{2k}(\zeta_{ij}))$. A priori, we cannot say whether it is even possible to allocate to $\zeta_{ij} = u_i^* v_j$ a joint Gaussian random vector of type $\operatorname{vec}(\mathbf{Z}_i, \mathbf{W}_j) \sim \mathbb{F} N_{2k}(0, \Sigma_{2k}(\zeta_{ij}))$. In fact, our next cornerstone result reveals that such a "joint Gaussian splitting of an inner product" is guaranteed if we assume that H is separable, $u_i \in S_H$ and $v_j \in S_H$. For this, we fix an arbitrary complete probability space $(\Omega, \mathscr{F}, \mathbb{P})$ and construct a suitable random field.

Proposition 3.8 (Inner Product Splitting) *Let $k \in \mathbb{N}$, $\mathbb{F} \in \{\mathbb{R}, \mathbb{C}\}$ and H be a separable $\mathbb{F}$-Hilbert space. There exists a family $\{\mathbf{Z}_x : x \in H\}$ of random vectors $\mathbf{Z}_x \equiv (Z_x^{(1)}, Z_x^{(2)}, \ldots, Z_x^{(k)})^\top$ in $\mathbb{F}^k$, such that*

$$\operatorname{vec}(\mathbf{Z}_x, \mathbf{Z}_y) \sim \mathbb{F} N_{2k}(0, C_{2k}(x, y)) \text{ for all } x, y \in H, \tag{3.5.1}$$

where

$$C_{2k}(x, y) := \begin{pmatrix} \|x\|_H^2 I_k & \langle x, y \rangle_H I_k \\ \langle y, x \rangle_H I_k & \|y\|_H^2 I_k \end{pmatrix}.$$

In particular,

$$\mathbb{E}[|Z_x^{(v)}|^2] = \|x\|_H^2 \text{ and } \langle x, y \rangle_H = \mathbb{E}[Z_x^{(v)} \overline{Z_y^{(v)}}] \text{ for all } v \in [k] \text{ and } x, y \in H.$$

If $w \in S_H$, then $\mathbf{Z}_w \sim \mathbb{F} N_k(0, I_k)$ and

$$\operatorname{vec}(\mathbf{Z}_u, \mathbf{Z}_v) \sim \mathbb{F} N_{2k}(0, \Sigma_{2k}(\langle u, v \rangle_H)) \text{ for all } u, v \in S_H. \tag{3.5.2}$$

3.5 The Gaussian Inner Product Splitting Property

If $e_1, e_2 \in S_H$ are orthogonal, then

$$\mathrm{vec}(\mathbf{Z}_\zeta, \mathbf{Z}_1) \sim \mathbb{F}N_{2k}(0, \Sigma_{2k}(\zeta)) \text{ for all } \zeta \in \overline{\mathbb{D}}, \tag{3.5.3}$$

where $\mathbf{Z}_\zeta := \mathbf{Z}_{\zeta e_1 + \sqrt{1-|\zeta|^2} e_2}$ and $\mathbf{Z}_1 := \mathbf{Z}_{e_1}$. In particular, $\mathrm{vec}(\mathbf{Z}_{\langle u, v \rangle_H}, \mathbf{Z}_1) \stackrel{d}{=} \mathrm{vec}(\mathbf{Z}_u, \mathbf{Z}_v)$ for all $u, v \in S_H$. $\mathbf{Z}_x \in L^2(\Omega)^k$ for all $x \in H$, and $Tx := \mathbf{Z}_x$ defines a bounded linear operator $T \in \mathfrak{L}(H, L^2(\Omega)^k)$, such that $\frac{1}{\sqrt{k}}T$ is an isometry. For any $\nu \in [k]$, the family $\{Z_x^{(\nu)} : x \in H\}$ is an H-isonormal process.

Proof Fix $k \in \mathbb{N}$, and let $n \in \mathbb{N}$. Consider the $n \times k$ random matrix

$$\Xi := (\xi_1 \vert \xi_2 \vert \ldots \vert \xi_k) = \begin{pmatrix} \xi_{11} & \xi_{12} & \cdots & \xi_{1k} \\ \xi_{21} & \xi_{22} & \cdots & \xi_{2k} \\ \vdots & \vdots & \vdots & \vdots \\ \xi_{n1} & \xi_{n2} & \cdots & \xi_{nk} \end{pmatrix},$$

where $\xi_\nu = \mathrm{vec}(\xi_{1\nu}, \ldots, \xi_{n\nu}) \sim \mathbb{F}N_n(0, I_n)$ for all $\nu \in [k]$ and the random vectors $\xi_1, \ldots, \xi_k$ are mutually independent, implying that $\xi_{11}, \ldots, \xi_{n1}, \ldots, \xi_{1k}, \ldots, \xi_{nk}$ are i.i.d. and standard normally distributed $\mathbb{F}$-valued random variables (i.e., with probability law $\mathbb{F}N_1(0, 1)$). Thus, the family

$$\{\xi_{i\nu} : (i, \nu) \in \mathbb{N} \times [k]\} \tag{3.5.4}$$

consists of i.i.d. and standard normally distributed $\mathbb{F}$-valued random variables. Since by assumption the Hilbert space H is separable, then H is either isometrically isomorphic to $\mathbb{F}_2^n$ for some $n \in \mathbb{N}$ or isometrically isomorphic to l_2. Firstly, we treat the finite-dimensional case: $H := \mathbb{F}_2^n$. Given an arbitrary vector $x \in H$, put

$$\mathbf{Z}_x := \Xi^* x = (x^\top \overline{\Xi})^\top = (x^\top \overline{\xi_1}, x^\top \overline{\xi_2}, \ldots, x^\top \overline{\xi_k})^\top, \tag{3.5.5}$$

respectively

$$Z_x^{(\nu)} := x^\top \overline{\xi_\nu} \quad (\nu \in [k]).$$

To prove (3.5.1) in the finite-dimensional case, fix $x, y \in H$. Let $a \in \mathbb{F}^{2k}$. Then $a = \mathrm{vec}(s, t)$, for some $s, t \in \mathbb{F}^k$, and

$$a^* \mathrm{vec}(\mathbf{Z}_x, \mathbf{Z}_y) = x^\top \overline{\Xi} \overline{s} + y^\top \overline{\Xi} \overline{t} = \sum_{\nu=1}^{k} \sum_{i=1}^{n} c_{\nu i} \overline{\xi_{i\nu}}$$

$$= \sum_{\nu=1}^{k} \sum_{i=1}^{n} \overline{c_{\nu i}} \xi_{i\nu} \sim \mathbb{F}N_1\left(0, \sum_{\nu=1}^{k} \sum_{i=1}^{n} |c_{\nu i}|^2\right),$$

where $c_{vi} := x_i \overline{s_v} + y_i \overline{t_v}$ (since all random variables $\xi_{iv} \sim \mathbb{F}N_1(0, 1)$ are mutually independent). A straightforward calculation shows that

$$0 \leq \sum_{v=1}^{k} \sum_{i=1}^{n} |c_{vi}|^2 = \|x\|^2 \|s\|^2 + 2\operatorname{Re}(\langle x, y \rangle \langle t, s \rangle) + \|y\|^2 \|t\|^2 = a^* C_{2k}(x, y) a.$$

Consequently, since $a \in \mathbb{F}^{2k}$ was chosen arbitrarily, $C_{2k}(x, y) \in \mathbb{P}_{2k}(\mathbb{F})$ is positive semidefinite, and [5, Theorem 2.8] implies that

$$\operatorname{vec}(\mathbf{Z}_x, \mathbf{Z}_y) \sim \mathbb{F}N_{2k}(0, C_{2k}(x, y)).$$

In particular, if $(u, v) \in S_H \times S_H$, then $C_{2k}(u, v) = \Sigma_{2k}(\langle u, v \rangle_H)$, and (3.5.2) follows at once. Next, we consider the infinite-dimensional case ($H := l_2$). Fix $x, y \in H$, and let $w \in \{x, y\}$. Put $\pi_n(w) := (w_1, w_2, \ldots, w_n)^\top \in \mathbb{F}_2^n =: H_n$. Since $Z_{\pi_n(w)}^{(v)} = \sum_{i=1}^{n} w_i \overline{\xi_{iv}}$ for all $v \in [k]$ (see (3.5.5)) and $\|w\|_H^2 = \sum_{i=1}^{\infty} |w_i|^2 < \infty$, the assumed independence structure of the Gaussians ξ_{iv} (cf. (3.5.4)) implies that for any $v \in [k]$ the sequence $(Z_{\pi_n(w)}^{(v)})_{n \in \mathbb{N}}$ is a Cauchy sequence in $L^2(\Omega)$. Thus, for any $v \in [k]$, the v'th component of the k-dimensional random vector $\mathbf{Z}_{\pi_n(w)}$ converges in $L^2(\Omega)$ to $\mathbf{Z}_w^{(v)} := \sum_{i=1}^{\infty} w_i \overline{\xi_{iv}} \in L^2(\Omega)$ (cf. also [20, Theorem 1.1.4.] for the real case). It follows that $s^* \mathbf{Z}_x + t^* \mathbf{Z}_y$ is the L^2-limit of the sequence $(s^* \mathbf{Z}_{\pi_n(x)} + t^* \mathbf{Z}_{\pi_n(y)})_{n \in \mathbb{N}}$ for all $s, t \in \mathbb{F}^k$. From the proof of the finite-dimensional case, we know that in $H_n = \mathbb{F}_2^n$

$$s^* \mathbf{Z}_{\pi_n(x)} + t^* \mathbf{Z}_{\pi_n(y)} \sim N_1(0, a^* C_{2k}(\pi_n(x), \pi_n(y)) a),$$

where $a := \operatorname{vec}(s, t)$. Since $\lim_{n \to \infty} \langle \pi_n(x'), \pi_n(y') \rangle_{H_n} = \langle x', y' \rangle_H$ for all $x', y' \in H$, we obtain

$$\lim_{n \to \infty} a^* C_{2k}(\pi_n(x), \pi_n(y)) a = a^* C_{2k}(x, y) a.$$

Consequently, since L^2-convergence implies convergence in probability, and hence convergence in distribution (cf. e.g. [74, Proposition E.1.5.]), we conclude that

$$\operatorname{vec}(\mathbf{Z}_x, \mathbf{Z}_y) \sim \mathbb{F}N_{2k}(0, C_{2k}(x, y)).$$

Finally, (3.5.3) is a particular case of (3.5.2), where $u := \zeta e_1 + \sqrt{1 - |\zeta|^2} e_2 \in S_H$ and $v := e_1 \in S_H$. □

Let $\Sigma = (\sigma_{ij})_{(i,j) \in [n] \times [n]} \in C(n; \mathbb{F})$ be an arbitrary correlation matrix. Then $\Sigma = \Gamma_{H_n}(w, w)$ for some $w \equiv (w_1, w_2, \ldots, w_n) \in S_{H_n}^n$, where $H_n := \mathbb{F}_2^n$ (due to Lemma 3.1), implying that $\sigma_{ij} = w_i^* w_j = \langle w_j, w_i \rangle_{H_n}$ for all $i, j \in [n]$. Consequently, (3.5.2), applied to all pairs $w_i, w_j \in S_H$, immediately results in

3.5 The Gaussian Inner Product Splitting Property

Corollary 3.6 (Correlation Matrix Splitting) *Let $n \in \mathbb{N}$ and $\Sigma = (\sigma_{ij})_{(i,j)\in[n]\times[n]} \in C(n; \mathbb{F})$. Let $k \in \mathbb{N}$. Then there exist n $\mathbb{F}^k$-valued random vectors $\mathbf{Z}_1, \mathbf{Z}_2, \ldots, \mathbf{Z}_n$ such that*

$$vec(\mathbf{Z}_i, \mathbf{Z}_j) \sim \mathbb{F}N_{2k}(0, \Sigma_{2k}(\sigma_{ij})) \text{ and } \frac{1}{\sqrt{k}}\mathbf{Z}_i \in S_{H_k}$$

for all $(i, j) \in [n] \times [n]$, where $H_k := L^2(\Omega)^k$. In particular,

$$\sigma_{ij} = \mathbb{E}[Z_i^{(\nu)} \overline{Z_j^{(\nu)}}] = \langle \frac{1}{\sqrt{k}}\mathbf{Z}_i, \frac{1}{\sqrt{k}}\mathbf{Z}_j \rangle_{H_k} \text{ for all } (i, j) \in [n] \times [n] \text{ and } \nu \in [k].$$

Moreover, for any $\mathbb{F}$-Hilbert space H, for any $u, v \in S_H$, there exist two joint Gaussian random vectors $\mathbf{W}_u$ and $\mathbf{W}_v$ in $\mathbb{F}^k$, such that $\frac{1}{\sqrt{k}}\mathbf{W}_u \in S_{H_k}$, $\frac{1}{\sqrt{k}}\mathbf{W}_v \in S_{H_k}$, $W_u^{(\nu)} \in S_{L^2(\Omega)}$, $W_v^{(\nu)} \in S_{L^2(\Omega)}$ and

$$\langle v, u \rangle_H = \mathbb{E}[W_v^{(\nu)} \overline{W_u^{(\nu)}}] = \langle \frac{1}{\sqrt{k}}\mathbf{W}_v, \frac{1}{\sqrt{k}}\mathbf{W}_u \rangle_{H_k} \text{ for all } \nu \in [k].$$

Chapter 4
Powers of Inner Products of Random Vectors, Uniformly Distributed on the Sphere

4.1 Gaussian Sign-Correlation

It seems to be the case that any rigorous proof of the Grothendieck inequality is built on two *equalities*, namely the Grothendieck equality (if $\mathbb{F} = \mathbb{R}$—cf. e.g. [49, 86], or the proof of [39, Prop. 4.4.2]) and the Haagerup equality (if $\mathbb{F} = \mathbb{C}$—see [49, 60, 86]). In fact, if we reveal the inherent *bivariate* Gaussian random structure, these equalities emerge as two special cases of the representation of a single Pearson correlation coefficient which applies likewise for the real case and the complex case (Corollary 4.1). Rewritten in terms of real Gaussian random vectors (if $\mathbb{F} = \mathbb{R}$) and complex Gaussian random vectors (if $\mathbb{F} = \mathbb{C}$) namely, we firstly obtain a representation of the two equalities, indicating an already lurking common underlying probabilistic structure for both fields, $\mathbb{R}$ and $\mathbb{C}$ (cf. also (4.2.8) and (4.2.9)). To this end, recall (cf., e.g., [6] and [10, Chapter II]) that for any $a, b \in \mathbb{C}$, any $c \in \mathbb{C} \setminus \{-n : n \in \mathbb{N}_0\}$ and any $z \in \mathbb{D}$ the well-defined power series

$$_2F_1(a, b, c; z) := \frac{\Gamma(c)}{\Gamma(a)\Gamma(b)} \sum_{n=0}^{\infty} \frac{\Gamma(a+n)\Gamma(b+n)}{\Gamma(c+n)} \frac{z^n}{n!}$$

$$= 1 + \frac{\Gamma(c)}{\Gamma(a)\Gamma(b)} \sum_{n=1}^{\infty} \frac{\Gamma(a+n)\Gamma(b+n)}{\Gamma(c+n)} \frac{z^n}{n!}$$

denotes the *Gaussian hypergeometric function*. If in addition $\text{Re}(c) > \text{Re}(a+b)$ then the series converges absolutely on $\mathbb{T}$ and satisfies $_2F_1(a, b, c; 1) = \frac{\Gamma(c)\Gamma(c-a-b)}{\Gamma(c-a)\Gamma(c-b)}$ (Gauss Summation Theorem). Recall that $\mathbb{S}_{\mathbb{C}^n} := \{w \in \mathbb{C}^n : \|w\|_{\mathbb{C}_2^n} = 1\} = J_2^{-1}(\mathbb{S}^{2n-1})$ denotes the unit sphere in $\mathbb{C}^n$ (where $n \in \mathbb{N}$, of course).

The Grothendieck Equality Let $n \in \mathbb{N}$ and $u, v \in \mathbb{S}^{n-1}$. Let $\mathbf{X} \equiv (X_1, \ldots, X_n)^\top \sim N_n(0, I_n)$ be a standard-normally distributed real Gaussian random vector. Then

$$\mathbb{E}[\text{sign}(u^\top \mathbf{X})\text{sign}(v^\top \mathbf{X})] = \frac{2}{\pi} \arcsin(u^\top v)$$

$$= \frac{2}{\pi} u^\top v \, _2F_1(\tfrac{1}{2}, \tfrac{1}{2}, \tfrac{3}{2}; (u^\top v)^2)$$

$$= \mathbb{E}[|X_1|]^2 \, u^\top v \, _2F_1(\tfrac{1}{2}, \tfrac{1}{2}, \tfrac{3}{2}; (u^\top v)^2). \quad (4.1.1)$$

The Haagerup Equality Let $n \in \mathbb{N}$ and $\mathbf{Z} \equiv (Z_1, \ldots, Z_n)^\top \sim \mathbb{C}N_n(0, I_n)$ be a standard-normally distributed complex Gaussian random vector. Then

$$\mathbb{E}[\text{sign}(u^*\mathbf{Z})\text{sign}(\overline{v^*\mathbf{Z}})] = \frac{\pi}{4} \text{sign}(u^*v)\Big(\frac{1}{\pi}\int_0^{2\pi} \arcsin(|u^*v|\cos(t))\cos(t)\,dt\Big)$$

$$= \frac{\pi}{4} \text{sign}(u^*v) \, |u^*v| \, _2F_1(\tfrac{1}{2}, \tfrac{1}{2}, 2; |u^*v|^2)$$

$$= \mathbb{E}[|Z_1|]^2 \, u^*v \, _2F_1(\tfrac{1}{2}, \tfrac{1}{2}, 2; |u^*v|^2). \quad (4.1.2)$$

Remark 4.1 The second equality in (4.1.2) is a particular case of the equality

$$\frac{1}{\pi}\int_0^{2\pi} \arcsin(x\cos(t))\cos(t)\,dt = x \, _2F_1(\tfrac{1}{2}, \tfrac{1}{2}, 2; x^2) \text{ for all } x \in [-1, 1], \quad (4.1.3)$$

implied by the Maclaurin series representation of the function arcsin and the well-known fact that $\int_0^{2\pi} \cos^{2(n+1)}(t)\,dt = \frac{2\sqrt{\pi}}{(n+1)!}\Gamma(n+\tfrac{3}{2}) = \frac{2\sqrt{\pi}}{\Gamma(n+2)}(n+\tfrac{1}{2})\Gamma(n+\tfrac{1}{2})$ for all $n \in \mathbb{N}_0$.

We can see clearly that both, (4.1.1) and (4.1.2) do not depend on the choice of the dimension n. The reason for this is Lemma 2.3, but not the use of the sign function. If we namely fix an arbitrary random vector $\mathbf{W} \sim \mathbb{F}N_n(0, I_n)$ and consider the matrix $A_{u,v} := \begin{pmatrix} u_1 & u_2 & \ldots & u_n \\ v_1 & v_2 & \ldots & v_n \end{pmatrix} \in \mathbb{M}_{2,n}(\mathbb{F})$, then $(u^*\mathbf{W}, v^*\mathbf{W})^\top = A_{u,v}\mathbf{W} \sim N_2(0, \Sigma_2(u^*v))$ (due to Lemma 2.3). Hence, if $(S_1, S_2)^\top \sim N_2(0, \Sigma_2(u^*v))$ is given, then $\mathbb{P}_{(S_1,S_2)^\top} = \mathbb{P}_{(u^*\mathbf{W}, v^*\mathbf{W})^\top} = (A_{u,v})_* \mathbb{P}_\mathbf{W}$. The change of variables formula therefore implies that (in particular) *for any* choice of a. e. bounded functions $f, g \in L^\infty(\mathbb{F})$, the following equality holds:

$$\mathbb{E}[f(u^*\mathbf{W})g(\overline{v^*\mathbf{W}})] = \mathbb{E}[(f \otimes \overline{g}) \circ A_{u,v}(\mathbf{W})]$$

$$= \int_{\mathbb{F}^2} f \otimes \overline{g}\,d((A_{u,v})_* \mathbb{P}_\mathbf{W}) = \mathbb{E}[f(S_1)g(\overline{S_2})]. \quad (4.1.4)$$

Consequently, if we also include Lemma 3.1 (or the obvious fact that the mapping $S_{\mathbb{F}^n} \times S_{\mathbb{F}^n} \ni (u, v) \mapsto u^*v \in \overline{\mathbb{D}} \cap \mathbb{F}$ is onto for any $n \in \mathbb{N}_2$) and Corollary 4.4, then (4.1.4), applied to $f := g :=$ sign implies that (4.1.1) and (4.1.2) are special cases of an equality which "just" involves the function sign : $\mathbb{R} \longrightarrow \{-1, 1\}$ and a 2-dimensional Gaussian random vector, where the latter consists of two arbitrarily correlated random variables, though. Remembering the fact (2.2.2), we obtain:

Corollary 4.1 (Gaussian Sign-Correlation Coefficient) *Fix $\mathbb{F} \in \{\mathbb{R}, \mathbb{C}\}$. Let $\Sigma \in C(2; \mathbb{F})$ and $(S_1, S_2) \sim \mathbb{F}N_2(0, \Sigma)$. Then $\Sigma = \Sigma_2(\zeta)$ for some $\zeta \in \mathbb{F} \cap \overline{\mathbb{D}}$, and the Pearson correlation coefficient between $sign(S_1)$ and $sign(S_2)$ is given by*

$$\mathbb{E}[sign(S_1)sign(\overline{S_2})] = \mathbb{E}[|S_1|]^2 \, \zeta \, {}_2F_1(\tfrac{1}{2}, \tfrac{1}{2}, \tfrac{d_{\mathbb{F}}+2}{2}; |\zeta|^2)$$
$$= \frac{1}{k_G^{\mathbb{F}}} \zeta \, {}_2F_1(\tfrac{1}{2}, \tfrac{1}{2}, \tfrac{d_{\mathbb{F}}+2}{2}; |\zeta|^2), \quad (4.1.5)$$

where $d_{\mathbb{R}} := 1$ and $d_{\mathbb{C}} := 2$.

4.2 Integration Over $\mathbb{S}^{n-1}$ and the Gamma Function

In fact, the Grothendieck equality as well as the Haagerup equality instantly unfold as a *special case* of a (much more general) result, where we explicitly describe all non-negative integer powers of an expectation of inner products of suitably correlated—real—random vectors, uniformly distributed on the unit sphere (cf. Theorem 4.1 and Proposition 6.7). In this regard, we possibly should point to the so-called "kernel trick", used also for the computation of inner products in high-dimensional feature spaces using simple functions defined on pairs of input patterns which is a crucial ingredient of support vector machines in statistical learning theory; i.e., learning machines that construct decision functions of sign type. This trick allows the formulation of nonlinear variants of any algorithm that can be cast in terms of inner products (cf. [151, Chapter 5.6]).

Firstly, it is quite helpful to understand the actual source of the values $\frac{2}{\pi}$ and $\frac{\pi}{4}$ (cf. (4.2.10), Corollary 4.4, Proposition 6.7 and [33, Chapter 8.7]).

Lemma 4.1 *Let $(b_n)_{n \in \mathbb{N}}$ be the sequence of real numbers, defined as*

$$b_n := \begin{cases} 1 & \text{if } n \text{ is even} \\ \sqrt{\pi/2} & \text{if } n \text{ is odd}. \end{cases} \quad (4.2.1)$$

Then

$$\Gamma\left(\frac{n}{2}\right) = \frac{(n-2)!!}{\sqrt{2^{n-2}}} b_n \quad \text{for all } n \in \mathbb{N}. \quad (4.2.2)$$

In particular, $b_n = b_{2m+n}$ for all $m, n \in \mathbb{N}$, and

$$_2F_1\left(\frac{k}{2}, \frac{l}{2}, \frac{m}{2}; z\right) = \frac{(m-2)!!}{(k-2)!!(l-2)!!} \sum_{n=0}^{\infty} \frac{(2(n-1)+k)!!(2(n-1)+l)!!}{(2n)!!(2(n-1)+m)!!} z^n \quad (4.2.3)$$

for all $k, l, m \in \mathbb{N}$ and $z \in \mathbb{D}$. If in addition $m > k+l$, then (4.2.3) holds for any $z \in \mathbb{T}$.

Proof Regarding the proof of (4.2.2), we have to distinguish two cases; namely the even case (i.e., $n = 2l$ for some $l \in \mathbb{N}$) and the odd case (i.e., $n = 2k+1$ for some $k \in \mathbb{N}_0$). However, since $(2l-2)!! = (2(l-1))!! = 2^{l-1}(l-1)!$ and $\Gamma(k+\frac{1}{2}) = \frac{(2k-1)!!}{2^k}\sqrt{\pi}$, (4.2.2) follows at once. Equipped with (4.2.2), the proof of the representation (4.2.3) just involves a few remaining basic algebraic transformations (including a multiple shortening of fractions), implied by the definition of Gaussian hypergeometric functions. □

We also need results about the real and complex Gaussian randomness structure, embedded in the Gamma function, which are of their own interest; built on an important link between the Gamma function and powers of absolute moments of standard normally distributed real random variables. To this end, we consider both, the real and the complex unit sphere as a probability space. Put

$$\sigma_{n-1}(A) := \frac{\omega_n(A)}{\omega_n} = \frac{\Gamma(n/2)}{2\pi^{n/2}} \omega_n(A) = \frac{\Gamma(n/2)}{2\pi^{n/2}} n \lambda_n(\{r\xi : 0 < r \leq 1 \text{ and } \xi \in A\})$$

$$= \frac{\Gamma(\frac{n}{2}+1)}{\pi^{n/2}} \lambda_n(\{r\xi : 0 < r \leq 1 \text{ and } \xi \in A\}),$$

where $A \in \mathcal{B}(\mathbb{S}^{n-1})$ and $\omega_n \equiv \omega_n(\mathbb{S}^{n-1}) = \frac{2\pi^{n/2}}{\Gamma(n/2)}$ denotes the surface area of the unit sphere $\mathbb{S}^{n-1} \subseteq \mathbb{R}^n$. σ_{n-1} denotes the rotation-invariant probability measure (Haar measure) on $\mathbb{S}^{n-1}$. Moreover, $\sigma_n^{\mathbb{C}} := (J_2^{-1})_* \sigma_{2n-1}$ denotes the surface area probability measure on the complex unit sphere $\mathbb{S}_{\mathbb{C}^n}$.

Proposition 4.1 Let $n \in \mathbb{N}$, $X \sim N_1(0, 1)$, $\mathbf{X} \sim N_n(0, I_n)$, $Z \sim \mathbb{C}N_1(0, 1)$, $\mathbf{Z} \sim \mathbb{C}N_n(0, I_n)$ and $\mathbf{Y} = \text{vec}(\mathbf{Y}_1, \mathbf{Y}_2) \sim N_{2n}(0, I_{2n})$. Let $p, q \in \mathbb{R}$ such that $p > -1$ and $q > 0$. Then

(i)

$$2\int_0^{\infty} s^p \gamma_1(ds) = \mathbb{E}[|X|^p] = \frac{2^{p/2}}{\sqrt{\pi}} \Gamma(\tfrac{p+1}{2}) \quad (4.2.4)$$

and

$$\Gamma(q) = \frac{\sqrt{2\pi}}{2^q} \mathbb{E}[|X|^{2q-1}].$$

4.2 Integration Over $\mathbb{S}^{n-1}$ and the Gamma Function

In particular,

$$\mathbb{E}[|X|^k] = \frac{(k-1)!!}{b_k} = \begin{cases} (k-1)!! & \text{if } k \text{ is even} \\ \sqrt{\frac{2}{\pi}}(k-1)!! & \text{if } k \text{ is odd} \end{cases} \quad \text{for all } k \in \mathbb{N}_0,$$
(4.2.5)

where b_k satisfies (4.2.1).

(ii) *Let $n \geq 2$ and $f : \mathbb{R}^n \longrightarrow \mathbb{R}$, such that*

$$f(rx) = r^p f(x) \text{ for all } (r, x) \in (0, \infty) \times \mathbb{R}^n.$$

Then $f \in L^1(\mathbb{R}^n, \gamma_n)$ if and only if $f\big|_{\mathbb{S}^{n-1}} \in L^1(\mathbb{S}^{n-1}, \sigma_{n-1})$, and

$$\mathbb{E}[f(\mathbf{X})] = \int_{\mathbb{R}^n} f(x)\gamma_n(dx) = 2^{p/2} \frac{\Gamma\left(\frac{n+p}{2}\right)}{\Gamma\left(\frac{n}{2}\right)} \int_{\mathbb{S}^{n-1}} f(\xi) \, d\sigma_{n-1}(\xi).$$
(4.2.6)

(iii) *Let $b : \mathbb{C}^n \longrightarrow \mathbb{C}$, such that*

$$b(rz) = r^p b(z) \text{ for all } (r, z) \in (0, \infty) \times \mathbb{C}^n.$$

Then $b \in L^1(\mathbb{C}^n, \gamma_n^{\mathbb{C}})$ if and only if $\text{Re}(b) \circ J_2^{-1}\big|_{\mathbb{S}^{2n-1}} \in L^1(\mathbb{S}^{2n-1}, \sigma_{2n-1})$ and $\text{Im}(b) \circ J_2^{-1}\big|_{\mathbb{S}^{2n-1}} \in L^1(\mathbb{S}^{2n-1}, \sigma_{2n-1})$, and

$$\mathbb{E}[b(\mathbf{Z})] = \int_{\mathbb{C}^n} b(z)\gamma_n^{\mathbb{C}}(dz) = \frac{\Gamma(n+\frac{p}{2})}{(n-1)!} \int_{S_{\mathbb{C}^n}} b(\zeta) \, d\sigma_n^{\mathbb{C}}(\zeta)$$

$$= \frac{\Gamma(n+\frac{p}{2})}{(n-1)!} \Big(\int_{\mathbb{S}^{2n-1}} \text{Re}(b(y_1 + i\, y_2)) \, d\sigma_{2n-1}((y_1, y_2))$$

$$+ i \int_{\mathbb{S}^{2n-1}} \text{Im}(b(y_1 + i\, y_2)) \, d\sigma_{2n-1}((y_1, y_2)) \Big)$$

$$= 2^{-p/2} \big(\mathbb{E}[\text{Re}(b(\mathbf{Y}_1 + i\, \mathbf{Y}_2))] + i\, \mathbb{E}[\text{Im}(b(\mathbf{Y}_1 + i\, \mathbf{Y}_2))] \big).$$
(4.2.7)

Proof

(i) Recall that $\Gamma(z) = \int_0^\infty t^{z-1} e^{-t} \, dt$, where $\text{Re}(z) > 0$. Consequently (if we substitute t through $\frac{s^2}{2}$), we obtain

$$\Gamma\left(\frac{p+1}{2}\right) = \frac{\sqrt{2\pi}}{(\sqrt{2})^{p-1}} \int_0^\infty s^p \, \gamma_1(ds),$$

and it follows that

$$\mathbb{E}[|X|^p] = \int_{-\infty}^0 (-s)^p \gamma_1(ds) + \int_0^\infty s^p \gamma_1(ds)$$

$$= 2\int_0^\infty s^p \gamma_1(ds) = \frac{2^{p/2}}{\sqrt{\pi}} \Gamma\left(\frac{p+1}{2}\right).$$

Equation (4.2.2) implies the particular case (4.2.5).

(ii) Since $\mathbf{X} \equiv (X_1, \ldots, X_n)^\top \sim N_n(0, I_n)$ is a (centered) standard Gaussian random vector, we have

$$\mathbb{E}[f(\mathbf{X})] = \mathbb{E}[f(X_1, \ldots, X_n)] = (2\pi)^{-n/2} \int_{\mathbb{R}^n} f(x) \exp\left(-\frac{1}{2}\|x\|^2\right) \lambda_n(dx).$$

It follows that for λ_n-almost all $r \in (0, \infty)$ we have

$$\int_{\mathbb{R}^n} f(x) \exp(-\tfrac{1}{2}\|x\|^2) \lambda_n(dx) = \int_0^\infty \Bigl(\int_{\mathbb{S}^{n-1}} f(r\xi) \exp(-\tfrac{1}{2}\|r\xi\|^2)\, d\omega_n(\xi)\Bigr) r^{n-1}\, dr$$

$$= \int_0^\infty \exp(-\tfrac{1}{2}r^2) r^{n+p-1}\, dr \cdot \int_{\mathbb{S}^{n-1}} f(\xi)\, d\omega_n(\xi).$$

Hence,

$$\mathbb{E}[f(\mathbf{X})] = (2\pi)^{-n/2} \kappa_{n+p} \int_{\mathbb{S}^{n-1}} f(\xi)\, d\omega_n(\xi),$$

where $\kappa_\nu := \int_0^\infty r^{\nu-1} \exp(-\tfrac{1}{2}r^2)\, dr = \sqrt{\tfrac{\pi}{2}} \mathbb{E}[|X_1|^{\nu-1}] = 2^{\frac{\nu-2}{2}} \Gamma(\tfrac{\nu}{2})$, $\nu \in \mathbb{N}$.

Consequently, since $\frac{2\pi^{n/2}}{\Gamma(\frac{n}{2})} = \omega_n(\mathbb{S}^{n-1})$ is the surface area of $\mathbb{S}^{n-1}$, it follows that

$$\mathbb{E}[f(\mathbf{X})] = (2\pi)^{-n/2} \frac{2\pi^{n/2}}{\Gamma(\frac{n}{2})} \kappa_{n+p} \int_{\mathbb{S}^{n-1}} f(\xi)\, d\sigma_{n-1}(\xi)$$

$$= 2^{p/2} \frac{\Gamma(\frac{n+p}{2})}{\Gamma(\frac{n}{2})} \int_{\mathbb{S}^{n-1}} f(\xi)\, d\sigma_{n-1}(\xi).$$

4.2 Integration Over $\mathbb{S}^{n-1}$ and the Gamma Function

(iii) That equality follows instantly from (ii) and the definition of the measure $\sigma_n^{\mathbb{C}}$. We only have to recall (2.1.9) and the representation $S_{\mathbb{C}^n} = \{\zeta \in \mathbb{C}^n : \|\zeta\|_{\mathbb{C}_2^n} = 1\} = J_2^{-1}(\mathbb{S}^{2n-1})$ of the unit sphere in $\mathbb{C}^n$. □

Corollary 4.2 Let $p \in (-1, \infty)$, $m, n \in \mathbb{N}$, $X \sim N_1(0, 1)$, $\mathbf{X} \sim N_n(0, I_n)$, $\mathbf{Y} \sim N_{2n}(0, I_{2n})$, $Z \sim \mathbb{C}N_1(0,1)$ and $\mathbf{Z} \sim \mathbb{C}N_n(0, I_n)$. Then

$$\mathbb{E}[\|\mathbf{X}\|_{\mathbb{R}_2^n}^p] = 2^{p/2} \frac{\Gamma(\frac{n+p}{2})}{\Gamma(\frac{n}{2})} = \frac{\mathbb{E}[|X|^{n-1+p}]}{\mathbb{E}[|X|^{n-1}]} \tag{4.2.8}$$

and

$$\mathbb{E}[\|\mathbf{Z}\|_{\mathbb{C}_2^n}^p] = \frac{\Gamma(n + \frac{p}{2})}{(n-1)!} = 2^{-p/2} \mathbb{E}[\|\mathbf{X}\|_{\mathbb{R}_2^{2n}}^p] = 2^{-p/2} \frac{\mathbb{E}[|X|^{2n-1+p}]}{\mathbb{E}[|X|^{2n-1}]}. \tag{4.2.9}$$

In particular,

$$\mathbb{E}[\|\mathbf{X}\|_{\mathbb{R}_2^n}^m] = a_n(m) \frac{(n-2+m)!!}{(n-2)!!} \quad \text{and} \quad \mathbb{E}[\|\mathbf{Z}\|_{\mathbb{C}_2^n}^m] = a_{2n}(m) \, 2^{-m/2} \frac{(2n-2+m)!!}{(2n-2)!!}$$

and

$$\mathbb{E}[|X|] = \sqrt{\frac{2}{\pi}} \quad \text{and} \quad \mathbb{E}[|Z|] = \sqrt{\frac{\pi}{4}} = \frac{\sqrt{\pi}}{2}, \tag{4.2.10}$$

where

$$a_n(m) := \frac{b_{n+m}}{b_n} = \begin{cases} \sqrt{\pi/2} & \text{if } n \text{ is even and } m \text{ is odd} \\ \sqrt{2/\pi} & \text{if } n \text{ is odd and } m \text{ is odd} \\ 1 & \text{if } m \text{ is even} \end{cases} \tag{4.2.11}$$

and b_n is defined as in Lemma 4.1.

Proof We just have to apply Proposition 4.1 to the function $\mathbb{F}^n \ni z \mapsto f_p(z) := \|z\|^p$ (and to recall that by definition $\|\cdot\|_{\mathbb{R}_2^1} := |\cdot|$). □

A further, very important special case (which allows an easy proof of Theorem 4.1) arises if we consider the function $\mathbb{R}^n \ni x \mapsto \langle u, x\rangle_{l_2^n}^m = (u^\top x)^m$, where $m \in \mathbb{N}_0$ and $u \in \mathbb{S}^{n-1}$ are given.

Corollary 4.3 *Let $p \in (-1, \infty), n \in \mathbb{N}_2, x \in \mathbb{S}^{n-1}$ and $Y \sim N_1(0,1)$. Let $f : \mathbb{R} \longrightarrow \mathbb{R}$ be a function such that $f \in L^1(\mathbb{R}, \gamma_1)$ and $f(ry) = r^p f(y)$ for all $(r, y) \in (0, \infty) \times \mathbb{R}$. Then $f(x^\top \cdot) \in L^1(\mathbb{S}^{n-1}, \sigma_{n-1})$, and*

$$\int_{\mathbb{S}^{n-1}} f(x^\top u)\, \sigma_{n-1}(du) = 2^{-p/2} \frac{\Gamma(\frac{n}{2})}{\Gamma(\frac{n+p}{2})} \mathbb{E}[f(Y)].$$

In particular,

$$\int_{\mathbb{S}^{n-1}} (x^\top u)^m\, \sigma_{n-1}(du) = \frac{1+(-1)^m}{2} \frac{\Gamma(\frac{m+1}{2})\Gamma(\frac{n}{2})}{\sqrt{\pi}\, \Gamma(\frac{m+n}{2})}$$

$$= \frac{1+(-1)^m}{2} \left(\frac{\Gamma(\frac{n}{2})}{\sqrt{\pi}\, \Gamma(\frac{n-1}{2})} B(\tfrac{m+1}{2}, \tfrac{n-1}{2}) \right) \quad (4.2.12)$$

for all $m \in \mathbb{N}_0$. Here, $(0, \infty) \times (0, \infty) \ni (x_1, x_2) \mapsto B(x_1, x_2) := \int_0^1 t^{x_1-1}(1-t)^{x_2-1}\, dt = \frac{\Gamma(x_1)\Gamma(x_2)}{\Gamma(x_1+x_2)}$ denotes the real beta function.

Proof Since $\|x\|_{l_n^2}^2 = 1$, it follows that $x^\top \mathbf{Y} \stackrel{d}{=} Y \sim N_1(0, 1)$ for any $\mathbf{Y} \sim N_n(0, I_n)$. Thus, $\mathbb{E}[f(Y)] = \mathbb{E}[f(x^\top \mathbf{Y})]$, so that we may apply (4.2.6) to the function $\mathbb{R}^n \ni x \mapsto f(x^\top \cdot)$. Regarding the particular case $\mathbb{R} \ni y \mapsto y^m$ (respectively, $\mathbb{R}^n \ni x \mapsto (x^\top \cdot)^m$), we only have to include Lemma 4.1 and the well-known fact that the m-th moment of $Y \sim N_1(0, 1)$ satisfies $\mathbb{E}[Y^m] = \frac{1+(-1)^m}{2}(m-1)!!$. □

Lemma 4.2 *Consider the sequence $(c_k)_{k \in \mathbb{N}}$, defined as*

$$c_k := \frac{1}{\sqrt{{}_2F_1(\frac{1}{2}, \frac{1}{2}, \frac{k+2}{2}; 1)}}.$$

Let $\mathbf{X} \sim N_k(0, I_k)$ and $\mathbf{Z} \sim \mathbb{C}N_k(0, I_k)$. Then

$$c_k = \sqrt{\frac{2}{k}} \frac{\Gamma(\frac{k+1}{2})}{\Gamma(\frac{k}{2})} = a_k \frac{1}{\sqrt{k}} \frac{(k-1)!!}{(k-2)!!} = \frac{1}{\sqrt{k}} \mathbb{E}[\|\mathbf{X}\|_{\mathbb{R}_2^k}] = \sqrt{\frac{2\pi}{k}} \frac{\omega_k}{\omega_{k+1}},$$
(4.2.13)

where $a_k := a_k(1)$ satisfies (4.2.11), $\omega_1 := 2$ and ω_m denotes the surface area of the unit sphere $\mathbb{S}^{m-1}$ ($m \in \mathbb{N}_2$). In particular, $c_1^2 = \frac{2}{\pi}$, $c_2^2 = \frac{\pi}{4}$ and

$$c_{2k} = \frac{1}{\sqrt{k}} \frac{\Gamma(k+\frac{1}{2})}{(k-1)!} = \frac{1}{\sqrt{k}} \sqrt{\frac{\pi}{4}} \frac{(2k-1)!!}{(2k-2)!!} = \frac{1}{\sqrt{k}} \mathbb{E}[\|\mathbf{Z}\|_{\mathbb{C}_2^k}]. \quad (4.2.14)$$

4.2 Integration Over $\mathbb{S}^{n-1}$ and the Gamma Function

Moreover, $0 < c_k < 1$ for all $k \in \mathbb{N}$, $\lim_{k \to \infty} c_k = 1$, and

$$\frac{1}{\sqrt{k}} \mathbb{E}[\|\mathbf{W}\|_{\mathbb{F}_2^k}] = c_{\nu_k^\mathbb{F}} = \frac{1}{\sqrt{{}_2F_1(\frac{1}{2}, \frac{1}{2}, \frac{\nu_k^\mathbb{F}+2}{2}; 1)}} \xrightarrow{k \to \infty} 1 \text{ for all } \mathbf{W} \sim \mathbb{F}N_k(0, I_k),$$

where $\nu_k^\mathbb{R} := k$ and $\nu_k^\mathbb{C} := 2k$.

Proof Fix $k \in \mathbb{N}$. Firstly, the Gauss Summation Theorem implies that

$$_2F_1(\tfrac{1}{2}, \tfrac{1}{2}, \tfrac{k}{2}+1; 1) = \frac{\Gamma(\frac{k}{2}+1)\Gamma(\frac{k}{2})}{\Gamma(\frac{k+1}{2})\Gamma(\frac{k+1}{2})} = \frac{k}{2} \cdot \frac{\Gamma^2(\frac{k}{2})}{\Gamma^2(\frac{k+1}{2})},$$

whence $c_k = \sqrt{\frac{2}{k}} \frac{\Gamma(\frac{k+1}{2})}{\Gamma(\frac{k}{2})} > 0$ (since $\Gamma(x) > 0$ for all $x > 0$). (4.2.13) now follows immediately from (4.2.8). Put $s_k := \frac{\Gamma(\frac{k+1}{2})}{\Gamma(\frac{k}{2})}$, where $k \in \mathbb{N}$. Recall that the function $\ln \circ \Gamma|_{(0,\infty)} : (0, \infty) \longrightarrow \mathbb{R}$ is convex, implying that in particular

$$\Gamma(\tfrac{k+1}{2}) = \Gamma(\tfrac{1}{2} x_k + \tfrac{1}{2} y_k) \leq \sqrt{\Gamma(x_k)}\sqrt{\Gamma(y_k)} = \sqrt{\Gamma(\tfrac{k}{2})}\sqrt{\Gamma(\tfrac{k+2}{2})},$$

where $x_k := \frac{k}{2}$ and $y_k := \frac{k+2}{2} = \frac{k}{2} + 1$. Thus, $s_k \leq s_{k+1}$ for all $k \in \mathbb{N}$. However, since $s_k s_{k+1} = \frac{k}{2}$ for all $k \in \mathbb{N}$, it consequently follows that for all $k \in \mathbb{N}_2$

$$\tfrac{k-1}{2} = s_{k-1} s_k \leq s_k^2 \leq s_k s_{k+1} = \tfrac{k}{2},$$

whence $\lim_{k \to \infty} c_k^2 = \lim_{k \to \infty} \frac{2}{k} s_k^2 = 1$. Finally, since

$$\frac{1}{c_k^2} - 1 = {}_2F_1(\tfrac{1}{2}, \tfrac{1}{2}, \tfrac{k+2}{2}; 1) - 1 = \frac{\Gamma(\frac{k+2}{2})}{\pi} \sum_{\nu=1}^{\infty} \frac{\Gamma^2(\nu + \frac{1}{2})}{\Gamma(\nu + \frac{k+2}{2})\nu!} > 0 \text{ for all } k \in \mathbb{N},$$

it follows that in fact $0 < c_k < 1$ for all $k \in \mathbb{N}$. $\square$

In the context of Theorem 1.2, the one-dimensional special cases of (4.2.8) and (4.2.9) disclose a unification of the real and complex Gaussian structure, encoded at least in the *little* Grothendieck constant:

Corollary 4.4 *Let $\mathbb{F} \in \{\mathbb{R}, \mathbb{C}\}$. Then the little Grothendieck constant $k_G^\mathbb{F}$ can be written as*

$$k_G^\mathbb{F} = {}_2F_1(\tfrac{1}{2}, \tfrac{1}{2}, \tfrac{d_\mathbb{F}+2}{2}; 1) = \begin{cases} \pi/2 & \text{if } \mathbb{F} = \mathbb{R} \text{ and } d_\mathbb{R} = 1 \\ 4/\pi & \text{if } \mathbb{F} = \mathbb{C} \text{ and } d_\mathbb{C} = 2 \end{cases}.$$

Corollary 4.5 (Krivine [92]) *Let $f \in C([-1, 1])$, $k \in \mathbb{N}_2$ and $u \in \mathbb{S}^{k-1}$. Then*

$$\int_{\mathbb{S}^{k-1}} f(\langle u, v\rangle_{\mathbb{R}_2^k}) d\sigma_{k-1}(v) = \int_{-1}^{1} f(t) \, d\mathbb{Q}_k(t)$$

$$= \sqrt{\frac{k-1}{2\pi}} \, c_{k-1} \int_{-1}^{1} f(t)(1-t^2)^{\frac{k-3}{2}} \, dt \, ,$$

where $\mathbb{Q}_k(ds) := \frac{\Gamma(k/2)}{\sqrt{\pi}\,\Gamma((k-1)/2)} (1-t^2)^{\frac{k-3}{2}} \, ds$ *is a probability measure on* $[-1, 1]$.

Proof We just have to link (4.2.13) with the (unindexed) equality on page 27 of [92]. □

Remark 4.2 (Absolutely p-Summing Operators and GT in Matrix Form)
Recall that $T \in \mathfrak{L}(E, F)$ between Banach spaces E and F is called absolutely p-summing ($1 \le p < \infty$) if there exists a constant $c \ge 0$ such that for all $n \in \mathbb{N}$ and $x_1, \ldots, x_n \in E$

$$\Big(\sum_{i=1}^{n} \|T x_i\|^p\Big)^{1/p} \le c \, w_p(x_1, \ldots, x_n) \, , \tag{4.2.15}$$

where $w_p(x_1, \ldots, x_n) := \sup_{\psi \in B_{E'}} \big(\sum_{i=1}^n |\langle x_i, \psi\rangle|^p\big)^{1/p}$. The p-summing norm $\|T\|_{\mathfrak{P}_p}$ is defined as the infimum of all constants $c \ge 0$ which satisfy (4.2.15) (cf., e.g., [33, Chapter 11] or [37, Chapter 2]). It is well-known that $(\mathfrak{P}_p, \|\cdot\|_{\mathfrak{P}_p})$ is a 1-Banach ideal. Expressed in the terminology of absolutely 1-summing operators, Grothendieck proved that his inequality in particular is *equivalent* to

$$\|T\|_{\mathfrak{P}_1} \le K_G^{\mathbb{F}} \|T\| \tag{4.2.16}$$

for all $\mathbb{F}$-Hilbert spaces H, $n \in \mathbb{N}$ and finite rank operators $T \in \mathfrak{L}(l_1^n, H)$ (cf. [102, 122] and [79, Theorem 10.7]). Actually, the proof of (4.2.16) in the finite rank case is quite simple. It is based on the following two facts. Firstly, since $(l_1^n)' \cong l_\infty^n$, it follows that for all $a_1, \ldots, a_m \in l_1^n$ ($a_i \equiv (a_{i1}, \ldots, a_{in})^\top$), $\|A\|_{\infty,1} = w_1(a_1, \ldots, a_m)$ and $\|A^*A\|_{\infty,1} = (w_2(a_1, \ldots, a_m))^2$, where

$$A := (a_1 | a_2 | \cdots | a_m)^\top \equiv (a_{ij}) \in \mathbb{M}_{m,n}(\mathbb{F}).$$

Secondly, since $H' \cong H$ (Riesz), we obtain that

$$\sum_{i=1}^{m} \|T a_i\|_H = \sum_{i=1}^{m} \Big\|\sum_{j=1}^{n} a_{ij} T e_j\Big\|_H = \operatorname{tr}(A^* \Gamma_H(u, z_T)) \le K_G^{\mathbb{F}} \|T\| \|A\|_{\infty,1}$$

$$= K_G^{\mathbb{F}} \|T\| \, w_1(a_1, \ldots, a_m),$$

4.2 Integration Over $\mathbb{S}^{n-1}$ and the Gamma Function

for some $u \in B_H^m$ and $z_T \in H^n$, where the latter is defined as $(z_T)_j := Te_j \in \|T\| B_H$ ($j \in [n]$). Similarly, we obtain the well-known $\mathfrak{P}_2$-representation of the little Grothendieck inequality ("little GT"), which even is *equivalent* to little GT in matrix form, since the use of $(\mathfrak{P}_2, \|\cdot\|_{\mathfrak{P}_2})$ in fact naturally implies the emergence of the *positive semidefinite* matrix $A^*A \in \mathbb{M}_n(\mathbb{F})^+$:

$$\sum_{i=1}^m \|Ta_i\|_H^2 = \sum_{l=1}^n \sum_{j=1}^n \Big(\sum_{i=1}^m \overline{a_{il}} a_{ij}\Big)\langle (z_T)_j, (z_T)_l\rangle_H = \mathrm{tr}((A^*A)\Gamma_H(z_T, z_T))$$

$$\leq k_G^{\mathbb{F}} \|T\|^2 \|A^*A\|_{\infty,1} = k_G^{\mathbb{F}} \|T\|^2 (w_2(a_1, \ldots, a_m))^2,$$

In other words,

$$\|T\|_{\mathfrak{P}_2} \leq \sqrt{k_G^{\mathbb{F}}} \|T\| \qquad (4.2.17)$$

for all $\mathbb{F}$-Hilbert spaces H, $n \in \mathbb{N}$ and (finite rank operators) $T \in \mathfrak{L}(l_1^n, H)$. It is surprising that no attention seems to have been paid to the equivalence of the psd matrix version of little GT and the absolutely 2-summing version of little GT so far. So, its worth to state it here. Moreover, if we combine [33, Theorem 11.10], (4.2.8) and (4.2.9), we can somewhat simplify the representation of the 1-summing norm of $Id_{\mathbb{F}_2^k}$ for any $k \in \mathbb{N}$. We namely have

$$\|Id_{\mathbb{R}_2^k}\|_{\mathfrak{P}_1} = \sqrt{\frac{\pi}{2}} \mathbb{E}[\|\mathbf{X}\|_{\mathbb{R}_2^k}] = \sqrt{\frac{\pi}{2}} a_k \frac{(k-1)!!}{(k-2)!!} \stackrel{(4.2.13)}{=} \sqrt{k}\Big(\sqrt{\frac{\pi}{2}} c_k\Big) \text{ for all } k \in \mathbb{N}$$

and

$$\|Id_{\mathbb{C}_2^k}\|_{\mathfrak{P}_1} = \sqrt{\frac{4}{\pi}} \mathbb{E}[\|\mathbf{Z}\|_{\mathbb{C}_2^k}] = \frac{2}{\pi} \|Id_{\mathbb{R}_2^{2k}}\|_{\mathfrak{P}_1}$$

$$= \frac{(2k-1)!!}{(2k-2)!!} \stackrel{(4.2.14)}{=} \sqrt{k}\Big(\sqrt{\frac{4}{\pi}} c_{2k}\Big) \text{ for all } k \in \mathbb{N},$$

where $a_k := a_k(1)$ satisfies (4.2.11). Consequently, Lemma 4.2 recovers [79, Proposition 8.8], respectively [33, Corollary 11.10] and reveals a link between norms of certain absolutely 1-summing operators and values of Gaussian hypergeometric functions (since $\lim_{\nu \to \infty} c_\nu = 1$):

$$\lim_{k\to\infty} \frac{1}{k} \|Id_{\mathbb{F}_2^k}\|_{\mathfrak{P}_1}^2 = k_G^{\mathbb{F}} = {}_2F_1\Big(\frac{1}{2}, \frac{1}{2}, \frac{d_{\mathbb{F}}+2}{2}; 1\Big).$$

That is,

$$\lim_{k\to\infty} \frac{1}{k}\|Id_{\mathbb{R}_2^k}\|_{\mathfrak{P}_1}^2 = \frac{\pi}{2} = {}_2F_1\left(\frac{1}{2}, \frac{1}{2}, \frac{3}{2}; 1\right) \text{ and } \lim_{k\to\infty} \frac{1}{k}\|Id_{\mathbb{C}_2^k}\|_{\mathfrak{P}_1}^2 = \frac{4}{\pi}$$
$$= {}_2F_1\left(\frac{1}{2}, \frac{1}{2}, 2; 1\right),$$

implying a further facet of a link between (Euclidean norms of) Gaussian random vectors and the 1-Banach ideal of absolutely 1-summing operators.

Proposition 4.1 also allows us to give a straightforward, simple proof of the following important well-known surface integral characterisation of the trace of a matrix:

Corollary 4.6 *Let* $\mathbb{F} \in \{\mathbb{R}, \mathbb{C}\}$, $m, n \in \mathbb{N}$, $A \in \mathbb{M}_{m,m}(\mathbb{F})$, $B \in \mathbb{M}_{m,n}(\mathbb{F})$, $C \in \mathbb{M}_{n,m}(\mathbb{F})$ *and* $D \in \mathbb{M}_{n,n}(\mathbb{F})$. *Let* $\mathbf{Z} \sim \mathbb{F}N_m(0, I_m)$. *Then the following properties hold:*

(i)
$$tr(\text{Re}(A)) + i\, tr(\text{Im}(A)) = tr(A) = \mathbb{E}[f_A(\mathbf{Z})],$$

where $\mathbb{F}^m \ni z \mapsto f_A(z) := z^*Az = tr(Azz^*)$.

(ii)
$$tr(A) = m\left(\int_{\mathbb{S}^{m-1}} u^\top \text{Re}(A)u\, d\sigma_{m-1}(u) + i \int_{\mathbb{S}^{m-1}} u^\top \text{Im}(A)u\, d\sigma_{m-1}(u)\right).$$

In particular,

$$\int_{\mathbb{S}^{m+n-1}} tr(B\Gamma_\mathbb{R}(u,v))\, d\sigma_{m+n-1}(vec(u,v))$$
$$= \int_{\mathbb{S}^{m+n-1}} u^\top Bv\, d\sigma_{m+n-1}(vec(u,v)) = 0.$$

Proof

(i) Since $\mathbf{Z} \sim \mathbb{F}N_m(0, I_m)$, it follows that $\mathbb{E}[\mathbf{ZZ}^*] = I_m$. Consequently, the linearity of $\mathbb{E}$ implies that

$$\text{tr}(A) = \text{tr}(A\mathbb{E}[\mathbf{ZZ}^*]) = \mathbb{E}[\text{tr}(A\mathbf{ZZ}^*)] = \mathbb{E}[\text{tr}(\mathbf{Z}^*A\mathbf{Z})] = \mathbb{E}[f_A(\mathbf{Z})].$$

(ii) Obviously, we may assume that $A \in \mathbb{M}_m(\mathbb{R})$. Since then $f_A(rx) = r^2 f_A(x)$ for all $(r, x) \in (0, \infty) \times \mathbb{R}^m$, we may apply Proposition 4.1 to $p = 2$, implying that

$$\mathrm{tr}(A) = 2 \frac{\Gamma(\frac{m}{2}+1)}{\Gamma(\frac{m}{2})} \int_{\mathbb{S}^{m-1}} f_A(u) \, d\sigma_{m-1}(u) = m \int_{\mathbb{S}^{m-1}} u^\top A u \, d\sigma_{m-1}(u)$$

(since $\Gamma(\frac{m}{2}+1) = \frac{m}{2}\Gamma(\frac{m}{2})$). Finally, if we put $w := \mathrm{vec}(u, v)$ and $\Delta(B) := \frac{1}{2}\begin{pmatrix} 0 & B \\ B^\top & 0 \end{pmatrix} \in \mathbb{M}_{m+n,m+n}(\mathbb{R})$, it obviously follows that $w^\top \Delta(B) w = u^\top B v$ and $\mathrm{tr}(\Delta(B)) = 0$. Consequently, we obtain

$$0 = \mathrm{tr}(\Delta(B)) = (m+n) \int_{\mathbb{S}^{m+n-1}} u^\top B v \, d\sigma_{m+n-1}(w).$$

$\square$

4.3 Integrating Powers of Inner Products of Random Vectors, Uniformly Distributed on $\mathbb{S}^{n-1}$

Let us recall the (real, respectively complex) correlation matrices

$$\Sigma_{2n}(z) := \begin{pmatrix} I_n & z I_n \\ \bar{z} I_n & I_n \end{pmatrix} = \begin{pmatrix} 1 & z \\ \bar{z} & 1 \end{pmatrix} \otimes I_n,$$

where $|z| \leq 1$ and $n \in \mathbb{N}$ (cf. (2.2.1) and Proposition 2.4). Moreover, if $\mathbf{X} \sim N_n(0, I_n)$ and $A \in \mathcal{B}(\mathbb{S}^{n-1})$ is an arbitrary Borel subset of the unit sphere $\mathbb{S}^{n-1} \subseteq \mathbb{R}^n$ ($n \geq 2$), (4.2.6), applied to the function $\mathbb{R}^n \setminus \{0\} \ni x \mapsto f_A(x) := \mathbb{1}_A(\frac{x}{\|x\|_{\mathbb{R}^n_2}})$ (and $p = 0$) implies that in particular

$$\mathbb{P}(\frac{\mathbf{X}}{\|\mathbf{X}\|_{\mathbb{R}^n}} \in A) = \mathbb{E}[f_A(\mathbf{X})] = \int_{\mathbb{S}^{n-1}} \mathbb{1}_A(\xi) \, d\sigma_{n-1}(\xi) = \sigma_{n-1}(A)$$

(since $\mathbf{X} \neq 0$ $\mathbb{P}$-a.s.). Thus, we get again the well-known fact that $\frac{\mathbf{X}}{\|\mathbf{X}\|_{\mathbb{R}^n}}$ is uniformly distributed on the unit sphere $\mathbb{S}^{n-1}$ if $\mathbf{X} \sim N_n(0, I_n)$. Note that, in addition, for any $X \sim N_1(0, 1)$, it is true that

$$\mathbb{P}(\frac{X}{|X|} = \varepsilon) = \mathbb{P}(X > 0 \text{ and } \varepsilon = 1) + \mathbb{P}(X \leq 0 \text{ and } \varepsilon = -1)$$

$$= \frac{1}{2} \text{ for all } \varepsilon \in \{-1, 1\},$$

so that we could also say that the random variable $\frac{X}{|X|}$ is uniformly distributed on the "unit sphere" $S^0 := \{-1, 1\} \subseteq \mathbb{R}^1$ if $X \sim N_1(0, 1)$.

Theorem 4.1 *Let $m, n \in \mathbb{N}$, $\rho \in (-1, 1)$ and $vec(\mathbf{X}, \mathbf{Y}) = vec(X_1, \ldots, X_n, Y_1, \ldots, Y_n) \sim N_{2n}(0, \Sigma_{2n}(\rho))$.*

(i) *If m is odd then*

$$\mathbb{E}\Big[\Big\langle \frac{\mathbf{X}}{\|\mathbf{X}\|_{\mathbb{R}^n_2}}, \frac{\mathbf{Y}}{\|\mathbf{Y}\|_{\mathbb{R}^n_2}} \Big\rangle_{\mathbb{R}^n_2}^m \Big] = c_{odd}(m, n)\, \rho\, (1 - \rho^2)^{\frac{n}{2}}$$

$$_3F_2\Big(\frac{n+1}{2}, \frac{n+1}{2}, \frac{m+2}{2}; \frac{3}{2}, \frac{m+n+1}{2}; \rho^2\Big),$$

where

$$c_{odd}(m, n) := \frac{m}{\sqrt{\pi}}\, \frac{\Gamma^2(\frac{n+1}{2})\, \Gamma(\frac{m}{2})}{\Gamma(\frac{n}{2})\Gamma(\frac{m+n+1}{2})}.$$

(ii) *If m is even then*

$$\mathbb{E}\Big[\Big\langle \frac{\mathbf{X}}{\|\mathbf{X}\|_{\mathbb{R}^n_2}}, \frac{\mathbf{Y}}{\|\mathbf{Y}\|_{\mathbb{R}^n_2}} \Big\rangle_{\mathbb{R}^n_2}^m \Big] = c_{even}(m, n)\, (1 - \rho^2)^{\frac{n}{2}}$$

$$_3F_2\Big(\frac{n}{2}, \frac{n}{2}, \frac{m+1}{2}; \frac{1}{2}, \frac{m+n}{2}; \rho^2\Big),$$

where

$$c_{even}(m, n) := \frac{1}{\sqrt{\pi}}\, \frac{\Gamma(\frac{n}{2})\, \Gamma(\frac{m+1}{2})}{\Gamma(\frac{m+n}{2})}.$$

In particular,

$$\mathbb{E}\Big[\Big\langle \frac{\mathbf{X}}{\|\mathbf{X}\|_{\mathbb{R}^n_2}}, \frac{\mathbf{Y}}{\|\mathbf{Y}\|_{\mathbb{R}^n_2}} \Big\rangle_{\mathbb{R}^n_2}^m \Big] = c_n^2\, (1 - \rho^2)^{\frac{n}{2}}\, \rho\, _2F_1\Big(\frac{n+1}{2}, \frac{n+1}{2}; \frac{n+2}{2}; \rho^2\Big)$$

$$= c_n^2\, \rho\, _2F_1\Big(\frac{1}{2}, \frac{1}{2}; \frac{n+2}{2}; \rho^2\Big), \tag{4.3.1}$$

where $c_n := c_{odd}(1, n) = \sqrt{\frac{2}{n}}\, \frac{\Gamma(\frac{n+1}{2})}{\Gamma(\frac{n}{2})}$.

Proof Fix $m \in \mathbb{N}$ and $\rho \in (-1, 1)$. Firstly, let $n = 1$. Observe that $\mathbb{R}^1_2 = (\mathbb{R}, |\cdot|)$, implying that the inner product on $\mathbb{R}^1_2$ is given by the standard product of real numbers. Moreover, note that $c_{odd}(m, 1) = \frac{2}{\pi}$ and $c_{even}(m, 1) = 1$. Thus,

4.3 Integrating Powers of Inner Products of Random Vectors, Uniformly...

if $m = 2l + 1$ is odd, where $l \in \mathbb{N}_0$, the Grothendieck equality (cf. Corollary 4.1) implies that

$$\mathbb{E}\Big[\Big(\frac{X}{|X|} \cdot \frac{Y}{|Y|}\Big)^m\Big] = \mathbb{E}[\text{sign}(X)\,\text{sign}(Y)] = \frac{2}{\pi}\rho\,{}_2F_1\Big(\frac{1}{2},\frac{1}{2};\frac{3}{2};\rho^2\Big) = \frac{2}{\pi}\arcsin(\rho).$$

On the other hand,

$$c_{\text{odd}}(m,1)\sqrt{1-\rho^2}\,\rho\,{}_3F_2\Big(1,1,\frac{m+2}{2};\frac{3}{2},\frac{m+2}{2};\rho^2\Big)$$

$$= \frac{2}{\pi}\rho\Big(\sqrt{1-\rho^2}\,{}_2F_1\Big(1,1;\frac{3}{2};\rho^2\Big)\Big)$$

$$= \frac{2}{\pi}\rho\,{}_2F_1\Big(\frac{1}{2},\frac{1}{2};\frac{3}{2};\rho^2\Big) = \frac{2}{\pi}\arcsin(\rho),$$

(where the penultimate equality follows from Euler's transformation formula of hypergeometric functions (cf., e.g., [6, Theorem 2.2.5, formula (2.2.7)]).

If $m = 2l$ is even ($l \in \mathbb{N}_0$), then

$$c_{\text{even}}(m,1)\sqrt{1-\rho^2}\,{}_3F_2\Big(\frac{1}{2},\frac{1}{2},\frac{m+1}{2};\frac{1}{2},\frac{m+1}{2};\rho^2\Big)$$

$$= \sqrt{1-\rho^2}\,{}_2F_1\Big(\frac{1}{2},\frac{1}{2};\frac{1}{2};\rho^2\Big) = 1$$

(since ${}_2F_1\big(\frac{1}{2},\frac{1}{2};\frac{1}{2};\rho^2\big) = \arcsin'(\rho) = \frac{1}{\sqrt{1-\rho^2}}$). This concludes the proof for the case $n = 1$.

So, let now $n \geq 2$ be given. We have to calculate the following well-defined (Lebesgue) integral

$$I := c_n^{(1)}(\rho) \int_{\mathbb{R}^n}\int_{\mathbb{R}^n} \Big\langle\frac{x}{\|x\|},\frac{y}{\|y\|}\Big\rangle_{\mathbb{R}_2^n}^m \exp\Big(-\frac{\|x\|^2 + \|y\|^2 - 2\rho\langle x,y\rangle_2}{2(1-\rho^2)}\Big)d^n x\, d^n y,$$

where $c_n^{(1)}(\rho) := \frac{1}{(2\pi)^n (1-\rho^2)^{n/2}}$ (cf. (2.2.3)). The main idea is to calculate I by a twofold application of the n-dimensional polar coordinates formula, implying the appearance of a double integral of type $\int_0^\infty \int_0^\infty F(r,s)dr\,ds$ and a decoupled spherical integral of type $\int_{\mathbb{S}^{n-1}} G(u,v)d\sigma_{n-1}(v)$, where the latter actually does not depend on u. Despite the entanglement of the radius integrand parts $r \in (0,\infty)$ and $s \in (0,\infty)$ and the spherical integrand parts $u \in \mathbb{S}^{n-1}$ and $v \in \mathbb{S}^{n-1}$ in the density function, that decoupling can be obtained, simply by making use of the Maclaurin series representation of the entangled part of the density function $(0,\infty)^2 \times (\mathbb{S}^{n-1})^2 \ni (r,s,u,v) \mapsto \exp\big(\frac{rs\rho\langle u,v\rangle_{\mathbb{R}_2^n}}{1-\rho^2}\big)$. Due to (4.2.12), the resulting integrals can then be calculated easily. We will also recognise the need for the even/odd case-distinction at this point. To concretise the calculation steps, put

$$c_n^{(2)}(\rho) := c_n^{(1)}(\rho)(\omega_n(\mathbb{S}^{n-1}))^2 = \frac{1}{(2\pi)^n (1-\rho^2)^{n/2}} \left(\frac{2\pi^{n/2}}{\Gamma(n/2)}\right)^2$$

$$= \frac{1}{2^{n-2}\,\Gamma^2(n/2)\,(1-\rho^2)^{n/2}}. \tag{4.3.2}$$

A twofold application of the n-dimensional polar coordinates, together with Fubini's theorem implies that

$$I = c_n^{(2)}(\rho) \int_0^\infty \int_0^\infty \int_{\mathbb{S}^{n-1}} \int_{\mathbb{S}^{n-1}} \langle u, v \rangle_{l_2^n}^m \exp\left(-\frac{1}{2}\left(\frac{r^2+s^2}{1-\rho^2}\right)\right)$$
$$\times \exp\left(\frac{rs\rho\langle u,v\rangle_2}{1-\rho^2}\right) r^{n-1} s^{n-1} \, \mathrm{d}\sigma_{n-1}(v)\, \mathrm{d}\sigma_{n-1}(u).$$

Since $\exp\left(\frac{rs\rho\langle u,v\rangle_2}{1-\rho^2}\right) = \sum_{\nu=0}^\infty \frac{1}{\nu!} \frac{\rho^\nu}{(1-\rho^2)^\nu} r^\nu s^\nu \langle u,v\rangle_{l_2^n}^\nu$, it consequently follows that

$$I = c_n^{(2)}(\rho) \sum_{\nu=0}^\infty \frac{1}{\nu!} \frac{\rho^\nu}{(1-\rho^2)^\nu} I_\nu(\rho) J_\nu(m), \tag{4.3.3}$$

where

$$I_\nu(\rho) := \int_0^\infty \int_0^\infty r^{\nu+n-1} s^{\nu+n-1} \exp\left(-\frac{1}{2}\left(\frac{r^2+s^2}{1-\rho^2}\right)\right) \mathrm{d}r\, \mathrm{d}s$$
$$= \left(\int_0^\infty r^{\nu+n-1} \exp\left(-\frac{1}{2}\left(\frac{r}{\sqrt{1-\rho^2}}\right)^2\right) \mathrm{d}r\right)^2$$

and

$$J_\nu(m) := \int_{\mathbb{S}^{n-1}} \int_{\mathbb{S}^{n-1}} \langle u, v\rangle_{l_2^n}^{m+\nu} \, \mathrm{d}\sigma_{n-1}(v)\, \mathrm{d}\sigma_{n-1}(u).$$

A simple change of variables in $I_\nu(\rho)$ $(r \mapsto \frac{r}{\sqrt{1-\rho^2}})$ therefore implies that

$$I_\nu(\rho) = \frac{\pi}{2}(1-\rho^2)^{n+\nu} \left(2\int_0^\infty x^{n-1+\nu} \gamma_1(\mathrm{d}x)\right)^2$$

$$\stackrel{(4.2.4)}{=} (2^{n-2}(1-\rho^2)^n)\, 2^\nu\, (1-\rho^2)^\nu\, \Gamma^2\left(\frac{\nu+n}{2}\right)$$

$$\stackrel{(4.3.2)}{=} \frac{(1-\rho^2)^{n/2}}{c_n^{(2)}(\rho)\,\Gamma^2(n/2)}\, 2^\nu\, (1-\rho^2)^\nu\, \Gamma^2\left(\frac{\nu+n}{2}\right).$$

4.3 Integrating Powers of Inner Products of Random Vectors, Uniformly... 101

Because of (4.2.12), it follows that

$$J_\nu(m) = \int_{\mathbb{S}^{n-1}} \langle u, v \rangle_{l_n^2}^{m+\nu} \, d\sigma_{n-1}(v) = \frac{1+(-1)^{m+\nu}}{2} \left(\frac{\Gamma(\frac{\nu+m+1}{2})\Gamma(\frac{n}{2})}{\sqrt{\pi}\,\Gamma(\frac{\nu+m+n}{2})} \right).$$

Hence,

$$I \stackrel{(4.3.3)}{=} \frac{1}{\Gamma(n/2)} (1-\rho^2)^{n/2} \sum_{\nu=0}^{\infty} \frac{1+(-1)^{m+\nu}}{2} \frac{2^\nu}{\nu!\sqrt{\pi}} \frac{\Gamma^2(\frac{\nu+n}{2})\Gamma(\frac{\nu+m+1}{2})}{\Gamma(\frac{\nu+m+n}{2})} \rho^\nu$$

It is no coincidence that the factor $\frac{2^\nu}{\nu!\sqrt{\pi}}$ emerges. If we namely apply Legendre's duplication formula to $\Gamma(2a)$, where $a := \frac{\nu+1}{2}$ (cf., e.g., [6, Theorem 1.5.1]), it follows that $\frac{2^\nu}{\nu!\sqrt{\pi}} = \frac{1}{\Gamma(\frac{\nu+1}{2})\Gamma(\frac{\nu+2}{2})}$, whence

$$I = \frac{1}{\Gamma(n/2)} (1-\rho^2)^{n/2} \sum_{\nu=0}^{\infty} \frac{1+(-1)^{m+\nu}}{2} \frac{\Gamma^2(\frac{\nu+n}{2})\Gamma(\frac{\nu+m+1}{2})}{\Gamma(\frac{\nu+1}{2})\Gamma(\frac{\nu+2}{2})\Gamma(\frac{\nu+m+n}{2})} \rho^\nu.$$

Obviously, we only have to consider the set of all $\nu \in \mathbb{N}_0$, such that $\nu + m$ is even. So, we need to distinguish between the odd case and the even case, relative to m. Firstly, let m be odd. Then $\nu \in \{2l+1 : l \in \mathbb{N}_0\}$. Due to the definition of the constant $c_{\text{odd}}(m, n)$ and the structure of the special function $_3F_2$, it follows immediately that

$$I = (1-\rho^2)^{n/2} \frac{1}{\Gamma(n/2)} \rho \sum_{l=0}^{\infty} \frac{\Gamma^2(l+\frac{n+1}{2})\Gamma(l+\frac{m+2}{2})}{\Gamma(l+\frac{m+n+1}{2})\Gamma(l+\frac{3}{2})} \frac{(\rho^2)^l}{l!}$$

$$= c_{\text{odd}}(m, n)\, (1-\rho^2)^{\frac{n}{2}}\, \rho\, _3F_2\Big(\frac{n+1}{2}, \frac{n+1}{2}, \frac{m+2}{2}; \frac{3}{2}, \frac{m+n+1}{2}; \rho^2\Big),$$

which concludes the proof of (i).

Similarly, the representation (ii) can be derived if m is even. Finally, the last equality in (4.3.1) again is an implication of Euler's transformation formula of hypergeometric functions (cf., e.g., [6, Theorem 2.2.5, formula (2.2.7)]). □

Remark 4.3 The special case (4.3.1) is contained in (the proof of) [24, Lemma 2.1].

Remark 4.4 Let $p \in (-1, \infty)$ and $f \in L^2(\gamma_1)$, satisfying $f(ry) = r^p f(y)$ for all $(r, y) \in (0, \infty) \times \mathbb{R}$. A natural question is, whether $I_f := \mathbb{E}\big[\langle f\big(\frac{X}{\|X\|_{\mathbb{R}_2^n}}, \frac{Y}{\|Y\|_{\mathbb{R}_2^n}}\big)\rangle_{\mathbb{R}_2^n}\big]$ can be similarly represented as $I = I_{f_m}$? In general, it seems that a closed-form representation of I_f is not possible. However, our proof of Theorem 4.1 clearly reveals that

$$I_f = \frac{\sqrt{\pi}}{2^{p/2}\,\Gamma(n/2)}\,(1-\rho^2)^{n/2} \sum_{\nu=0}^{\infty} \frac{\Gamma^2(\frac{\nu+n}{2})}{2^{\nu/2}\,\Gamma(\frac{\nu+1}{2})\Gamma(\frac{\nu+2}{2})\Gamma(\frac{\nu+p+n}{2})}\,\mathbb{E}[f(X)X^\nu]\,\rho^\nu$$

$$= \mathbb{E}\Big[f(X)\,\frac{\sqrt{\pi}}{2^{p/2}\,\Gamma(n/2)}\,(1-\rho^2)^{n/2} \sum_{\nu=0}^{\infty} \frac{\Gamma^2(\frac{\nu+n}{2})}{2^{\nu/2}\,\Gamma(\frac{\nu+1}{2})\Gamma(\frac{\nu+2}{2})\Gamma(\frac{\nu+p+n}{2})}\,(\rho X)^\nu\Big],$$

where $X \sim N_1(0,1)$. Observe that the factor $2^{\nu/2}$ in the denominator cancels out in the f_m-case (i.e., if $f = f_m$)! It originates from the integral

$$\int_{\mathbb{S}^{n-1}} f(\langle u,v\rangle_{l_n^2})\,\langle u,v\rangle_{l_n^2}^\nu\,\mathrm{d}\sigma_{n-1}(v) = 2^{-p/2}\,2^{-\nu/2}\,\frac{\Gamma(\frac{n}{2})}{\Gamma(\frac{\nu+p+n}{2})}\,\mathbb{E}[f(X)X^\nu]$$

(cf. Corollary 4.3). In general, that important reduction of the fraction cannot be maintained, though. Hence, if we represent the latter series as the sum of the even series part and the odd series part, we obtain

$$I_f = I_f^{\text{even}} + I_f^{\text{odd}},$$

where

$$I_f^{\text{even}} := \mathbb{E}\Big[f(X)\,\frac{\sqrt{\pi}}{2^{p/2}\,\Gamma(n/2)}\,(1-\rho^2)^{n/2} \sum_{l=0}^{\infty} \frac{\Gamma^2(l+\frac{n}{2})}{\Gamma(l+\frac{1}{2})\Gamma(l+\frac{p+n}{2})}\,\frac{(\frac{1}{2}\rho^2 X^2)^l}{l!}\Big]$$

and

$$I_f^{\text{odd}} := \mathbb{E}\Big[Xf(X)\,\frac{\sqrt{\pi}}{2^{(p+1)/2}\,\Gamma(n/2)}\,\rho(1-\rho^2)^{n/2}$$

$$\times \sum_{l=0}^{\infty} \frac{\Gamma^2(l+\frac{1+n}{2})}{\Gamma(l+\frac{3}{2})\Gamma(l+\frac{p+n+1}{2})}\,\frac{(\frac{1}{2}\rho^2 X^2)^l}{l!}\Big].$$

A straightforward calculation shows that

$$I_f^{\text{even}} = c_+(p,n)\sqrt{\frac{\pi}{2}}\,\frac{1}{(p-1)!!\,b_{p+1}}\,(1-\rho^2)^{n/2}\mathbb{E}\big[f(X)$$

$$\,_2F_2\big(\frac{n}{2},\frac{n}{2};\frac{1}{2},\frac{p+n}{2};(\tfrac{1}{\sqrt{2}}\rho X)^2\big)\big]$$

and

$$I_f^{\text{odd}} = c_-(p,n)\sqrt{\frac{\pi}{2}}\frac{1}{p!!\,b_{p+2}}\,\rho(1-\rho^2)^{n/2}\mathbb{E}\big[Xf(X)$$
$$\,_2F_2\big(\frac{n+1}{2},\frac{n+1}{2};\frac{3}{2},\frac{p+n+1}{2};(\tfrac{1}{\sqrt{2}}\rho X)^2\big)\big],$$

where b_n satisfies (4.2.1) ($n \in \{p+1, p+2\}$).

Chapter 5
Completely Correlation Preserving Functions

5.1 Completely Real Analytic Functions and the Entrywise Matrix Functional Calculus

Already while looking for the smallest upper bound of both, $K_G^{\mathbb{R}}$ and $K_G^{\mathbb{C}}$, we are lead to a deep interplay of different subfields of mathematics (both, pure and applied) including Gaussian harmonic analysis and building blocks of Malliavin calculus (Mehler kernel, Ornstein-Uhlenbeck semigroup, Hermite polynomials), integration over spheres in $\mathbb{R}^n$), complex analysis (analytic continuation and biholomorphic mappings, special functions), combinatorial analysis (inversion of Taylor series and ordinary partial Bell polynomials), matrix analysis (positive semidefinite matrices, block matrices) and multivariate statistics and high-dimensional Gaussian dependence modelling (correlation matrices, real and complex Gaussian random vectors, Gaussian measure).

In particular, we have to look for those functions which map correlation matrices of any size and any rank entrywise into a correlation matrix of the same size again, by means of the so-called *Hadamard product (or Schur product)* of matrices:

Definition 5.1 (Hadamard Product) Let $m, n \in \mathbb{N}$. Let $A = (a_{ij}) \in \mathbb{M}_{m,n}(\mathbb{F})$ and $B = (b_{ij}) \in \mathbb{M}_{m,n}(\mathbb{F})$. The Hadamard product $A * B \in \mathbb{M}_{m,n}(\mathbb{F})$ is defined as

$$(A * B)_{ij} := a_{ij} b_{ij} \quad ((i,j) \in [m] \times [n]).$$

The Hadamard product is sometimes called the *entrywise product*, for obvious reasons, or the *Schur product*, because of some early and basic results about the product obtained by Issai Schur (cf. [70]). Like the usual matrix product, the distributive law also holds for the Hadamard product: $A * (B + C) = A * B + A * C$. Unlike the usual matrix product, the Hadamard product is commutative: $A * B = B * A$.

Remark 5.1 Very often, the Hadamard product is denoted by the symbol ∘. However, given our view, the symbolic notation ∘ perhaps could lead to a minor ambiguity, since quite regularly, ∘ denotes composition of mappings. This is why we adopt the symbol ∗ instead, used for the definition of the Schur product in [120, p 29 ff].

Remark 5.2 (Hadamard Product as Subordinated Kronecker Product) Fix $m, n \in \mathbb{N}$ and $A, B \in \mathbb{M}_{m,n}(\mathbb{F})$. There exists an interesting link between the Hadamard product $A \ast B$ and the Kronecker product $A \otimes B$, induced by (1.2.2) and (3.4.6). In order to recognise this, let $(i, j) \in [m] \times [n]$ be given arbitrarily. Then

$$(A \ast B)_{ij} = (e_i^\top A e_j)(e_i^\top B e_j) \stackrel{(!)}{=} e_i^\top \otimes e_i^\top (A \otimes B) e_j \otimes e_j$$
$$\stackrel{(3.4.6)}{=} e_{(i-1)m+i}^{(m^2)\top}(A \otimes B) e_{(j-1)n+i}^{(n^2)} = (A \otimes B)_{\Psi_m(i,i), \Psi_n(j,j)}.$$

Consequently,

$$A \ast B = (A \otimes B)_\psi,$$

where $\psi(i, j) := (\Psi_m(i, i), \Psi_n(j, j)) = ((i-1)m + i, (j-1)n + j)$ for all $(i, j) \in [m] \times [n]$.

If we combine the latter fact and Proposition 3.5, we immediately obtain (cf. also (5.1.2)):

Proposition 5.1 *Let $m, n \in \mathbb{N}$, $S \in \mathbb{M}_{m,n}(\mathbb{F})$ and $R \in \mathbb{M}_{m,n}(\mathbb{F})$. If $S \in \mathcal{Q}_{m,n}(\mathbb{F})$ and $R \in \mathcal{Q}_{m,n}(\mathbb{F})$, then $S \ast R \in \mathcal{Q}_{m,n}(\mathbb{F})$. Note again that the Hadamard product is commutative.*

Although, the proof of the following two facts are just a consequent application of the definition of the Schur product, they are of interest on their own, and they help us strongly to support, inter alia, a quick proof of Theorem 5.1. In this context, diagonal matrices play an important role: if $a = (a_1, \ldots, a_p)^\top \in \mathbb{F}^p$, then $D_a \in \mathbb{M}_p(\mathbb{F})$ denotes the matrix, whose (i, j)'th entry is given by $\delta_{ij} a_j$. Moreover, we need the matrix $J^{(m,n)} \in \mathbb{M}_{m,n}(\mathbb{F})$, whose (i, j)'th entry is given by 1: $J^{(m,n)} := \sum_{i=1}^{m} \sum_{j=1}^{n} e_i e_j^\top$.

Lemma 5.1 *Let $m, n \in \mathbb{N}$, $B \in \mathbb{M}_{m,n}(\mathbb{F})$ and $(x, y) \in \mathbb{F}^m \times \mathbb{F}^n$. Then*

$$\Gamma_\mathbb{F}(x, y) \ast B = \overline{x} y^\top \ast B = D_x^\ast B D_y.$$

In particular,

$$\Gamma_\mathbb{F}(y, y) \ast A \in \mathbb{M}_n(\mathbb{F})^+ \text{ for all } A \in \mathbb{M}_n(\mathbb{F})^+.$$

5.1 Completely Real Analytic Functions and the Entrywise Matrix Functional... 107

Lemma 5.2 (Hadamard Product Factor Shifting) *Let $m, n \in \mathbb{N}$ and $A, B, C \in \mathbb{M}_{m,n}(\mathbb{F})$. Then*

$$\langle A * \overline{B}, J^{(m,n)} \rangle_F = \langle A, B \rangle_F = \langle J^{(m,n)}, \overline{A} * B \rangle_F . \quad (5.1.1)$$

In particular,

$$\langle A * B, C \rangle_F = \langle B, \overline{A} * C \rangle_F .$$

Remark 5.3 Lemma 5.2 implies that for any $A \in \mathbb{M}_n(\mathbb{F})$ the adjoint of the "Schur multiplier" $S_A : (\mathbb{M}_{m,n}(\mathbb{F}), \|\cdot\|_F) \longrightarrow (\mathbb{M}_{m,n}(\mathbb{F}), \|\cdot\|_F)$, $B \mapsto A * B$ coincides with the "Schur multiplier" $S_{\overline{A}}$ (cf. [120, p. 29 ff]):

$$S_A^* = S_{\overline{A}} .$$

Fix $m, n \in \mathbb{N}$. The proof of [47, Theorem 2.2] reveals that in fact *any* matrix $A \in \mathbb{M}_{m,n}(\mathbb{F})$ is a Schur multiplier:

$$\|S_A : \mathfrak{L}(l_2^n, l_2^m) \longrightarrow \mathfrak{L}(l_2^n, l_2^m)\| = \sup\{\|Auv^\top\|_* : (u, v) \in S_{l_2^n} \times S_{l_2^m}\} < \infty .$$

Here, $\|Auv^\top\|_*$ is the trace norm of the matrix $Auv^\top$ (cf. Remark 3.6). Consequently, if we link Proposition 3.3 and (3.3.1) with a further fundamental and deep result of Grothendieck, we obtain the following noteworthy representation of *arbitrary* $m \times n$ matrices (cf. [28, Theorem 4.2], [120, Theorem 8.7] and [125, Theorem 5.1 and Proposition 5.4]):

Proposition 5.2 *Let $m, n \in \mathbb{N}$ and $A \in \mathbb{M}_{m,n}(\mathbb{F})$. Put $\lambda_A(m, n) := \|A : l_1^n \longrightarrow l_\infty^m\|_{\mathfrak{L}_2}$. Then $\lambda_A(m, n) = \|S_A : \mathfrak{L}(l_2^n, l_2^m) \longrightarrow \mathfrak{L}(l_2^n, l_2^m)\|$, and*

$$A = \lambda_A(m, n) \, Q_A ,$$

where $Q_A \in \mathcal{Q}_{m,n}(\mathbb{F})$.

Combining the latter result with the (not numbered) remark before [120, Theorem 3.7], we obtain a further non-negligible

Corollary 5.1 *Let $n \in \mathbb{N}$ and $B = (b_{ij}) \in \mathbb{M}_n(\mathbb{F})^+$ be positive semidefinite. Then*

$$B = \max_{i \in [n]}(b_{ii}) \, Q_B ,$$

where $Q_B \in \mathcal{Q}_n(\mathbb{F})$. In particular, if $b_{ii} = 0$ for all $i \in [n]$, then $B = 0$.

The usefulness of the structure of the Hadamard product is reflected in the Schur product theorem which states that the (closed, convex and self-dual) cone of all positive semidefinite matrices is stable under Schur multiplication (cf. [138], Remark 3.8]). We give a short and completely self-contained proof:

Theorem 5.1 (Schur [138]) *Let $n \in \mathbb{N}$. Let $A \in \mathbb{M}_n(\mathbb{F})^+$ and $B \in \mathbb{M}_n(\mathbb{F})^+$ be positive semidefinite. Then $A * B \in \mathbb{M}_n(\mathbb{F})^+$. In particular,*

$$C(n; \mathbb{F}) * C(n; \mathbb{F}) \subseteq C(n; \mathbb{F}). \tag{5.1.2}$$

Proof Fix $y \in \mathbb{F}^n$, and put $M := \Gamma_{\mathbb{F}}(\overline{y}, \overline{y}) * \overline{B} = y y^* * \overline{B}$. Because of Lemma 5.1, $M^* = M \in \mathbb{M}_n(\mathbb{F})^+$ (since also $\overline{B} \in \mathbb{M}_n(\mathbb{F})^+$). Consequently, Lemma 5.2 implies that

$$\begin{aligned} y^*(A * B) y &= \mathrm{tr}(yy^*(A*B)) = \langle A*B, yy^* \rangle_F \\ &= \langle A, M \rangle_F = \mathrm{tr}(M^{1/2} M^{1/2} A^{1/2} A^{1/2}) \\ &= \mathrm{tr}((M^{1/2} A^{1/2})(A^{1/2} M^{1/2})) = \|A^{1/2} M^{1/2}\|_F^2 \geq 0, \end{aligned}$$

and the claim follows. □

Definition 5.2 (Entrywise Functional Calculus) Let $m, n \in \mathbb{N}$. Given $\emptyset \neq U \subseteq \mathbb{F}$, a function $f : U \longrightarrow \mathbb{F}$ and a matrix $A = (a_{ij}) \in \mathbb{M}_{m,n}(U)$ put

$$f[A] := (f(a_{ij})) \quad ((i,j) \in [m] \times [n]).$$

In particular, if $f(x) = \sum_{n=0}^{\infty} c_n x^n$, $x \in U$, where $c_n \in \mathbb{F}$ for all $n \in \mathbb{N}$, we have

$$f[A]_{ij} = \sum_{n=1}^{\infty} c_n a_{ij}^n \text{ for all } (i,j) \in [m] \times [n].$$

Functions of the latter type, where $c_n \geq 0$ for all $n \in \mathbb{N}$ play a significant role, also with respect to an analysis of $K_G^{\mathbb{R}}$ and $K_G^{\mathbb{C}}$. This is particularly reflected in the next two results with respect to the real field, which however also play a key role in the complex case (cf. Theorem 7.1 and Corollary 7.1). Recall that a function $\psi : I \longrightarrow \mathbb{R}$, defined on an open interval $I \subseteq \mathbb{R}$, is *absolutely monotonic*, if $\psi \in C^{\infty}(I)$ and $\psi^{(n)} \geq 0$ on I for all $n \in \mathbb{N}_0$ (cf. e.g. [83, Chapter 19] and [156, Chapter IV] regarding a rigorous reassessment of that crucial and very rich concept, coined by S. N. Bernstein in 1929). Regarding a refresher of complex analysis of functions of one complex variable, we recommend to place the rich source [142] next to our work.

Lemma 5.3 *Let $r \in (0, \infty)$ and $\psi : (-r, r) \longrightarrow \mathbb{R}$ be real analytic. Suppose that $b := \left(\frac{\psi^{(n)}(0)}{n!} r^n\right)_{n \in \mathbb{N}_0} \in l_1$. Then $\psi = \tilde{\psi}\big|_{(-r,r)}$, where the complex, bounded*

5.1 Completely Real Analytic Functions and the Entrywise Matrix Functional... 109

function $\widetilde{\psi} : r\overline{\mathbb{D}} \longrightarrow \|b\|_1 \overline{\mathbb{D}}$ is defined as

$$\widetilde{\psi}(z) := \sum_{n=0}^{\infty} \frac{\psi^{(n)}(0)}{n!} z^n \text{ for all } z \in r\overline{\mathbb{D}}.$$

The real function $\psi_{abs} : [-r, r] \longrightarrow [-\|b\|_1, \|b\|_1], x \mapsto \sum_{n=0}^{\infty} \frac{|\psi^{(n)}(0)|}{n!} x^n$ is continuous and bounded, such as the complex-valued function $\widetilde{\psi}$. $\psi_{abs}|_{(-r,r)}$ is real analytic, and $\psi_{abs}|_{(0,r)}$ is absolutely monotonic on $(0, r)$. $\|b\|_1 = \psi_{abs}(r)$ and

$$|\widetilde{\psi}(z)| \leq \psi_{abs}(|z|) \leq \psi_{abs}(r) \text{ for all } z \in r\overline{\mathbb{D}}. \tag{5.1.3}$$

In particular, ψ_{abs} can be extended to the continuous complex function $\widetilde{\psi}_{abs} : r\overline{\mathbb{D}} \longrightarrow \|b\|_1 \overline{\mathbb{D}}$, defined as

$$\psi_{abs}(z) \equiv \widetilde{\psi}_{abs}(z) := \sum_{n=0}^{\infty} \frac{|\psi^{(n)}(0)|}{n!} z^n \text{ for all } z \in r\overline{\mathbb{D}}.$$

$\widetilde{\psi}|_{r\mathbb{D}}$ (respectively $\psi_{abs}|_{r\mathbb{D}}$) is the unique holomorphic extension of ψ (respectively $\psi_{abs}|_{(-r,r)}$) on the domain $r\mathbb{D}$.

Proof Since ψ on $(-r, r)$ is real analytic, it follows that (locally, around 0)

$$\psi(x) = \sum_{n=0}^{\infty} b_n x^n \text{ for all } x \in (-s, s),$$

for some $0 < s \leq r$, where $b_n := \frac{\psi^{(n)}(0)}{n!} \in \mathbb{R}$ for all $n \in \mathbb{N}_0$. By assumption, $r > 0$, and $\sum_{n=0}^{\infty} |b_n| r^n = \|b\|_1 < \infty$, whence $\sum_{n=0}^{\infty} |b_n||x|^n \leq \|b\|_1 < \infty$ for all $x \in [-r, r]$. Thus, $(-r, r) \ni x \mapsto \sum_{n=0}^{\infty} b_n x^n$ is a well-defined real analytic function which coincides with the real analytic function ψ on the open subset $(-s, s)$ of the open interval $(-r, r)$. Consequently, both real analytic functions on $(-s, s)$ have to coincide on $(-r, r)$ (cf. [89, Corollary 1.2.4 and Corollary 1.2.6]). Clearly, $\widetilde{\psi}$ is well-defined and satisfies $\psi = \widetilde{\psi}|_{(-r,r)}$. Since $\|b\|_1 < \infty$ (by assumption), it follows that the series of continuous functions $\sum_{n=0}^{\infty} |b_n| x^n$ converges normally in $[-r, r]$ (with respect to the supremum norm) and hence uniformly in $[-r, r]$ (due to the majorant criterium (or M-test) of Weierstrass). Consequently, the function

$$[-r, r] \ni x \mapsto \psi_{abs}(x) = \sum_{n=0}^{\infty} |b_n| x^n$$

is well-defined, real analytic on $(-r, r)$ and continuous on $[-r, r]$. Similarly, it follows that the function $\widetilde{\psi}$ is continuous. Finally, given the construction of ψ_{abs}, the existence of $\widetilde{\psi}_{\text{abs}}$ is trivial. □

Since that class of real analytic functions plays a recurring and decisive role in our monograph (particularly for $r = 1$), and since real (and complex) analyticity of a function actually is a local property, it is justifiable to introduce the following definition:

Definition 5.3 Let $r \in (0, \infty)$. Put

$$W_+^\omega((-r, r)) := \{\psi : \psi \in C^\omega((-r, r)) \text{ and } \big(\frac{\psi^{(n)}(0)}{n!} r^n\big)_{n \in \mathbb{N}_0} \in l_1\},$$

where $C^\omega((-r, r))$ denotes the set of all real functions which are real analytic on $(-r, r)$. Any element in $\psi \in W_+^\omega((-r, r))$ is said to be *completely real analytic on* $(-r, r)$ at 0.

Observe that by definition any function in $W_+^\omega((-r, r))$ coincides with its own Taylor series at 0 "completely" (i.e., *everywhere*) on its domain of definition $(-r, r)$ (and not "locally, around" 0 only). Moreover, due to Lemma 5.3, it follows that for any $\psi \in W_+^\omega((-r, r))$ also $\psi_{\text{abs}}\big|_{(-r,r)} \in W_+^\omega((-r, r))$, and

$$\psi_{\text{abs}}^{(n)}(0) = |\psi^{(n)}(0)| \text{ for all } n \in \mathbb{N}_0. \tag{5.1.4}$$

Let us explicitly highlight three facts, implied by Lemma 5.3. To this end, if $\alpha \in \mathbb{T}$ is given, we consider the biholomorphic function $M_\alpha : \mathbb{C} \longrightarrow \mathbb{C}$, defined as $M_\alpha(z) := \alpha z$. Obviously, $M_\alpha^{-1} = M_{\overline{\alpha}}$ and $M_\alpha(\mathbb{D}) = \mathbb{D}$.

Remark 5.4 (Sign Condition) Let ψ and $\widetilde{\psi}$ be given as in Lemma 5.3. If ψ is odd and $\text{sign}(\psi^{(2n+1)}(0)) = (-1)^n$ for all $n \in \mathbb{N}_0$, then

$$\psi_{\text{abs}}(z) = \frac{1}{i}\widetilde{\psi}(iz) = (M_{-i} \circ \widetilde{\psi} \circ M_i)(z) \text{ for all } z \in r\overline{\mathbb{D}}$$

and

$$\psi_{\text{abs}}^{(2n+1)}(\zeta) = (-1)^n \widetilde{\psi}^{(2n+1)}(i\zeta) \text{ for all } \zeta \in r\mathbb{D} \text{ and } n \in \mathbb{N}_0$$

(since $i^{2n} = (-1)^n$ for all $n \in \mathbb{N}_0$).

Remark 5.5 (Wiener Algebra) Let ψ be given as in Lemma 5.3. Assume that $r = 1$. Then $\widetilde{\psi}\big|_{\mathbb{D}}$ and $\psi_{\text{abs}}\big|_{\mathbb{D}}$ are elements of

$$W^+(\mathbb{D}) := \{f : f(z) = \sum_{n=0}^\infty b_n z^n \text{ is holomorphic on } \mathbb{D} \text{ and satisfies } \sum_{n=0}^\infty |b_n| < \infty\}.$$

$W^+(\mathbb{D})$ is known as the *Wiener algebra*. It is a unital commutative (complex) Banach algebra with respect to the norm $\|\sum_{n=0}^{\infty} b_n z^n\|_{W^+(\mathbb{D})} := \sum_{n=0}^{\infty} |b_n|$, where multiplication is defined as that one of analytic functions, via the Cauchy product formula (cf. [105]). In particular, $\|\tilde{\psi}|_{\mathbb{D}}\|_{W^+(\mathbb{D})} = \psi_{\text{abs}}(1) = \|\psi_{\text{abs}}|_{\mathbb{D}}\|_{W^+(\mathbb{D})}$. Since by construction of $W^+(\mathbb{D})$ the series $f = \sum_{n=0}^{\infty} f_n$ is normally convergent in $\overline{\mathbb{D}}$ (with respect to the supremum norm), where $\overline{\mathbb{D}} \ni z \mapsto f_n(z) := b_n z^n$, the series f is uniformly convergent in $\overline{\mathbb{D}}$, and it follows that every element of $W^+(\mathbb{D})$ can be continuously extended to an element of the unital commutative Banach algebra $A(\mathbb{D})$, where

$$A(\mathbb{D}) := \{g : g \in C(\overline{\mathbb{D}}) \text{ such that } g|_{\mathbb{D}} \text{ is holomorphic on } \mathbb{D}\}$$

is equipped with the supremum norm and the usual pointwise algebraic operations. $A(\mathbb{D})$ denotes the well-known *disc algebra* (cf. [50, V.1., Example 4] and [157, Chapter III.E.]). Thus, $W^+(\mathbb{D})$ could be viewed as a subalgebra of the disc algebra $A(\mathbb{D})$. Since $A(\mathbb{D}) \subseteq H^\infty \subseteq H^2 \subseteq H^1$, it is likely that a further link to the very rich theory of Hardy spaces might open up here (cf., e.g., [50, 82, 130, 132]).

Remark 5.6 (Inversion and Complete Real Analyticity) Like a common thread, the following highly non-trivial problem—which is decisive for computing the upper bounds of $K_G^{\mathbb{F}}$—runs throughout the whole book. Very generally formulated, let $\psi \in W_+^\omega((-1, 1))$ be odd and completely real analytic on $(-1, 1)$ at 0. Assume that $\psi = \psi_{\text{abs}}|_{(-1,1)}$ and that $\psi : (-1, 1) \xrightarrow{\cong} (-1, 1)$ is bijective. Assume further that ψ^{-1} is real analytic on $(-1, 1)$. Does $\psi^{-1} \in W_+^\omega((-1, 1))$ already apply; i.e., is then also $(\psi^{-1})_{\text{abs}}$ well-defined (cf., e.g., Lemma 6.1, Corollary 6.8, Theorem 6.9, Example 6.7 and Example 7.1)? Both, [22] and [60] present multi-page proofs, each one built on rather advanced complex analysis (including intricate holomorphic extensions), to answer this question to the positive for key functions $\psi \in W_+^\omega((-1, 1))$ related to computing the value of $K_G^{\mathbb{C}}$ and $K_G^{\mathbb{R}}$, respectively!

5.2 Completely Correlation Preserving Functions and Schoenberg's Theorem

Let $0 < c \leq r$ and ψ be given as in Lemma 5.3. Assume that $\psi \neq 0$. Since $\psi \in W_+^\omega((-r, r))$, it follows that $\psi^{(n_0)}(0) \neq 0$ for some $n_0 \in \mathbb{N}_0$, implying that $\psi_{\text{abs}}(c) \geq \frac{|\psi^{(n_0)}(0)|}{n_0!} c^{n_0} > 0$. Thus, the function

$$(-1, 1) \ni \rho \mapsto \psi_c(\rho) := \frac{\psi(c\rho)}{\psi_{\text{abs}}(c)} \qquad (5.2.1)$$

is well-defined, continuous, bounded and satisfies $0 \neq \psi_c \in W_+^\omega((-1, 1))$. In Theorem 6.4 we will shed light on the hidden structure of these functions ψ_c. Let

$n \in \mathbb{N}$. Since $(\psi_c)_{\text{abs}}(1) = 1$, it even follows that $(\psi_c)_{\text{abs}} = \frac{\psi_{\text{abs}}(c \cdot)}{\psi_{\text{abs}}(c)} : [-1, 1] \longrightarrow [-1, 1]$ transforms any real $n \times n$-correlation matrix entrywise into a real $n \times n$-correlation matrix; i.e.,

$$(\psi_c)_{\text{abs}}[A] \in C(n; \mathbb{R}) \text{ for all } A \in C(n; \mathbb{R}), \text{ for all } n \in \mathbb{N} \qquad (5.2.2)$$

(due to Theorem 5.1). Within the scope of our research, functions $f \in C([-1, 1])$, satisfying $f|_{(-1,1)} \in W_+^\omega((-1, 1))$ and $f|_{(-1,1)}(\rho) = (f|_{(-1,1)})_{\text{abs}}(\rho)$ for all $\rho \in (-1, 1)$, are of particular importance, due to the following version of a fundamental result of I. J. Schoenberg (cf. [83, Theorem 16.2] and [137]):

Theorem 5.2 (Schoenberg [137]) *Let* $f : [-1, 1] \longrightarrow \mathbb{R}$ *be a continuous function. Then the following statements are equivalent:*

(i) $f[A]$ *is positive semidefinite for all* $A \in \bigcup_{n=1}^\infty \mathbb{M}_n([-1, 1])^+$.
(ii) $f[\Sigma]$ *is positive semidefinite for all* $\Sigma \in \bigcup_{n=1}^\infty C(n; \mathbb{R})$.
(iii) $f(x)$ *equals a convergent series* $\sum_{n=0}^\infty a_n x^n$ *for all* $x \in [-1, 1]$, *where* $a_n \geq 0$ *for all* $n \in \mathbb{N}_0$.
(iv) $f|_{(-1,1)} \in W_+^\omega((-1, 1))$ *and* $f = (f|_{(-1,1)})_{\text{abs}}$.
(v) $f|_{(-1,1)} \in W_+^\omega((-1, 1))$ *and* $f|_{(0,1)}$ *is absolutely monotonic.*
(vi) $f|_{(-1,1)}$ *can be extended to a holomorphic function* $\mathbb{D} \ni z \mapsto \tilde{f}(z) := \sum_{n=0}^\infty a_n z^n$, *where* $a_n \geq 0$ *for all* $n \in \mathbb{N}_0$ *and* $f(1) = \sum_{n=0}^\infty a_n < \infty$.

In particular, if (iii) or (vi) holds, then $a_n = \frac{f^{(n)}(0)}{n!} \geq 0$ *for all* $n \in \mathbb{N}_0$, *and the series* $\sum_{n=0}^\infty a_n x^n$ *converges uniformly on* $[0, 1]$.

Proof

(i) $\Rightarrow$ (ii): trivial (since $C(n; \mathbb{R}) \subseteq \mathbb{M}_n([-1, 1])^+$ for all $n \in \mathbb{N}$).
(ii) $\Leftrightarrow$ (iii): See [83, Theorem 16.24, p. 107 and remarks below].
(iii) $\Rightarrow$ (iv): Since $f|_{(-1,1)}$ coincides with a real analytic function on $(-1, 1)$ (namely the series $\sum_{n=0}^\infty a_n x^n$), it follows that $a_n = \frac{f^{(n)}(0)}{n!} \geq 0$ (cf. [89, Corollary 1.1.16]).
(iv) $\Rightarrow$ (v): Given an arbitrary $n \in \mathbb{N}_0$, the assumption (iv) implies that we only have to calculate the n-th derivative of the series on $(0, 1)$.
(v) $\Rightarrow$ (vi): Firstly, Bernstein's Theorem (cf. e. g. [156, Theorem 3a, p. 146]) implies that $f|_{(0,1)}$ has the holomorphic extension $\mathbb{D} \ni z \mapsto \tilde{f}(z) := \sum_{n=0}^\infty a_n z^n$, where $a_n := \frac{f^{(n)}(0)}{n!} \geq 0$ for all $n \in \mathbb{N}_0$. Since $f|_{(-1,1)} \in C^\omega((-1, 1))$ in particular is real analytic, it follows that $f|_{(-1,1)} = \tilde{f}|_{(-1,1)}$ (cf. [89, Corollary 1.1.16]). Since f is continuous (by assumption), it therefore follows that $f(1) = \lim_{x \uparrow 1} \tilde{f}(x) = \sum_{n=0}^\infty a_n$, where the latter equality follows from Abel's theorem on power series.

5.2 Completely Correlation Preserving Functions and Schoenberg's Theorem

(vi) $\Rightarrow$ (iii): Since $f(1) = \sum_{n=0}^{\infty} a_n < \infty$ and $a_n \geq 0$ for all $n \in \mathbb{N}_0$, it clearly follows that $\sum_{n=0}^{\infty}(-1)^n a_n \leq f(1) < \infty$. Since f is continuous (by assumption), Abel's theorem on power series therefore implies that $f(-1) = \lim_{x \downarrow -1} \tilde{f}(x) = \sum_{n=0}^{\infty}(-1)^n a_n$, and (iii) follows.

(iii) $\Rightarrow$ (i): Nothing is to show if $f = 0$. If $f \neq 0$, assumption (iii), together with Lemma 5.3 (applied to $r = 1 = c$) implies that $0 \neq f = (f|_{(-1,1)})_{\text{abs}}$ and $f(1) = f_{\text{abs}}(1) > 0$. Hence, we may apply Lemma 5.3 to the well-defined function $f_1 := \frac{f_{\text{abs}}}{f_{\text{abs}}(1)} = \frac{f}{f(1)}$, and (i) follows.

Finally, since every real analytic function has a unique power series representation (cf. [89, Corollary 1.1.16]), it follows that $a_n = \frac{f^{(n)}(0)}{n!}$ for all $n \in \mathbb{N}_0$. The uniform convergence of the series $\sum_{n=0}^{\infty} a_n x^n$ on $[0, 1]$ follows from [89, Proposition 1.1.3]. $\square$

Fix $\mathbb{F} \in \{\mathbb{R}, \mathbb{C}\}$. Looking for common sources of the real and complex cases, we put $D_{\mathbb{F}} := \{a \in \mathbb{F} : |a| < 1\}$, so that $D_{\mathbb{R}} = (-1, 1)$ and $D_{\mathbb{C}} = \mathbb{D}$. Property (5.2.2) deserves an autonomous and far-reaching

Definition 5.4 (Completely Correlation Preserving Function) Fix $\mathbb{F} \in \{\mathbb{R}, \mathbb{C}\}$. Let $h : \overline{D_{\mathbb{F}}} \longrightarrow \mathbb{F}$ be a function.

(i) Given $n \in \mathbb{N}$, h is n-correlation-preserving (short: n-CP) if for any $n \times n$ correlation matrix $\Sigma \in C(n; \mathbb{F})$ also $h[\Sigma] \in C(n; \mathbb{F})$ is an $n \times n$ correlation matrix.

(ii) h is called completely correlation-preserving (short: CCP) if h is n-correlation-preserving for all $n \in \mathbb{N}$.

Since every CCP function h is 2-CP, Definition 5.4 directly implies that $h(\overline{D_{\mathbb{F}}}) \subseteq \overline{D_{\mathbb{F}}}$ and $h(1) = 1$. If $h : \overline{D_{\mathbb{F}}} \longrightarrow \mathbb{F}$ satisfies $h(1) > 0$, then $h[\Sigma] \in \mathbb{M}_n(\overline{D_{\mathbb{F}}})^+$ for all $\Sigma \in C(n; \mathbb{F})$ if and only if $\frac{1}{h(1)} h$ is n-CP. If h_1 and h_2 are two CCP functions (respectively two n-CP functions), and if $\lambda \geq 0$, Definition 5.4 immediately implies that also $h_2 \circ h_1$, $h_1 \circ h_2$ and $\lambda h_1 + (1 - \lambda) h_2$ are CCP functions (respectively n-CP functions). Furthermore, since $h_1 h_2[A] = h_1[A] * h_2[A]$ for all matrices $A \in \mathbb{M}_{m,n}(\mathbb{F})$, $m, n \in \mathbb{N}$, it follows from (5.1.2) that also the product $h_1 h_2$ of two CCP functions (respectively two n-CP functions) again is CCP (respectively n-CP). Much less trivial is the fact (which involves the Grothendieck constant !) that in general, the inverse function h^{-1} of an invertible CCP function h is *not* a CCP function (cf. Corollary 6.4, Remark 6.7 and Theorem 6.9, such as the inverse of $h := \frac{2}{\pi} \arcsin$, given by $[-1, 1] \ni y \mapsto h^{-1}(y) = \sin(\frac{\pi}{2} y) = \sum_{n=0}^{\infty} b_n \frac{y^n}{n!}$, where $b_n := (-1)^{\lfloor n/2 \rfloor} \cdot \frac{1-(-1)^n}{2} \cdot (\frac{\pi}{2})^n$.

In the real case, Theorem 5.2 immediately leads to a full characterisation of continuous CCP functions, since:

Theorem 5.3 *Let $h : [-1, 1] \longrightarrow \mathbb{R}$ be a continuous function. Then the following statements are equivalent:*

(i) *h is CCP.*
(ii) *$h[A]$ is positive semidefinite for all $A \in \bigcup_{n=1}^{\infty} \mathbb{M}_n([-1, 1])^+$ and $h(1) = 1$.*
(iii) *$h(x)$ has the unique series representation $h(x) = \sum_{n=0}^{\infty} a_n x^n$ for all $x \in [-1, 1]$, where $a_n \geq 0$ for all $n \in \mathbb{N}_0$ and $\sum_{n=0}^{\infty} a_n = 1$.*
(iv) *$[-1, 1] \ni \rho \mapsto h(\rho) = \mathbb{E}_{\mathbb{P}}[\rho^X] = \sum_{n=0}^{\infty} \mathbb{P}(X = n)\rho^n$ is the probability generating function of some discrete random variable X.*
(v) *$h|_{(-1,1)} \in W_+^{\omega}((-1, 1))$, $h = \left(h|_{(-1,1)}\right)_{abs}$ and $h(1) = 1$.*
(vi) *$h|_{(-1,1)} \in W_+^{\omega}((-1, 1))$, $h|_{(0,1)}$ is absolutely monotonic and $h(1) = 1$.*
(vii) *$h|_{(-1,1)}$ can be extended to a holomorphic function $\mathbb{D} \ni z \mapsto \tilde{h}(z) := \sum_{n=0}^{\infty} a_n z^n$, where $a_n \geq 0$ for all $n \in \mathbb{N}_0$ and $\sum_{n=0}^{\infty} a_n = 1$.*
(viii) *$h|_{(-1,1)}$ can be extended to a complex function $\tilde{H} : \overline{\mathbb{D}} \longrightarrow \overline{\mathbb{D}}, z \mapsto \sum_{n=0}^{\infty} a_n z^n$, where $a_n \geq 0$ for all $n \in \mathbb{N}_0$ and $\sum_{n=0}^{\infty} a_n = 1$.*

If (iii) *or* (vii) *or* (viii) *is given, the series $\sum_{n=0}^{\infty} a_n z^n$ converges then absolutely on $\overline{\mathbb{D}}$ and $\sum_{n=0}^{\infty} a_n x^n$ converges uniformly on $[0, 1]$.*

Remark 5.7 (CCP Versus RKHS) Theorem 5.3 also reveals a direct link to representing kernel Hilbert spaces (RKHS). [121, Theorem 4.12.], together with the remark on complexification in [121, Chapter 5.1] namely implies that

$$(-1, 1) \times (-1, 1) \ni (s, t) \mapsto K_h(s, t) := \sum_{n=0}^{\infty} a_n (st)^n = \sum_{n=0}^{\infty} f_n(s) f_n(t)$$

is a well-defined kernel function for a real RKHS $H(K_h)$ of real-analytic functions on $(-1, 1)$ (cf. [121, Definition 2.12]), where $f_n(\rho) := \sqrt{a_n}\, \rho^n$. The set of functions f_n for which $a_n \neq 0$ is an orthonormal basis for $H(K_h)$.

It is quite instructive to analyse how n-CP functions in particular relate to properties of Schoenberg's kernel functions $\mathcal{P}(\mathbb{S}^d)$, $d \in \mathbb{N}$. Schoenberg's approach was revisited by the geostatistician T. Gneiting in his impressive paper [52], where $\mathcal{P}(\mathbb{S}^d)$ is denoted as Ψ_d. So, let us recall the concept of a positive definite function on a metric space (which in particular is a kernel function (cf. [121, Definition 2.12.]), originally coined by Schoenberg in [137]. Supported by Lemma 3.1, these kernel functions result in a further characterisation of real-valued n-CP functions. In order to recognise this observation, let (X, d) be a metric space. Fix $d \in \mathbb{N}_2$. A function $f : [0, \infty) \longrightarrow \mathbb{R}$ is positive definite on X if f is continuous, and if *for any* $n \in \mathbb{N}$ and any $(x_1, x_2, \ldots, x_n) \in X^n$, the matrix $f[d(x_i, x_j)]_{i,j=1}^n$ is positive semidefinite (cf. [83, Definition 16.15] and [52, 137]). Consequently, if we apply Schoenberg's definition to the unit sphere $\mathbb{S}^{d-1}$, where the metric on $\mathbb{S}^{d-1}$ is induced by the geodesic distance $\mathbb{S}^{d-1} \times \mathbb{S}^{d-1} \ni (x, y) \mapsto \arccos(\langle x, y \rangle)$, Lemma 3.1 implies that any continuous function $f : [-1, 1] \longrightarrow \mathbb{R}$ which satisfies $f \circ \cos : [0, \pi] \longrightarrow \mathbb{R} \in \mathcal{P}(\mathbb{S}^{d-1}) \equiv \Psi_{d-1}$ in particular is d-CP. Thus, [52,

5.2 Completely Correlation Preserving Functions and Schoenberg's Theorem

Table 1] shows us a wealth of non-trivial d-CCP functions, where $d \in [3]$. Similarly, we recognise that a continuous function $h : [-1, 1] \longrightarrow \mathbb{R}$ is CCP if and only if $h \circ \cos \in \Psi_\infty = \bigcap_{d=1}^\infty \Psi_d$.

If we combine these facts with [52, Theorem 7], we are rewarded with a large class of (even) CCP functions. To this end, recall that a continuous function $\psi : [0, \infty) \longrightarrow \mathbb{R}$ is called *completely monotonic* if $\psi|_{(0,\infty)} \in C^\infty((0, \infty))$ and $(-1)^n \psi^{(n)}(x) \geq 0$ for all $n \in \mathbb{N}_0$ and all $x > 0$ (cf., e.g., [83, Definition 27.18] and [156, Definition 2c]). Many explicit examples of completely monotone functions are listed in [104]. They play a significant role in various subfields of probability theory including theory and applications of Lévy processes and infinite divisibility.

Theorem 5.4 *Let $\psi : [0, \infty) \longrightarrow \mathbb{R}$ be completely monotonic and non-constant. Suppose that $\psi(0) = 1$, then*

$$\psi \circ \arccos : [-1, 1] \longrightarrow \mathbb{R} \text{ is a CCP function.}$$

In particular, $\psi([0, \pi]) \subseteq [-1, 1]$.

If we apply the complex analogue of Schoenberg's Theorem, coined by J.P.R. Christensen and P. Ressel in 1982 (cf. [83, Theorem 16.7]) to complex CCP functions, we obtain

Theorem 5.5 *Let $h : \overline{\mathbb{D}} \longrightarrow \mathbb{C}$ be a continuous complex function. Then the following statements are equivalent:*

(i) *h is CCP.*
(ii) *$h(z)$ has the unique series representation $h(z) = \sum_{k=0}^\infty \sum_{l=0}^\infty a_{kl} z^k \overline{z}^l$ for all $z \in \overline{\mathbb{D}}$, where $a_{kl} \geq 0$ for all $k, l \in \mathbb{N}_0$ and $\sum_{k=0}^\infty (\sum_{l=0}^\infty a_{kl}) = 1$.*

Observe that a significant implication of Theorem 5.5 is that in general *complex* CCP functions are not holomorphic, respectively analytic. Theorem 5.3, together with Lemma 5.3 immediately implies how one can easily construct complex CCP functions out of real ones:

Remark 5.8 Let $h = h_{f,f} : [-1, 1] \longrightarrow [-1, 1]$ be a real CCP function ($\|f\|_{\gamma_k} = 1$) and $\tilde{h} : \overline{\mathbb{D}} \longrightarrow \overline{\mathbb{D}}$ be as in Lemma 5.3. Then the following functions are all complex CCP functions: $\tilde{h}, \overline{\tilde{h}}$ and $\tilde{h} \cdot \overline{\tilde{h}} = |\tilde{h}|^2$.

Chapter 6
The Real Case: Towards Extending Krivine's Approach

6.1 Some Facts About Real Multivariate Hermite Polynomials

Another important estimation (even with upper bound 1) which might help to support our search for a "suitable" CCP function which is different from the CCP function $\frac{2}{\pi}\arcsin$ in the real case, is included in the next two results. We start with the real case first. To this end, recall e.g. from [20, Chapter 1.3] that for fixed $k \in \mathbb{N}$ $\{H_\alpha : \alpha \in \mathbb{N}_0^k\}$ is an ortho*normal* basis in $L^2(\gamma_k)$, where the k-variate Hermite polynomial $H_\alpha : \mathbb{R}^k \longrightarrow \mathbb{R}$ is defined by

$$H_\alpha(x_1, x_2, \ldots, x_k) := \prod_{i=1}^{k} H_{\alpha_i}(x_i) = \frac{(-1)^{|\alpha|}}{\sqrt{\alpha!}} \frac{D^\alpha \varphi_k(x)}{\varphi_k(x)} . \tag{6.1.1}$$

Here, $\varphi_k(x_1, x_2, \ldots, x_k) := \prod_{i=1}^{k} \varphi(x_i)$, $D^\alpha := (\frac{\partial}{\partial x_1})^{\alpha_1} \cdots (\frac{\partial}{\partial x_k})^{\alpha_k}$, $\alpha! := \prod_{i=1}^{k} \alpha_i!$ and $|\alpha| := \sum_{i=1}^{k} \alpha_i$ ($n \in \mathbb{N}_0, y \in \mathbb{R}, x = (x_1, \ldots, x_k)^\top \in \mathbb{R}^k$), and

$$H_n(y) := \frac{1}{\sqrt{n!}}(-1)^n \exp\left(\frac{y^2}{2}\right) \frac{d^n}{dy^n} \exp\left(-\frac{y^2}{2}\right) = \frac{(-1)^n}{\sqrt{n!}} \frac{\varphi^{(n)}(y)}{\varphi(y)}$$

$$= \frac{1}{\sqrt{n!}} \sum_{k=0}^{\lfloor n/2 \rfloor} H_{n,k} (-1)^k y^{n-2k} = (-1)^n H_n(-y). \tag{6.1.2}$$

H_n denotes the (probabilist's version of the) one-dimensional Hermite polynomial, where

$$H_{n,k} := \binom{n}{2k}(2k-1)!! = \frac{n!}{2^k k!(n-2k)!}.$$

Thus,

$$H_\alpha \varphi_k = \frac{(-1)^{|\alpha|}}{\sqrt{\alpha!}} \frac{D^\alpha \varphi_k}{\varphi_k}. \tag{6.1.3}$$

Since $(y+iz)^n = \sum_{l=0}^n \binom{n}{l} i^l z^l y^{n-l}$ and $\mathbb{E}[X^m] = \frac{(-1)^{m}+1}{2}(m-1)!!$ for all $m, n \in \mathbb{N}_0$, $y, z \in \mathbb{R}$ and $X \sim N_1(0,1)$, it follows that

$$\mathbb{E}[(y+iX)^n] = \sum_{l=0}^n \binom{n}{l} i^l \mathbb{E}[X^l] y^{n-l}$$

$$= \sum_{k=0}^{\lfloor n/2 \rfloor} \binom{n}{2k}(-1)^k (2k-1)!! \, y^{n-2k} = \sqrt{n!}\, H_n(y)$$

for all $n \in \mathbb{N}_0$, $y \in \mathbb{R}$ and $X \sim N_1(0,1)$. In particular, we reobtain the numbers

$$H_{2l}(0) = (-1)^l \frac{1}{\sqrt{(2l)!}}(2l-1)!! = (-1)^l \sqrt{\frac{(2l-1)!!}{(2l)!!}}$$

$$= (-1)^l \frac{\sqrt{(2l)!}}{2^l\, l!} \quad \text{and} \quad H_{2l+1}(0) = 0 \tag{6.1.4}$$

for all $l \in \mathbb{N}_0$ (cf. also (6.3.7)). Recall that the generating function of the one-dimensional Hermite polynomials is given by (cf. [20, (1.3.1)])

$$\mathbb{R} \times \mathbb{R} \ni (\lambda, x) \mapsto \exp\left(\lambda x - \frac{1}{2}\lambda^2\right) = \sum_{\nu=0}^\infty \frac{\lambda^\nu}{\sqrt{\nu!}} H_\nu(x)$$

implying that in the k-dimensional case the equality

$$\exp\left(\lambda^\top x - \frac{1}{2}\|\lambda\|^2\right) = \prod_{i=1}^k \exp\left(\lambda_i x_i - \frac{1}{2}\lambda_i^2\right)$$

$$= \sum_{m \in \mathbb{N}_0^k} \frac{1}{\sqrt{m!}} H_m(x)\lambda^m = \sum_{\nu=0}^\infty \Big(\sum_{m \in C(\nu,k)} \frac{H_\alpha(x)}{\sqrt{m!}}\Big)\lambda^m$$

$$\tag{6.1.5}$$

6.1 Some Facts About Real Multivariate Hermite Polynomials

holds for all $\lambda, x \in \mathbb{R}^k$, where $C(\nu, k) := \{m \in \mathbb{N}_0^k : |m| = \nu\}$. Namely, since $|H_m(x)| \leq (2\pi)^{k/4} \exp\left(\frac{1}{4}\|x\|_2^2\right)$ for all $m \in \mathbb{N}_0^k$ and $x \in \mathbb{R}^k$ (see [36, Lemma 1], respectively [75, Proposition 2.3 (i)]), the quotient test implies that $\sum_{\nu=0}^{\infty} \frac{1}{\sqrt{\nu!}} H_\nu(x) \lambda^\nu$ even converges absolutely. Hence, each one of the k families $\left(\frac{1}{\sqrt{\nu!}} H_\nu(x_i) \lambda_i^\nu\right)_{\nu \in \mathbb{N}_0}$ is summable ($i \in [k]$). Consequently, it follows that also the family $\left(\prod_{i=1}^{k} \frac{1}{\sqrt{m_i!}} H_{m_i}(x_i) \lambda_i^{m_i}\right)_{m \in \mathbb{N}_0^k} = \left(\frac{1}{\sqrt{m!}} H_m(x) \lambda^m\right)_{m \in \mathbb{N}_0^k}$ is summable. Equation (6.1.5) now follows from the reiteration of the double summation principle and the associativity formula for summable families. Consequently, we obtain (see also [20, Lemma 1.3.2 (iii)]):

$$\sqrt{m!}\, H_m(x) = D^m \exp(\lambda^\top x - \tfrac{1}{2}\|\lambda\|^2)\big|_{\lambda=0} = \prod_{\nu=1}^{k} \left(\tfrac{\partial}{\partial \lambda_\nu}\right)^{m_\nu} \exp(\lambda_\nu x - \tfrac{1}{2}\lambda_\nu^2)\big|_{\lambda_\nu=0}$$

(6.1.6)

for all $m \in \mathbb{N}_0^k$ and $x \in \mathbb{R}^k$.

In fact, (6.1.6) allows a further (purely analytic) proof of the Grothendieck *equality* and its multi-dimensional generalisation. As a by-product, we provide a closed-form analytical representation of the multvariate distribution function of a Gaussian random vector $\mathbf{X} \sim N_{2k}(0, \Sigma_{2k}(\rho))$ (cf. Proposition 6.8). In general, closed-form analytical representations of general multivariate Gaussian distribution functions are not available (cf., e.g., [64]). All that can be derived from the calculation of the following vital "Fourier-Hermite coefficients", which include sign as a particular case (cf. Corollary 6.1 below):

Theorem 6.1 *Let $k \in \mathbb{N}$, $m = (m_1, \ldots, m_k)^\top \in \mathbb{N}_0^k$, $a = (a_1, \ldots, a_k)^\top \in \mathbb{R}^k$ and $b = (b_1, \ldots, b_k)^\top \in \mathbb{R}^k$. Put $I_a := \prod_{i=1}^{k}[a_i, \infty)$ and $J_b := \prod_{i=1}^{k}(-\infty, b_i]$. Then*

$$\int_{I_a} H_m \, d\gamma_k = \langle \mathbb{1}_{I_a}, H_m \rangle_{\gamma_k} = \prod_{\substack{i=1 \\ m_i=0}}^{k}(1 - \Phi(a_i)) \prod_{\substack{i=1 \\ m_i \neq 0}}^{k} \frac{1}{\sqrt{m_i}} \varphi(a_i) H_{m_i-1}(a_i)$$

(6.1.7)

and

$$\int_{J_b} H_m \, d\gamma_k = \langle \mathbb{1}_{J_b}, H_m \rangle_{\gamma_k} = (-1)^{c(k,m)} \prod_{\substack{i=1 \\ m_i=0}}^{k} \Phi(b_i) \prod_{\substack{i=1 \\ m_i \neq 0}}^{k} \frac{1}{\sqrt{m_i}} \varphi(b_i) H_{m_i-1}(b_i),$$

(6.1.8)

where $c(k, m) := k - \sum_{i=1}^{k} \delta_{m_i\, 0} \in \{0, 1, \ldots, k\}$ counts the number of non-zero components of the vector m.

Proof Firstly, note that

$$\int_{\mathbb{M}} H_m \, d\gamma_k = \prod_{i=1}^{k} \int_{\mathbb{M}_i} H_{m_i} \, d\gamma_1 = \prod_{\substack{i=1 \\ m_i = 0}}^{k} \int_{\mathbb{M}_i} H_{m_i} \, d\gamma_1 \prod_{\substack{i=1 \\ m_i \neq 0}}^{k} \int_{\mathbb{M}_i} H_{m_i} \, d\gamma_1,$$

where $\mathbb{M} \in \{I_a, J_b\}$, $\mathbb{M}_i := [a_i, \infty)$ if $\mathbb{M} = I_a$ and $\mathbb{M}_i := (-\infty, b_i]$ if $\mathbb{M} = J_b$. Consequently, we only have to compute the one-dimensional integral factors

$$\int_{[a_i, \infty)} H_{m_i} \, d\gamma_1 = \delta_{m_i\, 0} - \int_{(-\infty, a_i]} H_{m_i} \, d\gamma_1.$$

To this end, fix $(m, a) \in \mathbb{N}_0 \times \mathbb{R}$. If $m = 0$, then $H_m = 1$, whence

$$\int_{[a, \infty)} H_m \, d\gamma_1 = \gamma_1([a, \infty)) = \int_a^{\infty} \varphi(x) dx = 1 - \Phi(a) = 1 - \int_{(-\infty, a]} H_m \, d\gamma_1.$$

If $m \neq 0$, (6.1.6), together with the Leibniz integral rule, applied to the parameter-dependent integral $\int_{[a,\infty)} (\frac{\partial}{\partial \lambda})^m \exp(\lambda x - \frac{1}{2}\lambda^2) \gamma_1(dx) = 2\pi \int_{[a,\infty)} (\frac{\partial}{\partial \lambda})^m \varphi(x - \lambda) dx$, imply that

$$-\int_{(-\infty, a]} H_m \, d\gamma_1 = \int_{[a, \infty)} H_m \, d\gamma_1 = \tfrac{1}{\sqrt{m!}} ((\tfrac{\partial}{\partial \lambda})^m \int_{[a, \infty)} \varphi(x - \lambda) dx)\big|_{\lambda=0}$$

$$= \tfrac{1}{\sqrt{m!}} (\tfrac{\partial}{\partial \lambda})^m (1 - \Phi(a - \lambda))\big|_{\lambda=0} = \tfrac{1}{\sqrt{m!}} (\tfrac{\partial}{\partial \lambda})^m \Phi(\lambda - a)\big|_{\lambda=0}$$

$$= \tfrac{1}{\sqrt{m!}} \varphi^{(m-1)}(-a) \stackrel{(6.1.2)}{=} \tfrac{1}{\sqrt{m}} \varphi(a) H_{m-1}(a).$$

$\square$

Corollary 6.1 *Let $k \in \mathbb{N}$, $m = (m_1, \ldots, m_k)^\top \in \mathbb{N}_0^k$, $a = (a_1, \ldots, a_k)^\top \in \mathbb{R}^k$ and $b = (b_1, \ldots, b_k)^\top \in \mathbb{R}^k$. Put $I_a := \prod_{i=1}^{k} [a_i, \infty)$ and $J_b := \prod_{i=1}^{k} (-\infty, b_i]$. Let $\chi_a : \mathbb{R}^k \longrightarrow \{-1, 1\}$ and $\psi_b : \mathbb{R}^k \longrightarrow \{-1, 1\}$ be defined as*

$$\chi_a(x) := 2\, \mathbb{1}_{I_a}(x) - 1 \text{ and } \psi_b(x) := 1 - 2\, \mathbb{1}_{J_b}(x) = -\chi_{-b}(-x).$$

$$\langle \chi_a, H_0 \rangle_{\gamma_k} = 2 \prod_{i=1}^{k} (1 - \Phi(a_i)) - 1 \text{ and } \langle \psi_b, H_0 \rangle_{\gamma_k} = 1 - 2 \prod_{i=1}^{k} \Phi(b_i). \quad (6.1.9)$$

6.1 Some Facts About Real Multivariate Hermite Polynomials

If $m \neq 0$, then

$$\langle \chi_a, H_m \rangle_{\gamma_k} = 2 \prod_{\substack{i=1 \\ m_i=0}}^{k} (1 - \Phi(a_i)) \prod_{\substack{i=1 \\ m_i \neq 0}}^{k} \frac{1}{\sqrt{m_i}} \varphi(a_i) H_{m_i-1}(a_i)$$

and

$$\langle \psi_b, H_m \rangle_{\gamma_k} = 2(-1)^{c(k,m)+1} \prod_{\substack{i=1 \\ m_i=0}}^{k} \Phi(b_i) \prod_{\substack{i=1 \\ m_i \neq 0}}^{k} \frac{1}{\sqrt{m_i}} \varphi(b_i) H_{m_i-1}(b_i), \quad (6.1.10)$$

where $c(k, m) := k - \sum_{i=1}^{k} \delta_{m_i\,0} \in \{0, 1, \ldots, k\}$. In particular,

$$\langle \mathrm{sign}, H_{2n} \rangle_{\gamma_1} = 0 \text{ and } \langle \mathrm{sign}, H_{2n+1} \rangle_{\gamma_1} = (-1)^n \sqrt{\frac{2}{\pi}} \frac{(2n-1)!!}{\sqrt{(2n+1)!}}$$

$$= \frac{(-1)^n}{2^n} \sqrt{\frac{2}{\pi}} \frac{(n+1)!}{\sqrt{(2n+1)!}} C_n \quad (6.1.11)$$

for any $n \in \mathbb{N}_0$, where $C_n := \frac{1}{n+1}\binom{2n}{n}$ denotes the n'th Catalan number.

Moreover, because of the fact that $H_\nu(-x) = (-1)^\nu H_\nu(x)$ (respectively, $H_\nu(|x|) = (\mathrm{sign}(x))^\nu H_\nu(x)$) for all $\nu \in \mathbb{N}_0$ and $x \in \mathbb{R}$, a k-fold multiplication of the values of one-dimensional Hermite polynomials implies that

$$H_n(-x) = (-1)^{|n|} H_n(x) \text{ for all } n \in \mathbb{N}_0^k \text{ and } x \in \mathbb{R}^k.$$

Hence,

$$\langle g_\rho, H_n \rangle_{\gamma_k} = \langle g, (H_n)_\rho \rangle_{\gamma_k} = \rho^{|n|} \langle g, H_n \rangle_{\gamma_k} \text{ for all } g \in L^2(\gamma_k) \text{ and } \rho \in \{-1, 1\}, \quad (6.1.12)$$

where $\mathbb{R}^k \ni y \mapsto f_\rho(y) := f(\mathrm{sign}(\rho)y) = f(\rho y)$ satisfies $\|f_\rho\|_{\gamma_k} = \|f\|_{\gamma_k}$ for $f \in \{g, H_n\}$ (which follows from a simple change of variables and the trivial fact that $|\rho^k| = 1$ for all $\rho \in \{-1, 1\}$). Obviously, dependent on the smoothness structure of $g \in L^2(\gamma_k)$, it's quite a challenge to calculate the "Fourier-Hermite coefficients" $\langle g, (H_n)_\rho \rangle_{\gamma_k} = \int_{\mathbb{R}^k} g(x) H_n(x) \gamma_k(\mathrm{d}x)$ explicitly. Regarding that particular problem, let us note a first important fact (cf. also Proposition 6.11):

Proposition 6.1 *Let $k \in \mathbb{N}$, $X \sim N_k(0, I_k)$ and $g \in L^2(\gamma_k)$. Then the function $\mathbb{R}^k \ni \lambda \mapsto \mathbb{E}[g(X + \lambda)]$ is smooth around 0. It satisfies*

$$D^\alpha \mathbb{E}[g(X + \lambda)]\big|_{\lambda=0} = \sqrt{\alpha!} \langle g, H_\alpha \rangle_{\gamma_k} = \sqrt{\alpha!}\, \mathbb{E}[g(X) H_\alpha(X)] \text{ for all } \alpha \in \mathbb{N}_0^k. \quad (6.1.13)$$

In particular, if in addition g is smooth, then

$$\langle g, H_\alpha \rangle_{\gamma_k} = \mathbb{E}[g(X)H_\alpha(X)] = \frac{1}{\sqrt{\alpha!}} \mathbb{E}[D^\alpha g(X)]$$

for all $\alpha \in \mathbb{N}_0^k$.

Proof Consider the function $\mathbb{R}^k \times \mathbb{R}^k \ni (x, \lambda) \mapsto F(x, \lambda) := \exp(\langle x, \lambda \rangle_2 - \frac{1}{2}\|\lambda\|_2)$. A straightforward application of the k-dimensional change of variables formula implies that $I_g(\lambda) := \mathbb{E}[g(\mathbf{X} + \lambda)] = \mathbb{E}[g(\mathbf{X})F(\mathbf{X}, \lambda)]$. Consequently, due to (6.1.5), it follows that

$$I_g(\lambda) = \sum_{\alpha \in \mathbb{N}_0^k} \frac{\mathbb{E}[g(\mathbf{X})H_\alpha^*(\mathbf{X})]}{\alpha!} \lambda^\alpha ,$$

where we put $H_\alpha^*(x) := \sqrt{\alpha!}\, H_\alpha(x)$ for all $\alpha \in \mathbb{N}_0^k$ and $x \in \mathbb{R}^k$. Thus, the multi-indices version of the k-dimensional Taylor formula (cf. [41, Theorem 2.8.4]) implies that

$$D^\alpha \mathbb{E}[g(\mathbf{X}+\lambda)]\big|_{\lambda=0} = D^\alpha I_g(\lambda)\big|_{\lambda=0} = \mathbb{E}[g(\mathbf{X})H_\alpha^*(\mathbf{X})] = \sqrt{\alpha!}\, \mathbb{E}[g(\mathbf{X})H_\alpha(\mathbf{X})].$$

□

6.2 Real CCP Functions and Covariances: A Fourier-Hermite Analysis Approach

Fix $f, g \in L^2(\mathbb{R}^k, \gamma_k)$, $\nu \in \mathbb{N}_0$ and $k \in \mathbb{N}$. Put $C(\nu, k) := \{n \in \mathbb{N}_0^k : |n| = \nu\}$ and

$$p_\nu(f, g) := \sum_{n \in C(\nu,k)} \langle f, H_n \rangle_{\gamma_k} \langle g, H_n \rangle_{\gamma_k} = p_\nu(g, f). \tag{6.2.1}$$

Since the inequality of arithmetic and geometric means in particular holds for any pair of elements, indexed by elements of the set of finitely many elements $C(\nu, k)$, it allows (the proof of) a direct transfer of Hölder's inequality by means of summation over $C(\nu, k)$, whence

$$|p_\nu(f, g)| \leq \sqrt{p_\nu(f, f)}\sqrt{p_\nu(g, g)} \text{ for all } \nu \in \mathbb{N}_0. \tag{6.2.2}$$

6.2 Real CCP Functions and Covariances: A Fourier-Hermite Analysis Approach

Since $\{C(\nu, k) : \nu \in \mathbb{N}_0\}$ obviously is a partition of the set $\mathbb{N}_0^k$ and $\{H_\alpha : \alpha \in \mathbb{N}_0^k\}$ is an orthonormal basis in $L^2(\gamma_k)$, Hölder's inequality again implies that

$$[-1, 1] \ni \rho \mapsto h_{f,g}(\rho) := \sum_{n \in \mathbb{N}_0^k} \langle f, H_n \rangle_{\gamma_k} \langle g, H_n \rangle_{\gamma_k} \rho^{|n|} = \sum_{\nu=0}^\infty p_\nu(f, g) \rho^\nu \tag{6.2.3}$$

converges absolutely, and

$$h_{f,g}(\rho) \le \sum_{\nu=0}^\infty |p_\nu(f, g)| \le \|f\|_{\gamma_k} \|g\|_{\gamma_k}. \tag{6.2.4}$$

Hence, $h_{f,g}$ is well-defined and bounded. Note also that by construction $h_{f,g} = h_{g,f}$, $h_{f,g} = \frac{1}{4}(h_{f+g,f+g} - h_{f-g,f-g})$ (due to the polarisation equality) and $|h_{p,p}(\rho)| \le \sum_{n \in \mathbb{N}_0^k} \langle p, H_n \rangle_{\gamma_k}^2 |\rho|^{|n|} \le \|p\|_{\gamma_k}^2$ for all $p \in L^2(\mathbb{R}^k, \gamma_k)$ and $\rho \in [-1, 1]$. Consequently, $h_{f,g}|_{(-1,1)} \in W_+^\omega((-1, 1))$.

Observe also that *for any* $f = \sum_{n \in \mathbb{N}_0^k} a_n H_n \in L^2(\gamma_k)$ and *any* $g = \sum_{n \in \mathbb{N}_0^k} b_n H_n \in L^2(\gamma_k)$, it follows that $a_n = x_n(f) := \langle f, H_n \rangle_{\gamma_k}$ and $b_n = x_n(g) = \langle g, H_n \rangle_{\gamma_k}$ for all $n \in \mathbb{N}_0^k$, implying that $p_\nu(f, g) = \sum_{n \in C(\nu,k)} a_n b_n$, whence

$$h_{f,g}(\rho) = \sum_{\nu=0}^\infty \Big(\sum_{n \in C(\nu,k)} a_n b_n \Big) \rho^\nu \text{ for all } \rho \in [-1, 1]$$

and $\sum_{\nu=0}^\infty \big(\sum_{n \in C(\nu,k)} a_n^2 \big) = \sum_{n \in \mathbb{N}_0^k} a_n^2 = \|f\|_{\gamma_k}^2$. In particular,

$$h_{f,f}(\rho) = \sum_{\nu=0}^\infty \Big(\sum_{n \in C(\nu,k)} a_n^2 \Big) \rho^\nu \text{ for all } \rho \in [-1, 1].$$

This immediately results in another important statement which should be compared with [92, Proposition 5 and Théorème 3] and (6.4.6):

Proposition 6.2 *Let* $k \in \mathbb{N}$ *and* $f, g \in S_{L^2(\gamma_k)}$. *If* H *is a real Hilbert space, then there exists a real Hilbert space* $\mathbb{H}$ *such that for any* $u, v \in S_H$,

$$h_{f,g}(\langle u, v \rangle_H) = \langle \psi_u(f), \psi_v(g) \rangle_\mathbb{H},$$

where $\psi_w : L^2(\gamma_k) \longrightarrow \mathbb{H}$ *is a mapping which satisfies* $\psi_w(S_{L^2(\gamma_k)}) \subseteq S_\mathbb{H}$ *for any* $w \in S_H$. *In particular, for any* $m, n \in \mathbb{N}$, *the following statements hold*:

$$h_{f,g}[S] \in \mathcal{Q}_{m,n} \text{ for all } S \in \mathcal{Q}_{m,n}$$

and

$$h_{f,g}[A] \in \mathbb{M}_n(\mathbb{R})^+ \text{ for all } A \in \mathbb{M}_n(\mathbb{R})^+.$$

Proof Firstly, based on the construction of the tensor product of Hilbert spaces (cf. S. K. Berberian's beautiful explanation of this concept in https://web.ma.utexas.edu/mp_arc/c/14/14-2.pdf), it follows that for any $\nu \in \mathbb{N}_0$, there exist mappings $w_\nu : H \longrightarrow H_\nu$, such that $w_\nu(S_H) \subseteq S_{H_\nu}$ and

$$\langle x, y \rangle_H^\nu = \langle w_\nu(x), w_\nu(y) \rangle_{H_\nu} \text{ for all } x, y \in \mathbb{R}^n, \tag{6.2.5}$$

where the Hilbert space $H_\nu := \bigotimes_{l=1}^\nu H$ denotes the ν-fold tensor product of H, $w_\nu(x) := \bigotimes_{l=1}^\nu x$ and $w_\nu(y) := \bigotimes_{l=1}^\nu y$. If $H \equiv \mathbb{R}_2^m$ is finite-dimensional, the structure of the (real) Gaussian measure in fact allows an explicit construction of such mappings $w_\nu : \mathbb{R}_2^m \longrightarrow L^2(\mathbb{R}_2^{\nu m}, \gamma_{\nu m})$ (without having to make use of abstract tensor products of Hilbert spaces), via

$$w_\nu(x)(\text{vec}(\xi_1, \ldots, \xi_\nu)) := \prod_{i=1}^\nu \langle x, \xi_i \rangle_{\mathbb{R}_2^m}$$

(cf. also [79, Lemma 10.4]). We only have to implement the well-known fact that $\gamma_{\nu m}$ coincides with the product measure $\bigotimes_{i=1}^\nu \gamma_m$. Hence,

$$h_{f,g}(\langle u, v \rangle_H) = \sum_{\nu=0}^\infty \Big(\sum_{n \in C(\nu,k)} x_n(f) x_n(g) \Big) \langle w_\nu(u), w_\nu(v) \rangle_{H_\nu}$$

$$= \sum_{\nu=0}^\infty \Big(\sum_{n \in C(\nu,k)} \langle x_n(f) w_\nu(u), x_n(g) w_\nu(v) \rangle_{H_\nu} \Big).$$

Consequently, the definition of the $(l_2\text{-})$direct product of Hilbert spaces leads us instantly to the desired Hilbert space $\mathbb{H} := \bigoplus_{\nu=0}^\infty \Big(\bigoplus_{n \in C(\nu,k)} H_\nu \Big)$ and the mappings ψ_u and ψ_v (cf. proof of [80, Theorem 17.2.9]). □

We will recognise soon that an *additional boundedness assumption* on $f = \sum_{n \in \mathbb{N}_0^k} a_n H_n \in L^2(\gamma_k)$ itself is of utmost importance in relation to an approximation of the smallest upper bound of $K_G^\mathbb{F}$ (cf. Theorems 6.8, 6.9, and 7.6). To perform this highly non-trivial task, we have to look strongly for "suitable" $f = \sum_{n \in \mathbb{N}_0^k} a_n H_n \in$

6.2 Real CCP Functions and Covariances: A Fourier-Hermite Analysis Approach

$L^2(\gamma_k)$ which in addition are bounded (a.s.); i.e., we have to look for some $M > 0$ such that (pointwise!) for (γ_k-almost) all $x \in \mathbb{R}^k$,

$$|f(x)| \leq M \tag{6.2.6}$$

(as is the case with (6.3.8)).

We are now fully prepared to extend these important facts to one of our key results in this book. In particular, we provide a multi-dimensional generalisation of the one-dimensional case $k = 1$ (cf. [18, Section 3.1]) and specify a non-obvious tightening of the upper bound of $h_{f,g}$. Here, it should be noted that the inclusion of Proposition 2.5 would allow to view Theorem 6.2 as a straightforward simple implication of the key results in [111, Chapter 11], including the consideration of [111, Definition 11.10] (cf. also [17, Chapter 5.6.1], [127, Lemma 2.2] and Remark 6.1). However, we give a self-contained proof, built on the well-established Ornstein-Uhlenbeck semigroup (whose construction is recalled in the proof) and which sheds light also on the impact of the negative correlation case $-1 < \rho < 0$ in shape of an alternating sign change in the related power series.

Theorem 6.2 *Let $k \in \mathbb{N}$, $f, g \in L^2(\mathbb{R}^k, \gamma_k)$, $S \sim N_k(0, I_k)$, $\rho \in [-1, 1]$ and $vec(X, Y) \sim N_{2k}(0, \Sigma_{2k}(\rho))$. Then the following properties hold:*

(i) $h_{f,g}\big|_{(-1,1)} \in W_+^\omega((-1, 1))$,

$$h_{f,g}(0) = \mathbb{E}[f(S)]\mathbb{E}[g(S)] \text{ and } h_{f,g}(\rho) = \mathbb{E}[f(X)g(Y)]$$
$$= \text{Cov}(f(X), g(Y)) + h_{f,g}(0). \tag{6.2.7}$$

In particular, $h_{f,g}(1) = \mathbb{E}[f(S)g(S)] = \langle f, g \rangle_{\gamma_k}$, $h_{f,f}(1) = \|f\|_{\gamma_k}^2$ and $h_{f,g}(-1) = \mathbb{E}[f(S)g(-S)] = \langle f, g_{-1} \rangle_{\gamma_k}$.

(ii) $h_{f,g} : [-1, 1] \longrightarrow \mathbb{R}$ *is bounded and satisfies*

$$|h_{f,g}(\rho)| \leq \big(h_{f,g}\big|_{(-1,1)}\big)_{abs}(|\rho|) \leq \big(h_{f,g}\big|_{(-1,1)}\big)_{abs}(1) \leq \|f\|_{\gamma_k} \|g\|_{\gamma_k} \tag{6.2.8}$$

for all $\rho \in [-1, 1]$.

(iii) *If $\rho \in (-1, 1)$, then*

$$h_{f,g}(\rho) = \frac{1}{(2\pi)^k (1 - \rho^2)^{k/2}}$$
$$\int_{\mathbb{R}^k} \int_{\mathbb{R}^k} f(x)g(y) \exp\left(-\frac{\|x\|^2 + \|y\|^2 - 2\rho\langle x, y\rangle}{2(1 - \rho^2)}\right) d^k x \, d^k y. \tag{6.2.9}$$

(iv) *If H is a real separable Hilbert space and $u, v \in S_H$, there exists a family $\{X_w : w \in S_H\}$ of $\mathbb{R}^k$-valued random vectors, such that $\mathrm{vec}(X_u, X_v) \sim N_{2k}(0, \Sigma_{2k}(\langle u, v \rangle_H))$ and*

$$h_{f,g}(\langle u, v \rangle_H) = \mathbb{E}[f(X_u)g(X_v)]. \tag{6.2.10}$$

(v) *If either f or g is odd, then $p_{2\nu}(f, g) = 0$ for all $\nu \in \mathbb{N}_0$ and*

$$\mathrm{cov}(f(X), g(Y)) = \mathbb{E}[f(X)g(Y)] = h_{f,g}(\rho) = \rho \sum_{\nu=0}^{\infty} p_{2\nu+1}(f, g)\rho^{2\nu}.$$

In particular, $h_{f,g}$ is odd.

Proof

(i) Firstly, since $H_0 = 1$, it follows that $h_{f,g}(0) = p_0(f, g) = \langle f, 1 \rangle_{\gamma_k} \langle g, 1 \rangle_{\gamma_k}$. Therefore, we have to consider the remaining cases $|\rho| = 1$, $0 < \rho < 1$ and $-1 < \rho < 0$. Firstly, let $|\rho| = 1$. Observe that in this case a probability density function does not exist since $\Sigma_{2k}(\rho)$ is singular (see Proposition 2.4). However, $\mathrm{vec}(S, \rho S) = A_\rho S \sim N_{2k}(0, \Sigma_{2k}(\rho))$, where $\mathbb{M}_{2k,k}(\mathbb{R}) \ni A_\rho := (I_k, \rho I_k)^\top$ (since $S \sim N_k(0, I_k)$ and $A_\rho A_\rho^\top = \Sigma_{2k}$). Consequently, $\mathbb{P}_{\mathrm{vec}(X,Y)} = \mathbb{P}_{\mathrm{vec}(S, \rho S)} = (A_\rho)_* \mathbb{P}_S$ (image measure), whence

$$\mathbb{E}[f(S)g(\rho S)] = \int_{\mathbb{R}^{2k}} f \otimes g \, \mathrm{d}((A_\rho)_* \mathbb{P}_S) = \int_{\mathbb{R}^k} (f \otimes g) \circ A_\rho \, \mathrm{d}\mathbb{P}_S$$

$$= \int_{\mathbb{R}^k} f(x) g(\rho x) \mathbb{P}_S(\mathrm{d}x) = \langle f, g_\rho \rangle_{\gamma_k},$$

where $g_\rho \in L^2(\gamma_k)$ is given as in (6.1.12). Thus, since $\{H_\nu : \nu \in \mathbb{N}_0^k\}$ is an orthonormal basis in the Hilbert space $L^2(\gamma_k)$, and $\mathbb{N}_0^k$ is partitioned according to $\mathbb{N}_0^k = \bigcup_{\nu=0}^{\infty} C(\nu, k)$, the associativity formula for summable families leads to

$$\mathbb{E}[f(X)g(Y)] = \mathbb{E}[f(S)g_\rho(S)] = \langle f, g_\rho \rangle_{\gamma_k} = \sum_{\nu=0}^{\infty} \sum_{n \in C(\nu,k)} \langle f, H_n \rangle_{\gamma_k} \langle g_\rho, H_n \rangle_{\gamma_k}$$

$$= \sum_{\nu=0}^{\infty} \Big(\sum_{n \in C(\nu,k)} \langle f, H_n \rangle_{\gamma_k} \langle g, H_n \rangle_{\gamma_k} \Big) \rho^\nu = h_{f,g}(\rho),$$

whereby the penultimate equality follows from (6.1.12). Next, let $0 < \rho < 1$. Then $\vartheta(\rho) := -\ln(\rho) = \ln(1/\rho) > 0$, implying that the linear operator $T_{\vartheta(\rho)} \in \mathfrak{L}(L^2(\gamma_k))$ is well-defined, where $(T_t)_{t \geq 0}$ denotes the strongly continuous Ornstein-Uhlenbeck semigroup of bounded non-negative symmetric

6.2 Real CCP Functions and Covariances: A Fourier-Hermite Analysis Approach

linear operators on $L^2(\gamma_k)$ (see [20, Theorem 1.4.1.]). Thus, by construction of the Ornstein-Uhlenbeck semigroup, and since $\mathbf{S} \sim N_k(0, I_k)$, it follows that

$$\mathbb{E}[f(\rho\, y + \sqrt{1-\rho^2}\,\mathbf{S}] = \mathbb{E}[f(e^{-\vartheta(\rho)}\, y + \sqrt{1 - e^{-2\vartheta(\rho)}}\,\mathbf{S}]$$
$$= T_{\vartheta(\rho)} f(y) \text{ for all } y \in \mathbb{R}^k. \quad (6.2.11)$$

A straightforward change of variables calculation leads to

$$\mathbb{E}[f(\rho\, y + \sqrt{1-\rho^2}\,\mathbf{S}] = \mathbb{E}[f(\mathbf{S}) M_\rho(\mathbf{S}, y; k)] \text{ for all } y \in \mathbb{R}^k,$$

where $\mathbb{R}^k \times \mathbb{R}^k \ni (x, y) \mapsto M_\rho(x, y; k)$ denotes the k-dimensional Mehler kernel (cf. (2.2.4)). The latter is well-defined, since $\rho^2 \neq 1$ by assumption. Consequently, we may apply (2.2.3), and obtain

$$\mathbb{E}[f(\mathbf{X})g(\mathbf{Y})] = \frac{1}{(2\pi)^k (1-\rho^2)^{k/2}}$$
$$\int_{\mathbb{R}^{2k}} f(x)g(y) \exp\left(-\frac{\|x\|^2 + \|y\|^2 - 2\rho\langle x, y\rangle}{2(1-\rho^2)}\right) d^{2k}(x, y)$$
$$= \int_{\mathbb{R}^k} \mathbb{E}[f(\mathbf{S}) M_\rho(\mathbf{S}, y; k)]\, g(y) \gamma_k(dy) = \langle T_{\vartheta(\rho)} f, g \rangle_{\gamma_k}. \quad (6.2.12)$$

Moreover, an application of [20, Theorem 1.4.4.] and the construction of ϑ imply that

$$T_{\vartheta(\rho)} f = \sum_{\nu=0}^{\infty} \rho^\nu \Big(\sum_{n \in C(\nu,k)} \langle f, H_n \rangle_{\gamma_k} H_n \Big)$$

on $L^2(\gamma_k)$. Consequently, it follows that

$$\mathbb{E}[f(\mathbf{X})g(\mathbf{Y})] = \Big\langle \sum_{\nu=0}^{\infty} \rho^\nu \Big(\sum_{n \in C(\nu,k)} \langle f, H_n \rangle_{\gamma_k} H_n \Big), g \Big\rangle_{\gamma_k}$$
$$= \sum_{\nu=0}^{\infty} \Big(\sum_{n \in C(\nu,k)} \langle f, H_n \rangle_{\gamma_k} \langle H_n, g \rangle_{\gamma_k} \Big) \rho^\nu = h_{f,g}(\rho), \quad (6.2.13)$$

which finishes the prove of (i) for the case $0 < \rho < 1$. So, let us consider the remaining case $-1 < \rho < 0$. Since $\text{vec}(\mathbf{X}, \mathbf{Y}) \sim N_{2k}(0, \Sigma_{2k}(\rho))$ by assumption, it follows that $\text{vec}(-\mathbf{X}, \mathbf{Y}) \sim N_{2k}(0, \Sigma_{2k}(|\rho|))$ (since $0 < |\rho| = -\rho < 1$). Thus,

$$\mathbb{E}[f(\mathbf{X})g(\mathbf{Y})] = \mathbb{E}[f_{-1}(-\mathbf{X})g(\mathbf{Y})] \stackrel{(6.2.13)}{=} h_{f_{-1},g}(|\rho|)$$

$$= \sum_{\nu=0}^{\infty} p_\nu(f_{-1}, g)|\rho|^\nu \stackrel{(6.1.12)}{=} \sum_{\nu=0}^{\infty} p_\nu(f, g)\rho^\nu,$$

which also finishes the proof of (i) for the remaining case $-1 < \rho < 0$. To sum up, given an arbitrary $\rho \in (-1, 1) \setminus \{0\}$ and any $\text{vec}(\mathbf{X}, \mathbf{Y}) \sim N_{2k}(0, \Sigma_{2k}(\rho))$, we have:

$$\mathbb{E}[f(\mathbf{X})g(\mathbf{Y})] = \frac{1}{(2\pi)^k(1-\rho^2)^{k/2}}$$

$$\int_{\mathbb{R}^{2k}} f(x)g(y) \exp\left(-\frac{\|x\|^2 + \|y\|^2 - 2\rho\langle x, y\rangle}{2(1-\rho^2)}\right) d^{2k}(x, y)$$

$$= h_{f,g}(\rho) = h_{f_\rho,g}(|\rho|) = \langle T_{\vartheta(|\rho|)} f_\rho, g\rangle_{\gamma_k}. \tag{6.2.14}$$

Actually, if we put $\vartheta(1) := 0$, we have shown that

$$\mathbb{E}[f(\mathbf{X})g(\mathbf{Y})] = h_{f,g}(\rho) = \langle T_{\vartheta(|\rho|)} f_\rho, g\rangle_{\gamma_k}$$

$$= \langle f_\rho, T_{\vartheta(|\rho|)} g\rangle_{\gamma_k} \text{ for all } \rho \in [-1, 1] \setminus \{0\}. \tag{6.2.15}$$

This proves (i), which in turn leads to (iii).

(ii) follows from (6.2.4).

(iv) Let $u, v \in S_H$. Consider the real $2k$-dimensional Gaussian random vector $\text{vec}(\mathbf{Z}_u, \mathbf{Z}_v) \sim N_{2k}(0, \Sigma_{2k}(\langle u, v\rangle_H))$ (which exists, due to Proposition 3.8). Hence, if we put $\rho_{u,v} := \langle u, v\rangle_H \in [-1, 1]$, it follows that $h_{f,g}(\rho_{u,v}) = \mathbb{E}[f(\mathbf{Z}_u)g(\mathbf{Z}_v)]$, which concludes claim (iv).

(v) Without loss of generality, we may assume that f is odd. Then

$$\langle f, H_n\rangle_{\gamma_k} = -\langle f_{-1}, H_n\rangle_{\gamma_k} \stackrel{(6.1.12)}{=} (-1)^{1+|n|}\langle f, H_n\rangle_{\gamma_k}$$

for all $n \in \mathbb{N}_0^k$, so that $(1 - (-1)^{1+|n|})\langle f, H_n\rangle_{\gamma_k} = 0$ for all $n \in \mathbb{N}_0^k$. Consequently, it follows that

$$\langle f, H_n\rangle_{\gamma_k} = 0 \text{ for all } n \in \{m \in \mathbb{N}_0^k : |m| \text{ is even}\}. \tag{6.2.16}$$

6.2 Real CCP Functions and Covariances: A Fourier-Hermite Analysis Approach   129

In particular, $\mathbb{E}[f(\mathbf{X})] = \langle f, H_0 \rangle_{\gamma_k} = 0$. Thus,

$$\operatorname{cov}(f(\mathbf{X}), g(\mathbf{Y})) = \mathbb{E}[f(\mathbf{X})g(\mathbf{Y})] = h_{f,g}(\rho) \stackrel{(6.2.16)}{=} \rho \sum_{\nu=0}^{\infty} p_{2\nu+1}(f, g) \rho^{2\nu}.$$

$\square$

Expressed in matrix notation, Theorem 6.2 leads directly to a further crucial result:

Corollary 6.2 *Let $k, m, n \in \mathbb{N}$ and $f, g \in L^2(\gamma_k)$. Then the following matrix representations hold:*

(i) *For any $S \equiv (s_{ij}) \in \mathcal{Q}_{m,n}$ there exist a random vector $\boldsymbol{P}_f = (f(\boldsymbol{X}_1), \ldots, f(\boldsymbol{X}_m))^\top$ in $\mathbb{R}^m$ and a random vector $\boldsymbol{Q}_g = (g(\boldsymbol{Y}_1), \ldots, g(\boldsymbol{Y}_n))^\top$ in $\mathbb{R}^n$, such that $(\boldsymbol{X}_i, \boldsymbol{Y}_j) \sim N_{2k}(0, \Sigma_{2k}(s_{ij}))$ for all $(i, j) \in [m] \times [n]$ and*

$$h_{f,g}[S] = \mathbb{E}[\boldsymbol{P}_f \boldsymbol{Q}_g^\top] = \left(\mathbb{E}[f(\boldsymbol{X}_i)g(\boldsymbol{Y}_j)]\right)_{ij}. \quad (6.2.17)$$

(ii) *For any correlation matrix $\Sigma \equiv (\sigma_{ij}) \in C(n; \mathbb{R})$ there exists a random vector $\boldsymbol{R}_f = (f(\boldsymbol{Z}_1), \ldots, f(\boldsymbol{Z}_n))^\top$ in $\mathbb{R}^n \setminus \{0\}$, such that $(\boldsymbol{Z}_i, \boldsymbol{Z}_j) \sim N_{2k}(0, \Sigma_{2k}(\sigma_{ij}))$ for all $(i, j) \in [m] \times [n]$ and*

$$h_{f,f}[\Sigma] = \mathbb{E}[\Theta_{f,f}] = \mathbb{E}[\boldsymbol{R}_f \boldsymbol{R}_f^\top] = \left(\mathbb{E}[f(\boldsymbol{Z}_i)f(\boldsymbol{Z}_j)]\right)_{ij}, \quad (6.2.18)$$

where $\Theta_{f,f} := \boldsymbol{R}_f \boldsymbol{R}_f^\top \in \mathbb{M}_n(\mathbb{R})^+$ is a positive semidefinite random matrix of rank 1.

Corollary 6.3 *Let $k \in \mathbb{N}$, $f, g \in L^2(\mathbb{R}^k, \gamma_k)$, $r \in [-1, 1]$, $\boldsymbol{S} \sim N_k(0, I_k)$ and $\operatorname{vec}(\boldsymbol{X}, \boldsymbol{Y}) \sim N_{2k}(0, \Sigma_{2k}(r))$. If $\|f\|_{\gamma_k} = 1$, $\|g\|_{\gamma_k} = 1$ and $\mathbb{E}[f(\boldsymbol{S})] = \mathbb{E}[g(\boldsymbol{S})] = 0$, then*

$$h_{f,g}(r) = \rho(f(\boldsymbol{X}), g(\boldsymbol{Y}))$$

coincides with the Pearson correlation coefficient between the real random variables $f(\boldsymbol{X})$ and $g(\boldsymbol{Y})$.

Very recently, Krivine generalised the Grothendieck equality and constructed a related function $\Phi_{F,G} : [-1, 1] \longrightarrow \{-1, 1\}$ in the real case (cf. [93]). In fact, if $\varepsilon > 0$ is fixed, Proposition 2.5 implies that $\Phi_{F,G} = h_{f,g}$, where

$$\mathbb{R}^3 \ni (x_0, x_1, x_2)^\top \mapsto f(x_0, x_1, x_2) := \operatorname{sign}\left(\left\langle \begin{pmatrix} \cos(\sqrt{2}\varepsilon\, H_2(x_0)) \\ \sin(\sqrt{2}\varepsilon\, H_2(x_0)) \end{pmatrix}, \begin{pmatrix} x_1 \\ x_2 \end{pmatrix} \right\rangle_{\mathbb{R}^2}\right)$$

$$= \operatorname{sign}(F(x_0, x_1, x_2))$$

and

$$\mathbb{R}^3 \ni (y_0, y_1, y_2)^\top \mapsto g(y_0, y_1, y_2) := \mathrm{sign}\left(\left\langle \begin{pmatrix} \cos(\sqrt{2}\varepsilon\, H_2(y_0)) \\ -\sin(\sqrt{2}\varepsilon\, H_2(y_0)) \end{pmatrix}, \begin{pmatrix} y_1 \\ y_2 \end{pmatrix} \right\rangle_{\mathbb{R}_2^2}\right)$$
$$= \mathrm{sign}(G(y_0, y_1, y_2)).$$

The function $\Phi_{F,G}$ in [93] namely proves to be a beautiful example that fulfils the following statement (if $k = 1$ and $n = 2$):

Proposition 6.3 *Let $k, n \in \mathbb{N}$, $\rho \in [-1, 1]$ and $vec(X_0, X, Y_0, Y)$ be a random vector, which maps into $\mathbb{R}^k \times \mathbb{R}^n \times \mathbb{R}^k \times \mathbb{R}^n \equiv \mathbb{R}^{2(k+n)}$, such that $vec(X_0, X, Y_0, Y) \sim N_{2(k+n)}(0, \Sigma_{2(k+n)}(\rho))$. Let $F : \mathbb{R}^k \longrightarrow \mathbb{S}^{n-1} \subseteq \mathbb{R}^n$ and $G : \mathbb{R}^k \longrightarrow \mathbb{S}^{n-1} \subseteq \mathbb{R}^n$ be two arbitrary measurable functions. Put $\mathbb{R}^k \times \mathbb{R}^n \ni (x_0, x) \mapsto f(x_0, x) := \mathrm{sign}(\langle F(x_0), x\rangle_{\mathbb{R}_2^n})$ and $\mathbb{R}^k \times \mathbb{R}^n \ni (y_0, y) \mapsto g(y_0, y) := \mathrm{sign}(\langle G(y_0), y\rangle_{\mathbb{R}_2^n})$. Then*

$$h_{f,g}(\rho) = \frac{2}{\pi} \mathbb{E}\left[\arcsin\left(\rho\, \langle F(X_0), G(Y_0)\rangle_{\mathbb{R}_2^n}\right)\right]$$
$$= \frac{2}{\pi} \int_{\mathbb{R}^k} \arcsin\left(\rho\, \langle F(x_0), G(y_0)\rangle_{\mathbb{R}_2^n}\right)$$
$$\exp\left(-\frac{\|x_0\|^2 + \|y_0\|^2 - 2\rho\langle x_0, y_0\rangle}{2(1-\rho^2)}\right) \frac{d^k x_0\, d^k y_0}{(2\pi)^k (1-\rho^2)^{k/2}}.$$

Proof Since $vec(X_0, X, Y_0, Y) \sim N_{2(k+n)}(0, \Sigma_{2(k+n)}(\rho))$, the structure of the characteristic functions $\phi_{vec(X_0, X, Y_0, Y)}$, $\phi_{vec(X_0, Y_0)}$ and $\phi_{vec(X, Y)}$ clearly implies that $vec(X_0, Y_0) \sim N_{2k}(0, \Sigma_{2k}(\rho))$, $vec(X, Y) \sim N_{2n}(0, \Sigma_{2n}(\rho))$ and $\phi_{vec(X_0, X, Y_0, Y)} = \phi_{vec(X_0, Y_0)} \phi_{vec(X, Y)}$ (similar to the proof of Corollary 2.2). Hence, the Gaussian random vectors $vec(X_0, Y_0)$ and $vec(X, Y)$ are independent. Consequently,

$$h_{f,g}(\rho) = \mathbb{E}_{\mathbb{P}_{vec(X_0, Y_0)}}[\Psi(X_0, Y_0)],$$

where

$$\Psi(x_0, y_0) := \mathbb{E}_{vec(X, Y)}\left[f(x_0, X)\, g(y_0, Y)\right]$$
$$= \mathbb{E}_{vec(X, Y)}\left[\mathrm{sign}(\langle F(x_0), X\rangle_{\mathbb{R}_2^n})\, \mathrm{sign}(\langle F(y_0), Y\rangle_{\mathbb{R}_2^n})\right].$$

Since $\begin{pmatrix} \langle F(x_0), X\rangle_{\mathbb{R}_2^n} \\ \langle F(y_0), Y\rangle_{\mathbb{R}_2^n} \end{pmatrix} \sim \mathbb{F}N_2(0, \Sigma_2(\rho\, \langle F(x_0), G(y_0)\rangle_{\mathbb{R}_2^n}))$ (due to Lemma 2.3), it follows that we may apply the Grothendieck equality to $\Psi(x_0, y_0)$ for any $x_0, y_0 \in \mathbb{R}^k$, and the claim follows. □

6.2 Real CCP Functions and Covariances: A Fourier-Hermite Analysis Approach

In case of $f = g$ some important extra analytical facts emerge; particularly if in addition f is odd (which is the relevant case for the topic of this book):

Theorem 6.3 *Let $k \in \mathbb{N}$ and $f \in L^2(\mathbb{R}^k, \gamma_k)$ be odd, such that $r := \|f\|_{\gamma_k}^2 > 0$. Then the following properties hold:*

(i) $h_{f,f}|_{(-1,1)} \in W_+^\omega((-1,1))$.
(ii) $h_{f,f}$ *is a CCP function if and only if* $r = 1$.
(iii) $h_{f,f} : [-1,1] \longrightarrow [-r,r]$ *is an odd strictly increasing homeomorphism, which satisfies* $h_{f,f}((-1,0)) = (-r,0)$ *and* $h_{f,f}((0,1)) = (0,r)$. *Moreover,*

$$0 \leq h'_{f,f}(0) = \sum_{i=1}^{k} \left(\int_{\mathbb{R}^k} f(x) x_i \gamma_k(dx) \right)^2$$
$$= \int_{\mathbb{R}^{2k}} f(x) f(y) \langle x, y \rangle_{\mathbb{R}_2^k} \gamma_{2k}(d(x,y)). \qquad (6.2.19)$$

(iv) *If $h'_{f,f}(0) > 0$, then $h'_{f,f}(\rho) > 0$ for all $\rho \in (-1,1)$. In particular,*
$h_{f,f}^{-1}|_{(-r,r)} = \left(h_{f,f}|_{(-r,r)} \right)^{-1}$ *is real analytic on $(-r,r)$ if and only if $h'_{f,f}(0) > 0$.*

(v) *If $h_{f,f}^{-1}|_{(-r,r)} \in W_+^\omega((-r,r))$, then*

$$\left(h_{f,f}^{-1}|_{(-r,r)} \right)_{abs}(y) \geq \frac{1}{h'_{f,f}(0)} y \text{ for all } y \in [0,r]. \qquad (6.2.20)$$

In particular, $s(r,f) := \left(h_{f,f}^{-1}|_{(-r,r)} \right)_{abs}(r) > 0$. If $\alpha > 0$ and $h'_{f,f}(0) < \frac{r}{\alpha}$, then $s(r,f) > \alpha$. Moreover,

$$[-1,1] \ni t \mapsto \psi(t) := \frac{1}{s(r,f)} \left(h_{f,f}^{-1}|_{(-r,r)} \right)_{abs}(rt)$$

is an odd, strictly increasing and homeomorphic CCP function. In particular, the function $\left(h_{f,f}^{-1}|_{(-r,r)} \right)_{abs} : [-r,r] \xrightarrow{\cong} [-s(r,f), s(r,f)]$ is strictly increasing, odd and homeomorphic, as well as its inverse. $\left(h_{f,f}^{-1}|_{(-r,r)} \right)_{abs}(y) = s(r,f) \psi(\frac{y}{r})$ for all $y \in [-r,r]$ and $\left(\left(h_{f,f}^{-1}|_{(-r,r)} \right)_{abs} \right)^{-1}(x) = r \psi^{-1}(\frac{x}{s(r,f)})$ for all $x \in [-s(r,f), s(r,f)]$.

Proof

(ii) The claim follows from the power series representation of $h_{f,f}$ and condition (iii) of Theorem 5.3. Regarding the latter, we only have to perform Parseval's equality, applied to the Hilbert space $L^2(\gamma_k)$, since

$$r = \sum_{n \in \mathbb{N}_0^k} \langle f, H_n \rangle_{\gamma_k}^2 = \sum_{v=0}^{\infty} \Big(\sum_{n \in C(v,k)} \langle f, H_n \rangle_{\gamma_k}^2 \Big) = h_{f,f}(1) = \sum_{v=0}^{\infty} a_v,$$

(6.2.21)

where $0 \leq a_v := p_v(f, f) = \sum_{n \in C(v,k)} \langle f, H_n \rangle_{\gamma_k}^2$ for all $v \in \mathbb{N}_0$.

From now on, we may assume without loss of generality that $r = 1$ in the remaining part of the proof (since $h_{f,f} = r\, h_{f^\circ, f^\circ}$, where $f^\circ := \frac{1}{\sqrt{r}} f$).

(iii) Since $h_{f,f}$ is odd, it follows that $0 < 1 = h_{f,f}(1) = \sum_{v=0}^{\infty} p_{2v+1}(f, f)$. Hence, $p_{2\mu+1}(f, f) > 0$ for some $\mu \in \mathbb{N}_0$. Let $-1 \leq \rho_1 < \rho_2 \leq 1$. Recall that $\mathbb{R} \ni s \mapsto s^{2v+1}$ is strictly increasing for all $v \in \mathbb{N}_0$. Thus, $p_{2\mu+1}(f, f)\rho_1^{2\mu+1} < p_{2\mu+1}(f, f)\rho_2^{2\mu+1}$. Hence,

$$h_{f,f}(\rho_1) = p_{2\mu+1}(f, f)\rho_1^{2\mu+1} + \sum_{v \neq \mu} p_{2v+1}(f, f)\rho_1^{2v+1} < p_{2\mu+1}(f, f)\rho_2^{2\mu+1}$$

$$+ \sum_{v \neq \mu} p_{2v+1}(f, f)\rho_1^{2v+1} \leq h_{f,f}(\rho_2).$$

Due to (6.2.21), we may apply the M-test of Weierstrass, and it consequently follows that the strictly increasing odd function $h_{f,f} : [-1, 1] \longrightarrow [1, 1]$ is continuous and hence invertible (due to Bolzano's intermediate value theorem of calculus). Finally, since $[-1, 1]$ is compact, it follows that the bijective continuous function $h_{f,f}$ already is a homeomorphism which satisfies $h_{f,f}(0) = 0$, $h_{f,f}(1) = 1$, and hence $h_{f,f}((-1, 0)) = (-1, 0)$ and $h_{f,f}((0, 1)) = (0, 1)$. Since $h'(0) = p_1(f, f) = \sum_{i=1}^{k} \langle f, H_{e_i} \rangle_{\gamma_k}^2$ and $H_1(a) = a$ for all $a \in \mathbb{R}$, it follows that $h'_{f,f}(0) = \sum_{i=1}^{k} \big(\int_{\mathbb{R}^k} f(x) x_i \gamma_k(dx) \big)^2$. Moreover, since $\gamma_{2k} = \gamma_k \otimes \gamma_k$, the latter integral obviously coincides with $\int_{\mathbb{R}^{2k}} f(x) f(y) \langle x, y \rangle_{\mathbb{R}_2^k} \gamma_{2k}(d(x, y))$, which implies (6.2.19).

(iv) Since $h_{f,f}$ is odd, it follows that

$$h_{f,f}(\rho) = \sum_{v=0}^{\infty} p_{2v+1}(f, f)\rho^{2v+1} \quad \text{for all } \rho \in [-1, 1].$$

Thus,

$$h'_{f,f}(\rho) = p_1(f,f) + \sum_{\nu=1}^{\infty}(2\nu+1)p_{2\nu+1}(f,f)\rho^{2\nu} \geq p_1(f,f)$$

$$= h'_{f,f}(0) \text{ for all } \rho \in (-1,1).$$

Since $h'_{f,f}(0) > 0$ (by assumption), $h'_{f,f}$ is nonvanishing on $(-1,1)$: $h'_{f,f}(\rho) > 0$ for all $\rho \in (-1,1)$. Therefore, we may apply the Real Analytic Inverse Function Theorem (cf. [89, Theorem 1.5.3]), implying that also $\left(h_{f,f}|_{(-1,1)}\right)^{-1}$ is real analytic on $(-1,1)$. The converse implication follows from the simple fact that $h'_{f,f}(0) \geq 0$ and $0 < 1 = (h_{f,f} \circ h_{f,f}^{-1})'(0) = h'_{f,f}(0) \cdot (h_{f,f}^{-1})'(0)$.

(v) Let $0 \leq y \leq 1$. Since $\left(\left(h_{f,f}|_{(-1,1)}\right)^{-1}_{\mathrm{abs}}\right)'(0) = \left|(h_{f,f}^{-1})'(0)\right| = \left|\frac{1}{h'_{f,f}(0)}\right| = \frac{1}{h'_{f,f}(0)}$, it follows that

$$\left(h_{f,f}|_{(-1,1)}\right)^{-1}_{\mathrm{abs}}(y) = \left(\left(h_{f,f}|_{(-1,1)}\right)^{-1}_{\mathrm{abs}}\right)'(0)\, y + \sum_{\nu=1}^{\infty} b_{2\nu+1}\, y^{2\nu+1} \geq \frac{1}{h'_{f,f}(0)}\, y.$$

Theorem 5.3 implies that the function ψ is an odd CCP function, so that Lemma 5.3 and a similar approach as in the proof of (iii) conclude the proof of statement (v) (since $s_1 = \left(h_{f,f}^{-1}|_{(-1,1)}\right)_{\mathrm{abs}}(1) > 0$). $\square$

A first implication for *odd* CCP functions is a strong improvement of the boundedness condition in Theorem 6.2, induced by the Schwarz lemma from complex analysis:

Proposition 6.4 *Let $k \in \mathbb{N}$ and $f \in S_{L^2(\gamma_k)}$ be odd. Assume that $h_{f,f}(r) \neq r$ for some $r \in (0,1)$. Then*

$$h_{f,f}(\rho) < \rho \text{ for all } \rho \in (0,1) \text{ and } h_{f,f}(\tau) > \tau \text{ for all } \tau \in (-1,0). \tag{6.2.22}$$

Moreover, $0 \leq h'_{f,f}(0) < 1$. If in addition, $h_{f,f}^{-1}|_{(-1,1)} \in W_+^{\omega}((-1,1))$, then

$$\left(h_{f,f}^{-1}|_{(-1,1)}\right)_{\mathrm{abs}}(1) > 1. \tag{6.2.23}$$

Proof Due to Theorem 6.3, $[-1,1] \ni \rho \mapsto h_{f,f}(\rho) = \sum_{\nu=0}^{\infty} p_{2\nu+1}(f,f)\,\rho^{2\nu+1}$ is an odd, strictly increasing CCP function, which satisfies $h_{f,f}|_{(-1,1)} = \tilde{h}|_{(-1,1)}$, where $\mathbb{D} \ni z \mapsto \tilde{h}(z) := \sum_{\nu=0}^{\infty} p_{2\nu+1}(f,f) z^{2\nu+1}$ is holomorphic (cf. Lemma 5.3). Thus, for any $z \in \mathbb{D}$ it follows that

$$|\tilde{h}(z)| \leq h_{f,f}(|z|) < h_{f,f}(1) = 1,$$

whence $\tilde{h}(\mathbb{D}) \subseteq \mathbb{D}$. Since $\tilde{h}(0) = h_{f,f}(0) = 0$, we consequently may apply the Schwarz lemma to the holomorphic function $\tilde{h}$, implying that

$$|h_{f,f}(\rho)| = |\tilde{h}(\rho)| \leq |\rho| \text{ for all } \rho \in (-1, 1)$$

and $h'_{f,f}(0) \leq 1$. Assume by contradiction that $|\tilde{h}(\rho_0)| = |h_{f,f}(\rho_0)| = |\rho_0|$ for some $\rho_0 \in (-1, 1) \setminus \{0\}$ or $h'_{f,f}(0) = 1$. In each case, the Schwarz lemma implies that there exists $q \in \mathbb{Z}$, such that $h_{f,f}(\rho) = (-1)^q \rho$ for all $\rho \in (-1, 1)$ (since $e^{in\pi} = \cos(n\pi) = (-1)^n$ for all $n \in \mathbb{N}_0$). However, since e.g. $0 \leq h_{f,f}(1/2) \neq -1/2$, it follows that $h_{f,f}(\rho) = \rho$ for all $\rho \in [-1, 1]$, which contradicts the assumption. Consequently, an application of Theorem 6.3, (v) and the assumption implies that in fact $\left(h_{f,f}^{-1}|_{(-1,1)}\right)_{\mathrm{abs}}(1) > 1$ (since $\left(h_{f,f}^{-1}|_{(-1,1)}\right)_{\mathrm{abs}}$ is well-defined). □

All of a sudden we end up with an important general and non-obvious structural statement about *odd* CCP functions; namely:

Corollary 6.4 *Let* $k \in \mathbb{N}$ *and* $f \in S_{L^2(\gamma_k)}$ *be odd. Assume that* $h_{f,f}(r) \neq r$ *for some* $r \in (0, 1)$. *Then* $h_{f,f}^{-1}$ *is not a CCP function.*

Proof We just have to link Theorem 5.3, (v) and (6.2.23). □

We will recognise soon that Theorem 6.3 is strongly linked with the value of the Grothendieck constant $K_G^{\mathbb{F}}$ (cf. Theorem 6.9). We namely obtain in a natural way another significant new definition; the so-called "hyperbolic CCP transform". To make this concept understandable, recall Lemma 5.3 and Remark 5.4 and reconsider the odd CCP function $\psi := h_{\text{sign,sign}} = \frac{2}{\pi} \arcsin$; i.e., the Grothendieck function. Since $\psi^{-1} = \sin(\frac{\pi}{2} \cdot)$ on $[-1, 1]$, it follows that $\left(\psi^{-1}|_{(-1,1)}\right)_{\mathrm{abs}}(\tau) = \sinh(\frac{\pi}{2}\tau) = \frac{1}{i}\sin(\frac{\pi}{2}i\tau)$ for all $\tau \in [-1, 1]$. Note that $\left(\psi^{-1}|_{(-1,1)}\right)_{\mathrm{abs}}(1) = \sinh(\pi/2) > 1$, implying that $[-1, 1] \subseteq (-\sinh(\pi/2), \sinh(\pi/2))$. Thus,

$$\frac{2}{\pi} \ln(y + \sqrt{y^2 + 1}) = \frac{2}{\pi} \sinh^{-1}(y) = \left(\left(\psi^{-1}|_{(-1,1)}\right)_{\mathrm{abs}}\right)^{-1}(y)$$

$$= \frac{2}{\pi}\left(\frac{1}{i}\sin^{-1}(i\,y)\right) = \frac{1}{i}\tilde{\psi}(iy) \quad (6.2.24)$$

6.2 Real CCP Functions and Covariances: A Fourier-Hermite Analysis Approach   135

for all $y \in [-1, 1]$. The Maclaurin series representation of the real CCP function ψ therefore implies the Maclaurin series of $y \mapsto \frac{1}{i}\tilde{\psi}(iy)$ on (the whole of) $\mathbb{R}$ is given by

$$\frac{1}{i}\tilde{\psi}(iy) \stackrel{(6.3.7)}{=} \frac{2}{\pi} \sum_{v=0}^{\infty} (-1)^v \frac{((2v-1)!!)^2}{(2v+1)!} y^{2v+1}$$

$$= -\frac{1}{3\pi} + \sum_{\substack{v=0 \\ v \neq 1}}^{\infty} (-1)^v \frac{((2v-1)!!)^2}{(2v+1)!} y^{2v+1}.$$

Consequently, it follows that

$$\left((\psi^{-1}|_{(-1,1)})_{\mathrm{abs}}\right)^{-1}(\rho) = \frac{1}{i}\tilde{\psi}(i\rho) < \psi(\rho) < \rho \text{ for all } \rho \in (0, 1).$$

Theorem 6.3 and Proposition 6.4, together with Remark 5.4 imply that the Grothendieck function is a special case of

Lemma 6.1 (Hyperbolic CCP Transform) *Let $k \in \mathbb{N}$ and $\psi = h_{f,f}$, where $f \in S_{L^2(\gamma_k)}$ is odd. Assume that $\psi^{-1}|_{(-1,1)} \in W_+^\omega((-1,1))$ and $\psi(r) \neq r$ for some $r \in (0,1)$. Let the complex function $F := \widetilde{\psi|_{(-1,1)}} : \overline{\mathbb{D}} \longrightarrow \overline{\mathbb{D}}$ be defined as in Lemma 5.3. Then $F(\mathbb{D}) \subseteq \mathbb{D}$ and*

$$F(z) = \sum_{v=0}^{\infty} p_{2v+1}(f, f) z^{2v+1} \text{ for all } z \in \overline{\mathbb{D}},$$

where each $p_{2v+1}(f, f) \in [0, \infty)$ satisfies (6.2.1). Put $s^* := (\psi^{-1}|_{(-1,1)})_{\mathrm{abs}}(1)$. Then $s^* > 1$ and

$$h_{f,f}^{hyp} := ((\psi^{-1}|_{(-1,1)})_{\mathrm{abs}})^{-1} : [-s^*, s^*] \stackrel{\cong}{\longrightarrow} [-1, 1]$$

is an odd, strictly increasing homeomorphism, which satisfies $[-1, 1] \subseteq (-s^*, s^*)$, $\psi^{hyp}((0, 1]) \subseteq (0, 1)$ and $\psi^{hyp}([-1, 0)) \subseteq (-1, 0)$. Moreover,

$$0 < \psi^{hyp}(1) \leq \psi'(0) < 1, \tag{6.2.25}$$

and the following two statements hold:

(i)

$$|\psi^{hyp}(\rho)| < \psi(|\rho|) < |\rho| \text{ for all } \rho \in (-1, 1) \setminus \{0\}.$$

(ii) *If* $sign((\psi^{-1}|_{(-1,1)})^{(2n+1)}(0)) = (-1)^n$ *for all* $n \in \mathbb{N}_0$, *then*

$$\psi^{hyp}(x) = \sum_{\nu=0}^{\infty} (-1)^{\nu} p_{2\nu+1}(f, f) x^{2\nu+1}$$

$$= \frac{1}{i} F(ix) = (M_{-i} \circ F \circ M_i)(x) \text{ for all } x \in [-1, 1].$$

In particular,

$$\psi^{hyp}(1) = \frac{1}{i} F(i). \tag{6.2.26}$$

Proof Since $\psi^{hyp} = ((\psi^{-1}|_{(-1,1)})_{abs})^{-1}$ is strictly increasing and odd (such as $(\psi^{-1}|_{(-1,1)})_{abs}$), Proposition 6.4 implies that $\psi^{hyp}((0, 1]) \subseteq (0, 1)$ and $\psi^{hyp}([-1, 0)) \subseteq (-1, 0)$ (since $s^* > 1$). Recall from Proposition 6.4 that $0 < \psi'(0) < 1$. To verify (6.2.25), it therefore suffices to show that

$$(\psi^{-1}|_{(-1,1)})_{abs}(\psi'(0)) \overset{!}{\geq} 1. \tag{6.2.27}$$

Remember also that $|(\psi^{-1})'(0)| = \frac{1}{\psi'(0)}$. Thus, $(\psi^{-1}|_{(-1,1)})_{abs}(\psi'(0)) = 1 + \sum_{l=1}^{\infty} \alpha_{2l+1}(\psi'(0))^{2l+1}$, where $\alpha_{2l+1} \geq 0$ for all $l \in \mathbb{N}$, and (6.2.27) follows at once.

(i) Since also ψ^{hyp} is odd, we only have to verify that the claim holds for any $\rho \in (0, 1)$. So, fix $0 < \rho < 1$. Due to Proposition 6.4, it is sufficient to show that

$$(\psi^{-1}|_{(-1,1)})_{abs}(\psi(\rho)) \overset{!}{>} \rho \tag{6.2.28}$$

To this end, put $\sigma := \psi(\rho)$. Then $\sigma > \psi(0) = 0$. Since ψ is an odd CCP function, it follows that

$$\psi(x) = \psi'(0)x + \sum_{l=1}^{\infty} a_{2l+1} x^{2l+1} \text{ for all } x \in [-1, 1],$$

where $a_{2l+1} \geq 0$ for all $m \in \mathbb{N}$. Assume by contradiction that $a_{2l+1} = 0$ for all $l \in \mathbb{N}$. Then $\psi(x) = \psi'(0)x$ for all $x \in [-1, 1]$. In particular, $1 = \psi(1) = \psi'(0) < 1$, which is absurd. Thus, an $l_0 \in \mathbb{N}$ exists, such that $a_{2l_0+1} > 0$, implying that in particular $\sigma = \psi(\rho) > \psi'(0)\rho$. Given the assumption on $\psi^{-1}|_{(-1,1)}$, we therefore obtain that

$$(\psi^{-1}|_{(-1,1)})_{abs}(\sigma) = \frac{\sigma}{\psi'(0)} + \sum_{m=1}^{\infty} b_{2m+1} \sigma^{2m+1} > \rho + \sum_{m=1}^{\infty} b_{2m+1} \sigma^{2m+1} \geq \rho,$$

6.2 Real CCP Functions and Covariances: A Fourier-Hermite Analysis Approach   137

where $b_{2m+1} \geq 0$ for all $m \in \mathbb{N}$. Since $\rho \in (0, 1)$ was arbitrarily chosen, (6.2.28) and hence (i) follows.

(ii) Let $G := \widetilde{\psi^{-1}|_{(-1,1)}} : \overline{\mathbb{D}} \longrightarrow \overline{\mathbb{D}}$ be defined as in Lemma 5.3. Since ψ is strictly increasing, (5.1.3) implies that $|F(z)| \leq \psi(|z|) < \psi(1) = 1$ for all $z \in \mathbb{D}$, whence $F(\mathbb{D}) \subseteq \mathbb{D}$. Similarly, we obtain that $G(\mathbb{D}) \subseteq \mathbb{D}$ (since $\psi^{-1}|_{(-1,1)} \in W_+^\omega((-1,1))$ by assumption). $F|_\mathbb{D}$ is the unique holomorphic extension of $\psi|_{(-1,1)}$ (since $\mathbb{D}$ is a domain; i.e., a non-empty connected open set). $F|_\mathbb{D}$ is invertible, and $(F|_\mathbb{D})^{-1} = G|_\mathbb{D}$ is the unique holomorphic extension of $\psi^{-1}|_{(-1,1)}$. Let $x \in (-1, 1)$. Then $ix \in \mathbb{D}$, implying that

$$(M_{-i} \circ F \circ M_i)^{-1}(x) = (M_{-i} \circ G \circ M_i)(x).$$

However, since $\mathrm{sign}((\psi^{-1}|_{(-1,1)})^{(2n+1)}(0)) = (-1)^n$ for all $n \in \mathbb{N}_0$ (by assumption), Remark 5.4 implies that

$$(M_{-i} \circ G \circ M_i)(x) = (\psi^{-1}|_{(-1,1)})_{\mathrm{abs}}(x).$$

Hence, $\psi^{\mathrm{hyp}} = ((\psi^{-1}|_{(-1,1)})_{\mathrm{abs}})^{-1} = M_{-i} \circ F \circ M_i$ on $(-1, 1)$. Since ψ^{hyp} is continuous on $[-1, 1]$ and the holomorphic function $F|_\mathbb{D}$ in particular is continuous on $(-1, 1) \subseteq \mathbb{D} = M_i(\mathbb{D})$, a standard limit argument implies that $\psi^{\mathrm{hyp}}(1) = (M_{-i} \circ F \circ M_i)(1)$ and $\psi^{\mathrm{hyp}}(-1) = (M_{-i} \circ F \circ M_i)(-1)$. □

Next, consider for example, the function $a := \frac{1}{\sqrt{2}}(1 + H_2) : \mathbb{R} \longrightarrow \mathbb{R}$. Then $a \in S_{L^2(\gamma_1)}$ and $[-1, 1] \ni \rho \mapsto h_{a,a}(\rho) = \frac{1}{2}(1+\rho^2)$ defines an *even* CCP function, such that $h_{a,a}(0) = \frac{1}{2} > 0$. With this example in mind, the assumption in the second part of the following result is not empty.

Proposition 6.5 *Let $k \in \mathbb{N}$, $\rho \in [-1, 1]$ and $a, b, f, g \in L^2(\mathbb{R}^k, \gamma_k)$. Then $a \otimes f \in L^2(\mathbb{R}^{2k}, \gamma_{2k})$ and $b \otimes g \in L^2(\mathbb{R}^{2k}, \gamma_{2k})$, and*

$$h_{a \otimes f, b \otimes g} = h_{a,b} \cdot h_{f,g}.$$

In particular, the product $h^{ev} h^{odd}$ of a continuous, even CCP function and a continuous, odd CCP function is a continuous, odd, strictly increasing homeomorphic CCP function. If $(h^{odd})'(0) > 0$ and $h^{ev}(0) > 0$, then $(h^{ev} h^{odd})^{-1}|_{(-1,1)}$ is real analytic. A product of two odd CCP functions is an even CCP function.

Proof Let $\rho \in [-1, 1]$ and $\mathrm{vec}(\mathbf{X}_1, \mathbf{Y}_1, \mathbf{X}_2, \mathbf{Y}_2) = \mathrm{vec}(\mathbf{S}_1, \mathbf{S}_2) \sim N_{4k}(0, \Sigma_{4k}(\rho))$, where $\mathbf{S}_i := \mathrm{vec}(\mathbf{X}_i, \mathbf{Y}_i)$, $i = 1, 2$. Put $\mathbf{X} := \mathrm{vec}(\mathbf{X}_1, \mathbf{X}_2)$ and $\mathbf{Y} := \mathrm{vec}(\mathbf{Y}_1, \mathbf{Y}_2)$. Then Corollary 2.2 implies that $\mathbf{X} \stackrel{d}{=} \mathbf{Y} \sim N_{2k}(0, \Sigma_{2k}(\rho))$ are independent. Let $G \in O(4n)$ be the matrix, introduced in (1.2.6). Then $\mathrm{vec}(x_1, y_1, x_2, y_2) =$

$G \operatorname{vec}(x_1, x_2, y_1, y_2)$ for all $x_1, x_2, y_1, y_2 \in \mathbb{R}^k$. Thus, $\mathbb{P}_{\operatorname{vec}(\mathbf{S}_1, \mathbf{S}_2)} = G_* \mathbb{P}_{\operatorname{vec}(\mathbf{X}, \mathbf{Y})}$ (image measure), and

$$\begin{aligned}
h_{a \otimes f, b \otimes g}(\rho) &= \mathbb{E}\big[a \otimes f(\mathbf{S}_1)\, b \otimes g(\mathbf{S}_2)\big] = \mathbb{E}_{\mathbb{P}_{\operatorname{vec}(\mathbf{S}_1, \mathbf{S}_2)}}\big[(a \otimes f) \otimes (b \otimes g)\big] \\
&= \mathbb{E}_{\mathbb{P}_{\operatorname{vec}(\mathbf{X}, \mathbf{Y})}}\big[((a \otimes f) \otimes (b \otimes g)) \circ G\big] \\
&= \mathbb{E}_{\mathbb{P}_{\operatorname{vec}(\mathbf{X}, \mathbf{Y})}}\big[(a \otimes b) \otimes (f \otimes g)\big] \\
&= \mathbb{E}\big[(a \otimes b)(\mathbf{X})\, (f \otimes g))(\mathbf{Y})\big] = \mathbb{E}\big[a \otimes b(\mathbf{X})\big]\mathbb{E}\big[f \otimes g(\mathbf{Y})\big] \\
&= h_{a,b}(\rho) h_{f,g}(\rho).
\end{aligned}$$

Since $(h^{\mathrm{ev}} h^{\mathrm{odd}})' = (h^{\mathrm{ev}})' h^{\mathrm{odd}} + h^{\mathrm{ev}}(h^{\mathrm{odd}})'$, $(h^{\mathrm{odd}})'(0) > 0$ and $h^{\mathrm{ev}}(0) > 0$ (by assumption), it follows that $(h^{\mathrm{ev}} h^{\mathrm{odd}})'(0) > 0$. The claim now follows from Theorem 6.3. □

If we apply Theorem 6.2 to a pair of k-dimensional Hermite polynomials and recall that $\{H_\alpha : \alpha \in \mathbb{N}_0^k\}$ actually is an orthonormal basis in $L^2(\gamma_k)$ (cf. (6.1.1)), we immediately reobtain another remarkable property of Hermite polynomials (cf. [110, Lemma 1.1.1] and [111, Proposition 11.33]).

Corollary 6.5 *Let $k \in \mathbb{N}$, $\rho \in [-1, 1]$, $\alpha = (\alpha_1, \ldots, \alpha_k) \in \mathbb{N}_0^k$ and $\beta = (\beta_1, \ldots, \beta_k)^\top \in \mathbb{N}_0^k$. If $\operatorname{vec}(\mathbf{X}, \mathbf{Y}) \sim N_{2k}(0, \Sigma_{2k}(\rho))$, then*

$$h_{H_\alpha, H_\beta}(\rho) = \mathbb{E}[H_\alpha(\mathbf{X}) H_\beta(\mathbf{Y})] = \delta_{\alpha, \beta} \rho^{|\alpha|} = \prod_{i=1}^{k} \delta_{\alpha_i, \beta_i}\, \rho^{\alpha_i}.$$

Remark 6.1 (Noise Stability) Fix $k \in \mathbb{N}$. Let $f \in L^2(\gamma_k)$ and A, B Borel sets in $\mathbb{R}^k$. Let $\rho \in [-1, 1] \setminus \{0\}$ and $\operatorname{vec}(\mathbf{X}, \mathbf{Y}) \sim N_{2k}(0, \Sigma_{2k}(\rho))$. Then $\operatorname{sign}(\rho)\mathbf{X} \sim N_k(0, I_k)$. Consequently, (6.2.11) implies that

$$\begin{aligned}
T_{\vartheta(|\rho|)} f_\rho(y) &= \mathbb{E}[f(\rho\, y + \sqrt{1-\rho^2}\, \mathbf{X})] \\
&= \int_{\mathbb{R}^k} f(\rho\, y + \sqrt{1-\rho^2}\, x) \gamma_k(\mathrm{d}x) = U_\rho f(y) \quad (6.2.29)
\end{aligned}$$

for all $y \in \mathbb{R}^k$, where U_ρ is the Gaussian noise operator (cf. [111, Definition 11.12]). U_ρ also is well-defined for $\rho = 0$, with constant value $U_0 f = \mathbb{E}[f(\mathbf{X})]$. Observe that (6.2.29) implies that $U_\rho = T_{\vartheta(|\rho|)} M_\rho$, where the isometry $M_\rho = M_\rho^{-1} \in \mathfrak{L}(L^2(\gamma_k), L^2(\gamma_k))$ is given by $M_\rho f := f_\rho$. A special case of (6.2.14) is given by

$$\mathbb{P}(\mathbf{X} \in A, \mathbf{Y} \in B) = \mathbb{E}[\mathbb{1}_A(\mathbf{X}) \mathbb{1}_B(\mathbf{Y})] = h_{\mathbb{1}_A, \mathbb{1}_B}(\rho) = \langle \mathbb{1}_A, U_\rho \mathbb{1}_B \rangle_{\gamma_k}.$$

Therefore, Theorem 6.2 (under inclusion of Proposition 2.5) encompasses the key concept of Gaussian noise stability, most commonly introduced as **Stab**$_\rho[f] :=$

$\langle f, U_\rho f \rangle_{\gamma_k}$ ($\rho \in [-1, 1]$, $f \in L^2(\gamma_k)$). Gaussian noise stability also comprises deep connections to geometry of minimal surfaces, hypercontractivity, isoperimetric inequalities, communication complexity and Gaussian copulas. A very comprehensive introductory processing of these topics can be found in [111, Chapter 11] and the references therein, including the seminal results of E. Mossel and J. Neeman.

In fact, it can be verified that T_t (and hence $U_\rho = T_{\vartheta(|\rho|)} M_\rho$) is even a *nuclear* operator, implying that each T_t (and each U_ρ) in particular is a *compact* Hilbert-Schmidt operator! More precisely, we have

Proposition 6.6 *Let $k \in \mathbb{N}$, $t \geq 0$ and $\rho \in (-1, 1) \setminus \{0\}$. Then both, $T_t \in \mathfrak{L}(L^2(\gamma_k), L^2(\gamma_k))$, and $U_\rho \in \mathfrak{L}(L^2(\gamma_k), L^2(\gamma_k))$ are Hilbert-Schmidt operators, satisfying*

$$\|T_t\|_{\mathfrak{S}_2} = \frac{1}{(1 - e^{-2t})^{k/2}} \text{ and } \|U_\rho\|_{\mathfrak{S}_2} = \frac{1}{(1 - \rho^2)^{k/2}}.$$

T_t *as well as* U_ρ *are even nuclear, and*

(i) $\|T_t\|_{\mathfrak{N}} \leq \frac{1}{(1-e^{-t})^k}$.
(ii) $\|U_\rho\|_{\mathfrak{N}} \leq \frac{1}{(1-|\rho|)^k}$.

Proof Fix $t \geq 0$. Put $K := L^2(\gamma_k)$ and $\rho(t) := e^{-t}$. Then $t = \vartheta(\rho(t))$. Since $\|M_\rho\| = 1 = \|M_\rho^{-1}\|$ and $U_\rho = T_{\vartheta(|\rho|)} M_\rho$ (respectively, $M_\rho^{-1} U_\rho = T_{\vartheta(|\rho|)}$), the ideal property of the Banach ideal $(\mathfrak{S}_2, \|\cdot\|_{\mathfrak{S}_2})$ implies that we only have to verify the claim for T_t. For this purpose, we use the orthonormal basis $\{H_\alpha : \alpha \in \mathbb{N}_0^k\}$ of k-dimensional Hermite polynomials in $L^2(\gamma_k)$ (cf. (6.1.1)). The semigroup property implies that $T_t = T_{t/2}^2 = T_{t/2}^* T_{t/2}$. Corollary 6.5, together with (6.2.15) therefore leads to

$$\sum_{\alpha \in \mathbb{N}_0^k} \|T_{t/2} H_\alpha\|_K^2 = \sum_{\alpha \in \mathbb{N}_0^k} \langle T_t H_\alpha, H_\alpha \rangle_K = \sum_{\alpha \in \mathbb{N}_0^k} h_{H_\alpha, H_\alpha}(\rho(t)) = \sum_{\alpha \in \mathbb{N}_0^k} \rho(t)^{|\alpha|}$$

$$= \sum_{\alpha \in \mathbb{N}_0^k} \prod_{i=1}^k \rho(t)^{\alpha_i} = \prod_{i=1}^k \sum_{\alpha_i=0}^\infty \rho(t)^{\alpha_i} = \frac{1}{(1 - \rho(t))^k}.$$

Hence, $\left(\|T_{t/2} H_\alpha\|_K\right)_{\alpha \in \mathbb{N}_0^k} \in l^2(\mathbb{N}_0^k)$. So, we may apply [80, Proposition 20.2.7] (which holds in particular for the index set $I := \mathbb{N}_0^k$), and it follows that for any $r \geq 0$, T_r is a Hilbert-Schmidt operator with Hilbert-Schmidt norm

$$\|T_r\|_{\mathfrak{S}_2} = \Big(\sum_{\alpha \in \mathbb{N}_0^k} \|T_r H_\alpha\|_K^2 \Big)^{1/2} = \frac{1}{(1 - \rho(2r))^{k/2}}.$$

Consequently, $T_t = T_{t/2}^2$ is the composition of two Hilbert-Schmidt operators, which allows us to apply [80, Proposition 20.2.8]. It follows that $T_t \in \mathfrak{N}(K, K)$ and

$$\|T_t\|_{\mathfrak{N}} \leq \|T_{t/2}\|_{\mathfrak{S}_2}^2 = \frac{1}{(1-\rho(t))^k}.$$

□

6.3 Examples of Real CCP Functions, Gaussian Copulas and an Extension of Stein's Lemma

As was to be expected, Theorems 6.2 and 6.3 give us first non-trivial examples, such as

$$h_{\mathrm{sign,sign}} : [-1, 1] \longrightarrow [-1, 1], x \mapsto \frac{2}{\pi} \arcsin(x) = \frac{2}{\pi} x \,{}_2F_1\left(\frac{1}{2}, \frac{1}{2}, \frac{3}{2}; x^2\right)$$

in the one-dimensional real case (implying the Grothendieck equality) and

$$h_{f_2, f_2} : [-1, 1] \longrightarrow [-1, 1], \tau \mapsto \frac{\pi}{4} \tau \,{}_2F_1\left(\frac{1}{2}, \frac{1}{2}, 2; \tau^2\right)$$

in the one-dimensional complex case (implying the Haagerup equality), where

$$\mathbb{R}^2 \ni x \mapsto f_2(x) := \begin{cases} \sqrt{2}\,\frac{x_1}{\|x\|_2} & \text{if } x \neq 0 \\ 0 & \text{if } x = 0 \end{cases}.$$

(cf. [60, Lemma 3.2. and Proof of Theorem 3.1] and Example 7.1). If we namely apply Theorem 4.1 to $m = 1$ and arbitrary $k \in \mathbb{N}$, we will recognise that Theorems 6.2 and 6.3 lead us to CCP functions $h_{f,f}$, where $f \in S_{L^2(\gamma_k)}$ is bounded a.e.. These examples also include [24, Lemma 2.1] as a special case. More precisely, we have:

Proposition 6.7 *Let $k \in \mathbb{N}$, $\rho \in [-1, 1]$ and $\mathrm{vec}(\mathbf{X}, \mathbf{Y}) \equiv (X_1, \ldots, X_k, Y_1, \ldots, Y_k)^\top \sim N_{2k}(0, \Sigma_{2k}(\rho))$. Consider the odd function*

$$\mathbb{R}^k \ni x \mapsto f_k(x) := \begin{cases} \sqrt{k}\,\frac{x_1}{\|x\|_{\mathbb{R}_2^k}} & \text{if } x \neq 0 \\ 0 & \text{if } x = 0 \end{cases}.$$

Then $f_k \in S_{L^2(\gamma_k)} \cap L^\infty(\gamma_k)$ and $\|f_k\|_\infty = \sqrt{k}$. The function $h_{f_k, f_k} : [-1, 1] \longrightarrow [-1, 1]$ is an odd strictly increasing homeomorphism which is CCP and satisfies

6.3 Examples of Real CCP Functions, Gaussian Copulas and an Extension of... 141

$h_{f_k,f_k}|_{(-1,1)} \in W_+^\omega((-1,1))$. Let c_k be defined as in Lemma 4.2. Then $0 < h'_{f_k,f_k}(0) = c_k^2 < 1$. $h_{f_k,f_k}^{-1}|_{(-1,1)}$ is real analytic, and

$$h_{f_k,f_k}(\rho) = k\,\mathbb{E}\Big[\frac{X_1 Y_1}{\|\mathbf{X}\|_{\mathbb{R}_2^k}\|\mathbf{Y}\|_{\mathbb{R}_2^k}}\Big] = \mathbb{E}\Big[\Big\langle\frac{\mathbf{X}}{\|\mathbf{X}\|_{\mathbb{R}_2^k}},\frac{\mathbf{Y}}{\|\mathbf{Y}\|_{\mathbb{R}_2^k}}\Big\rangle_{\mathbb{R}_2^k}\Big]$$

$$= c_k^2\,\rho\,{}_2F_1\Big(\frac{1}{2},\frac{1}{2},\frac{k+2}{2};\rho^2\Big) = c_k^2\,k!!\sum_{n=0}^{\infty}\frac{((2n-1)!!)^2}{(2n)!!\,(2n+k)!!}\rho^{2n+1}$$

(6.3.1)

In particular, the function $[-1,1] \ni \rho \mapsto h_{f_k,f_k}(\rho) - c_k^2\rho$ *is CCP as well. If* $|\rho| < 1$, *then*

$$h_{f_k,f_k}(\rho) = \sqrt{\frac{2k}{\pi}}\,c_k\,\rho\int_0^1 \frac{(\sqrt{1-t^2})^{k-1}}{\sqrt{1-\rho^2 t^2}}\,dt. \tag{6.3.2}$$

Moreover,

$$\mathbb{E}\Big[\frac{X_i Y_i}{\|\mathbf{X}\|_{\mathbb{R}_2^k}\|\mathbf{Y}\|_{\mathbb{R}_2^k}}\Big] = \frac{c_k^2}{k}\,\rho\,{}_2F_1\Big(\frac{1}{2},\frac{1}{2},\frac{k+2}{2};\rho^2\Big) \text{ and } \mathbb{E}\Big[\frac{X_i^2}{\|\mathbf{X}\|_{\mathbb{R}_2^k}^2}\Big] = \frac{1}{k}$$

for all $i \in [k]$.

Proof If $k = 1$, (6.3.1) precisely coincides with the Grothendieck equality (Corollary 4.1). Thus, in order to prove (6.3.1), we may assume that $k \geq 2$. We just have to prove the second equality only (due to (4.3.1) and (4.2.3)). To this end, fix $i \in [k] \setminus \{1\}$. Consider the partitioned random vector $\mathrm{vec}(\mathbf{X}', \mathbf{Y}')$, defined as $\mathbf{X}' := P_i^1 \mathbf{X}$ and $\mathbf{Y}' := P_i^1 \mathbf{Y}$, where $P_i^1 := e_1 e_i^\top + e_i e_1^\top + \frac{1}{2}\sum_{j\in[k]\setminus\{1,i\}} e_j e_j^\top = (P_i^1)^\top = (P_i^1)^{-1}$. So, if we swap the 1'st row and the i'th row of the equality matrix I_k, we obtain P_i^1, implying that $X'_1 = X_i$, $X'_i = X_1$ and $X'_j = X_j$ for all $j \in [k] \setminus \{1,i\}$. Due to the construction of the random vector $\mathrm{vec}(\mathbf{X}', \mathbf{Y}')$ and the structure of the correlation matrix $\Sigma_{2k}(\rho)$, it follows that

$$\begin{pmatrix}\mathbf{X}'\\\mathbf{Y}'\end{pmatrix} = \begin{pmatrix}P_i^1 & 0\\0 & P_i^1\end{pmatrix}\begin{pmatrix}\mathbf{X}\\\mathbf{Y}\end{pmatrix}$$

and

$$\begin{pmatrix}P_i^1 & 0\\0 & P_i^1\end{pmatrix}\begin{pmatrix}I_k & \rho I_k\\\rho I_k & I_k\end{pmatrix}\begin{pmatrix}P_i^1 & 0\\0 & P_i^1\end{pmatrix} = \begin{pmatrix}I_k & \rho I_k\\\rho I_k & I_k\end{pmatrix}.$$

Hence, $\text{vec}(\mathbf{X}', \mathbf{Y}') \stackrel{d}{=} \text{vec}(\mathbf{X}, \mathbf{Y}) \sim N_{2k}(0, \Sigma_{2k}(\rho))$. Consequently, we obtain

$$\mathbb{E}\Big[\frac{X_1 Y_1}{\|\mathbf{X}\|_{\mathbb{R}_2^k}\|\mathbf{Y}\|_{\mathbb{R}_2^k}}\Big] = \mathbb{E}\Big[\frac{X_i' Y_i'}{\|\mathbf{X}'\|_{\mathbb{R}_2^k}\|\mathbf{Y}'\|_{\mathbb{R}_2^k}}\Big] = \mathbb{E}\Big[\frac{X_i Y_i}{\|\mathbf{X}\|_{\mathbb{R}_2^k}\|\mathbf{Y}\|_{\mathbb{R}_2^k}}\Big].$$

Since $h_{f_k,f_k}(\rho) = c_k^2 \rho + c_k^2 k !! \sum_{n=1}^{\infty} \frac{((2n-1)!!)^2}{(2n)!!(2n+k)!!} \rho^{2n+1}$ for all $\rho \in [-1,1]$, it follows that $h'_{f_k,f_k}(0) = c_k^2 > 0$, implying that $h^{-1}_{f_k,f_k}\big|_{(-1,1)}$ is real analytic (due to Theorem 6.3). Euler's integral representation of the function $(-1, 1) \ni \rho \mapsto {}_2F_1(\frac{1}{2}, \frac{1}{2}, \frac{k+2}{2}; \rho^2)$ (cf. e.g. [6, Theorem 2.2.1]). Theorem 6.3 now completes the proof. □

Remark 6.2 Let $k \in \mathbb{N}$ and c_k be defined as in Lemma 4.2. Then [23, Theorem 1] in fact implies that

$$\frac{\pi}{2} c_d^2 \le K_G^{\mathbb{R}}(d) \text{ for all } d \in \mathbb{N}_3.$$

Remark 6.3 (Krivine Rounding Scheme Reconsidered) Fix $k \in \mathbb{N}$. Due to Proposition 6.7, the following set of real-valued functions is non-empty:

$$\mathcal{G}_k := \{f : f \in S_{L^2(\gamma_k)}, f \text{ is odd}, h'_{f,f}(0) > 0\}. \tag{6.3.3}$$

Let $f \in \mathcal{G}_k$. If in addition, $|f| = 1$ and $h^{-1}_{f,f}\big|_{(-1,1)} \in W_+^{\omega}((-1,1))$ Lemma 6.1 implies that $\{f \circ \sqrt{2}, f \circ \sqrt{2}\}$ is a Krivine rounding scheme, introduced in [22, Definition 2.1] (since $h_{f,f}$ coincides with the function $H_{f \circ \sqrt{2}, f \circ \sqrt{2}}$, introduced in [22, Definition 2.1] and $c(f \circ \sqrt{2}, f \circ \sqrt{2}) = h_{f,f}^{\text{hyp}}(1) \in (0, 1))$. The latter fact should be compared with Theorem 6.9 !

Now it is no longer surprising that Proposition 6.7 encompasses the Grothendieck equality and the Haagerup equality as particular cases.

Example 6.1 ($k = 1$ (Grothendieck)) Fix $\rho \in [-1, 1]$. It is well-known that $\arcsin(\rho) = \rho \, {}_2F_1(\frac{1}{2}, \frac{1}{2}, \frac{3}{2}; \rho^2)$, whence $h_{f_1, f_1}(\rho) = \frac{2}{\pi} \arcsin(\rho)$.

Example 6.2 ($k = 2$ (Haagerup)) Let $n \in \mathbb{N}$, $u, v \in \mathbb{C}^n$ such that $\|u\| = 1$ and $\|v\| = 1$. Let $Z \sim \mathbb{C}N_n(0, I_n)$. The Haagerup equality (4.1.2), together with (4.1.3) implies that

$$\mathbb{E}[\text{sign}(u^*Z)\text{sign}(\overline{v^*Z})] = \text{sign}(u^*v) \, h_{f_2, f_2}(|u^*v|).$$

In the following, we shed some light on the underlying structure of the function h_{ψ_a, ψ_b}, where $\psi_p := 1 - 2\mathbb{1}_{J_p} \in S_{L^2(\gamma_k)}$ and $J_p := \prod_{i=1}^{k}(-\infty, p_i]$ for all $p \equiv (p_1, \ldots, p_k)^\top \in \mathbb{R}^k$ (introduced in Corollary 6.1). Recall here that for any $n \in \mathbb{N}$, $\Sigma \in \mathbb{M}_n(\mathbb{R})^+$ and $\mathbf{X} \equiv (X_1, \ldots, X_n)^\top \sim N_n(0, \Sigma)$, $\Phi_{0,\Sigma} : \mathbb{R}^n \longrightarrow [0, 1]$ denotes

6.3 Examples of Real CCP Functions, Gaussian Copulas and an Extension of... 143

the n-variate distribution function of $\mathbf{X}$, i.e.,

$$\Phi_{0,\Sigma}(x) := F_{\mathbf{X}}(x) = \mathbb{P}\big(\bigcap_{i=1}^{n}\{X_i \le x_i\}\big) \text{ for all } x \equiv (x_1,\ldots,x_n)^\top \in \mathbb{R}^n.$$

Proposition 6.8 *Let* $k \in \mathbb{N}$, $\rho \in [-1,1]$ *and* $\mathbf{X} \equiv (X_1,\ldots,X_{2k})^\top \sim N_{2k}(0, \Sigma_{2k}(\rho))$. *Let* $a, b \in \mathbb{R}^k$ *and* $x \equiv (x_1,\ldots,x_{2k})^\top = \mathrm{vec}(a,b)$. *Then*

$$h_{\psi_a,\psi_b}(\rho) = 4\Phi_{0,\Sigma_{2k}(\rho)}(x) + 1 - 2\Phi_{0,I_k}(a) - 2\Phi_{0,I_k}(b). \tag{6.3.4}$$

Furthermore,

$$\Phi_{0,\Sigma_{2k}(\rho)}(x) = \Phi_{0,I_k}(a)\Phi_{0,I_k}(b) + \sum_{\nu=1}^{\infty} d_\nu(x;k)\rho^\nu,$$

where

$$d_\nu(x;k) := \sum_{m \in C(\nu,k)} \Big(\prod_{\substack{i=1\\m_i=0}}^{2k} \Phi(x_i) \prod_{\substack{i=1\\m_i\ne 0}}^{2k} \frac{1}{\sqrt{m_i}}\varphi(x_i)H_{m_i-1}(x_i)\Big).$$

Proof Fix $\mathbf{X} \equiv (X_1,\ldots,X_{2k})^\top \sim N_{2k}(0, \Sigma_{2k}(\rho))$ and $x = \mathrm{vec}(a,b)$. Since $\psi_a(x_1)\psi_b(x_2) = (1 - 2\mathbb{1}_{J_a}(x_1))(1 - 2\mathbb{1}_{J_b}(x_2)) = 1 - 2\mathbb{1}_{J_a}(x_1) - 2\mathbb{1}_{J_b}(x_2) + 4\mathbb{1}_{J_a}(x_1)\mathbb{1}_{J_b}(x_2)$ for all $x_1, x_2 \in \mathbb{R}^k$ and $\mathbf{X} = \mathrm{vec}(\mathbf{X}_1, \mathbf{X}_2)$, where $\mathbf{X}_1 := (X_1,\ldots,X_k)^\top \sim N_k(0, I_k)$ and $\mathbf{X}_2 := (X_{k+1},\ldots,X_{2k})^\top \sim N_k(0, I_k)$, an application of Theorem 6.2 to the function h_{ψ_a,ψ_b} implies that

$$F_{\mathbf{X}}(x) = \mathbb{P}\big(\bigcap_{i=1}^{2k}\{X_i \le x_i\}\big) = \mathbb{E}[\mathbb{1}_{J_a}(\mathbf{X}_1)\mathbb{1}_{J_b}(\mathbf{X}_2)]$$

$$= \frac{1}{4}(h_{\psi_a,\psi_b}(\rho) - 1 + 2\Phi_{0,I_k}(a) + 2\Phi_{0,I_k}(b))$$

$$= \frac{1}{4}\Big((1 - 2\Phi_{0,I_k}(a))(1 - 2\Phi_{0,I_k}(b))$$

$$+ \sum_{\nu=1}^{\infty} p_\nu(\psi_a,\psi_b)\rho^\nu - 1 + 2\Phi_{0,I_k}(a) + 2\Phi_{0,I_k}(b)\Big)$$

$$= \Phi_{0,I_k}(a)\Phi_{0,I_k}(b) + \frac{1}{4}\sum_{\nu=1}^{\infty}\Big(\sum_{m \in C(\nu,k)}\langle \psi_a, H_m\rangle_{\gamma_k}\langle \psi_b, H_m\rangle_{\gamma_k}\Big)\rho^\nu.$$

The third equality is equivalent to (6.3.4). Corollary 6.1 clearly finishes the proof. □

Proposition 6.8 immediately gives us another significant example—which contains the Grothendieck equality as a special case again. Due to the famous Theorem of Sklar, it emerges from a lurking multivariate Gaussian copula; i.e., from a certain finite-dimensional distribution function with uniformly distributed marginals (cf., e.g., [116] and the references therein).

Example 6.3 (A Lurking $\Sigma_{2k}(\rho)$-Gaussian Copula) Let $k \in \mathbb{N}$, $m = (m_1, \ldots, m_k)^\top \in \mathbb{N}_0^k$, $\alpha = (\alpha_1, \ldots, \alpha_k)^\top \in (0, 1)^k$ and $\beta = (\beta_1, \ldots, \beta_k)^\top \in (0, 1)^k$. Let $a = (a_1, \ldots, a_k)^\top \in \mathbb{R}^k$ and $b = (b_1, \ldots, b_k)^\top \in \mathbb{R}^k$, where $a_i := \Phi^{-1}(\alpha_i)$ and $b_i := \Phi^{-1}(\beta_i)$ for all $i \in [k]$. Then

$$h_{\psi_a, \psi_b}(\rho) = 1 - 2\left(\prod_{i=1}^k \alpha_i + \prod_{i=1}^k \beta_i\right) + 4 c_{\Sigma_{2k}(\rho)}(\alpha, \beta) \text{ for all } \rho \in [-1, 1],$$

where

$$(0, 1)^k \ni (u_1, \ldots, u_k) \mapsto c_{\Sigma_{2k}(\rho)}(u_1, \ldots, u_k)$$
$$:= \Phi_{0, \Sigma_{2k}(\rho)}(\Phi^{-1}(u_1), \ldots, \Phi^{-1}(u_k))$$

denotes the $2k$-dimensional Gaussian copula with respect to the correlation matrix $\Sigma_{2k}(\rho)$ (cf., e.g., [116]). Consequently, if $(\alpha, \beta) = (\frac{1}{2}, \frac{1}{2})$, it follows that $a = 0$, $b = 0$ and

$$h_{\psi_0, \psi_0}(\rho) = 1 - \tfrac{1}{2^{k-2}} + 4\Phi_{0, \Sigma_{2k}(\rho)}(0) = 1 - \tfrac{1}{2^{k-2}} + 4\mathbb{P}\left(\bigcap_{i=1}^{2k} \{X_i \leq 0\}\right) \quad (6.3.5)$$

for all $\rho \in [-1, 1]$ and $\mathbf{X} = (X_1, \ldots, X_{2k})^\top \sim N_{2k}(0, \Sigma_{2k}(\rho))$ (since $\Phi(0) = \frac{1}{2}$). In fact, if $(\alpha, \beta) = (\frac{1}{2}, \frac{1}{2})$ and $k = 1$, then $\psi_0 = \text{sign} - \mathbb{1}_{\{0\}}$, implying that (6.3.5) even reduces to the Grothendieck equality. In order to recognise this, recall first that $(2l)! = 2^l \, l! \, (2l - 1)!!$ for any $l \in \mathbb{N}_0$, whence

$$((2l - 1)!!)^2 = \binom{2l}{l} \frac{(2l)!}{4^l}. \quad (6.3.6)$$

Consequently, the power series representation of the function h_{ψ_0, ψ_0}, together with Corollary 6.1 and (6.1.4) implies that

$$4\mathbb{P}(X_1 \leq 0, X_2 \leq 0) - 1 = h_{\text{sign,sign}}(\rho) = h_{\psi_0, \psi_0}(\rho)$$

$$= \sum_{n=0}^\infty (\langle \psi_0, H_n \rangle_{\gamma_1})^2 \rho^n = \frac{2}{\pi} \sum_{l=0}^\infty \frac{1}{2l+1} H_{2l}^2(0) \, \rho^{2l+1}$$

$$\stackrel{(6.1.4)}{=} \frac{2}{\pi} \sum_{l=0}^\infty ((2l-1)!!)^2 \frac{\rho^{2l+1}}{(2l+1)!} \stackrel{(6.3.6)}{=} \frac{2}{\pi} \arcsin(\rho) \quad (6.3.7)$$

6.3 Examples of Real CCP Functions, Gaussian Copulas and an Extension of... 145

for all $\rho \in [-1, 1]$ and $(X_1, X_2)^\top \sim N_2(0, I_2)$ (cf. also [64, 103]). Again, we recognise that the Hermite expansion of the function sign—in $L^2(\gamma_1)$—is given by

$$\text{sign} = \psi_0 = \sqrt{\frac{2}{\pi}} \sum_{l=0}^{\infty} (-1)^l \frac{(2l-1)!!}{\sqrt{(2l+1)!}} H_{2l+1} \qquad (6.3.8)$$

(cf. (6.1.11)).

A further interesting one-dimensional example ($k = 1$) is given by the function $h_{\Phi,\Phi}$, where $\Phi = F_X$ is the (continuous) distribution function of a standard normally distributed random variable $X \sim N_1(0, 1)$. To this end, recall that the (continuous) random variable $U := \Phi(X) \sim U(0, 1)$ is uniformly distributed on $[0, 1]$, implying that $\Phi \in L^2(\gamma_1)$, with $\langle \Phi, 1 \rangle_{\gamma_1} = \mathbb{E}[U] = \frac{1}{2}$ and $\|\Phi\|_{\gamma_1}^2 = \mathbb{E}[U^2] = Var(U) + \mathbb{E}^2[U] = \frac{1}{12} + \frac{1}{4} = \frac{1}{3}$ (cf. [116, Remark 2.17.]). In particular, $\mathbb{E}[\kappa(X)] = 0$, where $\kappa := 2\sqrt{3}\Phi - \sqrt{3} = \sqrt{3}(2\Phi - 1)$. Now, we are ready to prove

Proposition 6.9 *Let* $X \sim N_1(0, 1)$ *and* $\kappa := \sqrt{3}(2\Phi - 1)$. *Then the following properties hold:*

(i)

$$\frac{d^n}{dt^n} \mathbb{E}[\Phi(X + t)] = 2^{-n/2} \varphi^{(n-1)}\left(\frac{t}{\sqrt{2}}\right) \text{ for all } n \in \mathbb{N} \text{ and } t \in \mathbb{R}.$$

(ii)

$$\mathbb{E}[\Phi(X + t)] = \Phi\left(\frac{t}{\sqrt{2}}\right) \text{ for all } n \in \mathbb{N} \text{ and } t \in \mathbb{R}.$$

(iii) $\mathbb{E}[\Phi(X) H_{2n+1}(X)] = (-1)^n \sqrt{\frac{1}{2\pi}} \sqrt{\frac{1}{4^n} \frac{1}{2n+1} \binom{2n}{n}} \sqrt{(\frac{1}{2})^{2n+1}}$ for all $n \in \mathbb{N}_0$.
In particular, $\mathbb{E}[\kappa(X) H_{2n+1}(X)] = (-1)^n \sqrt{\frac{6}{\pi}} \sqrt{\frac{1}{4^n} \frac{1}{2n+1} \binom{2n}{n}} \sqrt{(\frac{1}{2})^{2n+1}}$ for all $n \in \mathbb{N}_0$.

(iv) $h_{\Phi,\Phi}(\rho) = \frac{1}{4} + \frac{1}{2\pi} \arcsin(\frac{\rho}{2})$ and $h_{\kappa,\kappa}(\rho) = \frac{6}{\pi} \arcsin(\frac{\rho}{2})$ for all $\rho \in [-1, 1]$.

(v) $h_{\sqrt{3}\Phi,\sqrt{3}\Phi} = 3h_{\Phi,\Phi} = \frac{3}{4}(1 + \frac{2}{\pi}\arcsin(\frac{1}{2}\cdot))$ is a strictly increasing homeomorphic CCP function which maps $[-1, 1]$ onto $[\frac{1}{2}, 1]$ and is neither odd nor even. $h_{\kappa,\kappa}$ is an odd, strictly increasing homeomorphic CCP function. $h^{-1}_{\sqrt{3}\Phi,\sqrt{3}\Phi}(t) = -2\cos(\frac{2\pi}{3}t)$ for all $t \in [\frac{1}{2}, 1]$, $h^{-1}_{\kappa,\kappa}(s) = 2\sin(\frac{\pi}{6}s)$ and $\left(h^{-1}_{\kappa,\kappa}\right)_{abs}(s) = 2\sinh(\frac{\pi}{6}s)$ for all $s \in [-1, 1]$. In particular, $\left(h^{-1}_{\kappa,\kappa}\right)_{abs}(1) > 1$. $h^{hyp}_{\kappa,\kappa}(\rho) = \frac{6}{\pi} \sinh^{-1}(\frac{\rho}{2})$ for all $\rho \in [-1, 1]$.

(vi) κ is odd, $\|\kappa\|_{\gamma_1} = 1$ and $h'_{\kappa,\kappa}(0) = \frac{3}{\pi} > 0$. Moreover, $\kappa \in L^\infty(\gamma_1)$, and $\|\kappa\|_\infty = \sqrt{3}$.

Proof

(i) Fix $t \in \mathbb{R}$. We prove this result by induction on n. Obviously, we have to verify the base case only ($n = 1$). To this end, we apply a few Fourier transform techniques including the well-known fact that $\widehat{\varphi} = \varphi$ is its own Fourier transform. So, let $n = 1$. Then

$$\frac{d}{dt}\mathbb{E}[\Phi(X+t)] = \int_{\mathbb{R}}(\tau_{-t}\varphi)(x)\widehat{\varphi}(x)dx = \int_{\mathbb{R}}\widehat{(\tau_{-t}\varphi)}(x)\varphi(x)dx,$$

where $\tau_{-t}\varphi(x) := \varphi(x-(-t))$ denotes the translation operator (here, applied to φ). Since $\widehat{(\tau_{-t}\varphi)}(x) = \widehat{\varphi}(x)\exp(itx) = \varphi(x)\exp(itx)$ for all $x \in \mathbb{R}$, it follows that

$$\frac{d}{dt}\mathbb{E}[\Phi(X+t)] = \frac{1}{2\pi}\int_{\mathbb{R}}\exp(-x^2)\exp(itx)dx$$

$$= \frac{1}{\sqrt{2}}\frac{1}{\sqrt{2\pi}}\int_{\mathbb{R}}\varphi(y)\exp\left(i\frac{t}{\sqrt{2}}y\right)dy$$

$$= \frac{1}{\sqrt{2}}\widehat{\varphi}\left(-\frac{t}{\sqrt{2}}\right) = \frac{1}{\sqrt{2}}\varphi\left(-\frac{t}{\sqrt{2}}\right) = \frac{1}{\sqrt{2}}\varphi^{(0)}\left(\frac{t}{\sqrt{2}}\right),$$

which finishes the verification of the base case.

(ii) We just have to integrate the base case on both sides (from 0 to t, say).

(iii) Since the Taylor series representation of φ obviously is given by

$$\varphi(x) = \frac{1}{\sqrt{2\pi}}\exp\left(-\frac{x^2}{2}\right) = \frac{1}{\sqrt{2\pi}}\sum_{m=0}^{\infty}\frac{(-1)^m}{2^m m!}x^{2m}$$

for all $x \in \mathbb{R}$, it follows that $\varphi^{(2m)}(0) = (2m)!\left(\frac{1}{\sqrt{2\pi}}\frac{(-1)^m}{2^m m!}\right) = \frac{1}{\sqrt{2\pi}}\frac{(-1)^m m!}{2^m}\binom{2m}{m}$ and $\varphi^{(2m+1)}(0) = 0$ for all $m \in \mathbb{N}_0$. (iii) now follows directly from (i) and (the one-dimensional case of) the equality (6.1.13), respectively Proposition 6.1.

(iv) follows directly from (iii).

(v) follows from (iv), Theorem 6.2, Lemma 6.1 and the fact that for example, $h_{\sqrt{3}\Phi,\sqrt{3}\Phi}(-1) = \frac{1}{2} \notin \{-1, 1\}$.

(vi) Since $\Phi(x) + \Phi(-x) = 1$ for all $x \in \mathbb{R}$, it follows that κ is odd. (iv), applied to $\rho = 1$ implies that $\|\kappa\|^2_{\gamma_k} = h_{\kappa,\kappa}(1) = 1$. Moreover, (iv) clearly implies that $h'_{\kappa,\kappa}(0) = \frac{3}{\pi} > 0$. Finally, since $\Phi : \mathbb{R} \longrightarrow [0, 1]$ is a distribution function, it follows that κ is bounded and $|\kappa(x)| \leq \sqrt{3} = \sqrt{3}\lim_{x\to\infty}(2\Phi(x) - 1) = \lim_{x\to\infty}\kappa(x)$ for all $x \in \mathbb{R}$. $\square$

6.3 Examples of Real CCP Functions, Gaussian Copulas and an Extension of... 147

We will soon realise that Theorem 6.2 actually reflects a characterisation of the class of all real continuous CCP functions (see Theorem 6.5).

Within the scope of our analysis of the Grothendieck constants, we need *invertible* CCP functions, implying that we may ignore even CCP functions, defined on $[-1, 1]$ (since these are non-injective). However, thanks to Theorem 5.3, $[-1, 1] \ni \rho \mapsto \rho h(\rho)$ is an odd CCP function for any continuous, even CCP function h. In fact, we will recognise next that in particular any CCP function $h_{f,f}$, which satisfies $h_{f,f}(-1) < 1$ induces canonically an odd CCP function $h_{f^\times, f^\times}$, where $f^\times \in S_H$ is odd, $H := L^2(\mathbb{R}^k, \gamma_k)$. In the complex case it seems that we even have to use odd functions, to avoid calculations with non-positive semidefinite entrywise absolute values of correlation matrices (cf. (7.1.17) and Theorem 7.6, (ii)).

Proposition 6.10 (Odd CCP Transform) *Let $k \in \mathbb{N}$ and $f \in L^2(\mathbb{R}^k, \gamma_k)$. Consider the odd function $\mathbb{R}^k \ni x \mapsto f^{odd}(x) := \frac{f(x) - f(-x)}{2}$ (the odd part of f). Then f is odd if and only if $f^{odd} = f$. In particular, $(f^{odd})^{odd} = f^{odd}$. f is even if and only if $f^{odd} = 0$. Moreover, $f^{odd} \in L^2(\mathbb{R}^k, \gamma_k)$ and*

(i)
$$\|f^{odd}\|^2_{\gamma_k} = \frac{h_{f,f}(1) - h_{f,f}(-1)}{2} \geq 0.$$

(ii) *f is even λ_k-a.s. if and only if $h_{f,f}(-1) = h_{f,f}(1)$.*
(iii) *f is odd λ_k-a.s. if and only if $h_{f,f}(-1) = -h_{f,f}(1)$.*
(iv) *Assume that $h_{f,f}(1) > h_{f,f}(-1)$. Put*

$$f^\times := \sqrt{\frac{2}{h_{f,f}(1) - h_{f,f}(-1)}} \, f^{odd}. \tag{6.3.9}$$

Then f is not even, $\|f^\times\|_{\gamma_k} = 1$, and $h_{f^\times, f^\times} = \frac{2}{h_{f,f}(1) - h_{f,f}(-1)} h_{f^{odd}, f^{odd}}$ is an odd CCP function.

Proof Due to Theorem 6.2, respectively Theorem 6.3, we just have to verify the non-trivial parts of the claims (i), (ii) and (iii).

(i) Fix $\mathbf{X} \sim N_k(0, I_k)$. Then

$$\|f^{odd}\|^2_{\gamma_k} = \frac{1}{4} \mathbb{E}[(f^\circ)^2(\mathbf{X})] = \frac{1}{4}\left(\mathbb{E}[f^2(\mathbf{X})] - 2\mathbb{E}[f(\mathbf{X})f(-\mathbf{X})] + \mathbb{E}[f^2(-\mathbf{X})]\right).$$

Since $-\mathbf{X} \stackrel{d}{=} \mathbf{X} \sim N_k(0, I_k)$ and $\text{vec}(\mathbf{X}, -\mathbf{X}) \sim N_{2k}(0, \Sigma_{2k}(-1))$, it follows from Theorem 6.2 that

$$\|f^{odd}\|^2_{\gamma_k} = \frac{1}{2}\left(\|f\|^2_{\gamma_k} - \mathbb{E}[f(\mathbf{X})f(-\mathbf{X})]\right) = \frac{1}{2}(h_{f,f}(1) - h_{f,f}(-1)).$$

(ii) follows directly from (i).

(iii) Consider the even function $f^{\text{ev}} := f - f^{\text{odd}} \in L^2(\mathbb{R}^k, \gamma_k)$; i.e., the even part of f. By construction, $f^{\text{ev}}(x) = \frac{f(x)+f(-x)}{2}$ for all $x \in \mathbb{R}^k$. Similarly, as in the proof of (i), we therefore obtain

$$\|f^{\text{ev}}\|_{\gamma_k}^2 = \frac{h_{f,f}(1) + h_{f,f}(-1)}{2},$$

wherefrom claim (iii) obviously follows. □

Example 6.4 Since $\arcsin(\frac{1}{2}) = \frac{\pi}{6}$, Proposition 6.9 implies that $h_{\Phi,\Phi}(1) = \frac{1}{3}$ and $h_{\Phi,\Phi}(-1) = \frac{1}{6}$, whence

$$\Phi^\times = \sqrt{3}(2\Phi - 1) = \kappa.$$

Example 6.5 Let $[-1,1] \ni \rho \mapsto C^{\text{Ga}}(\frac{1}{2}, \frac{1}{2}; \rho)$ denote the bivariate Gaussian copula with Pearson's correlation coefficient ρ as parameter, evaluated at $(\frac{1}{2}, \frac{1}{2})$. Then

$$C^{\text{Ga}}(\frac{1}{2}, \frac{1}{2}; \rho) = \Phi_{\Sigma_2(\rho)}(\Phi^{-1}(\frac{1}{2}), \Phi^{-1}(\frac{1}{2})) = \mathbb{P}(X \leq 0, Y \leq 0) = h_{g,g}(\rho),$$

where $\Phi_{\Sigma_2(\rho)}$ denotes the bivariate distribution function of the random vector $(X,Y)^\top \sim N_2(0, \Sigma_2(\rho))$ and $g := \mathbb{1}_{(-\infty,0]}$. Thus, $g^{\text{odd}} = -\frac{1}{2}\text{sign}$, $h_{g,g}(1) = \mathbb{P}(X \leq 0) = \Phi(0) = \frac{1}{2}$ and $h_{g,g}(-1) = \mathbb{P}(X \leq 0, -X \leq 0) = \mathbb{P}(X = 0) = 0$. Consequently,

$$g^\times = -\text{sign} = 2\mathbb{1}_{(-\infty,0)} - 1 + \mathbb{1}_{\{0\}},$$

and it follows that $h_{g^\times,g^\times} = \frac{2}{\pi}\arcsin$ (on $[-1,1]$).

Regarding an explicit calculation of the correlation coefficients $\mathbb{E}[f(\mathbf{X})g(\mathbf{Y})]$, induced by a pair of non-linearly transformed Gaussian random vectors and given "sufficiently smooth" functions f and g, we implement a few facts from the theory of distributions and test function spaces. A detailed in-depth introduction to these "generalised functions" and test function spaces and their analysis is provided by e.g. [63, 78, 131]. To this end, let $k \in \mathbb{N}$, $f \in L^1_{\text{loc}}(\mathbb{R}^k)$ (i.e., f is locally integrable) and $\psi \in \mathcal{S}_k$ be an arbitrary test function, where $\mathcal{S}_k$ denotes the Schwartz space of rapidly decreasing test functions on $\mathbb{R}^k$. Put

$$\langle \psi, \Lambda_f \rangle := \int_{\mathbb{R}^k} f(x)\psi(x) \mathrm{d}^k x.$$

Observe that the latter symbol $\langle \cdot, \cdot \rangle$ denotes the duality bracket on $\mathcal{S}_k \times \mathcal{S}'_k$ and not an inner product of Hilbert space elements. When there is no ambiguity, we adopt the common habit to identify the tempered distribution $\Lambda_f \in \mathcal{S}'_k$ with $f \in L^1_{\text{loc}}(\mathbb{R}^k)$

6.3 Examples of Real CCP Functions, Gaussian Copulas and an Extension of... 149

itself. More generally, recall that tempered distributions and their derivatives are elements of the dual space $\mathcal{S}'_k$ of $\mathcal{S}_k$, such as the Dirac delta distribution $\delta_0 = \delta$, defined via $\langle \psi, \delta \rangle := \psi(0)$ for any $\psi \in \mathcal{S}_k$. Observe that there is no $g \in L^1_{\text{loc}}(\mathbb{R}^k)$ such that $\delta = \Lambda_g$.

Let us quickly recall how differentiation of tempered distributions is defined. It originates from a reiteration of the integration by parts formula and is also known as "weak differentiation" (cf. e. g. [131, Chapter 6.12]):

$$\langle \psi, D^n u \rangle := (-1)^{|n|} \langle D^n \psi, u \rangle = (-1)^{|n|} \langle \psi, D^{n-1}(Du) \rangle \qquad (6.3.10)$$

for all $\psi \in \mathcal{S}_k, u \in \mathcal{S}'_k$ and $n \in \mathbb{N}_0^k$. Consequently, since $H_n \varphi_k \in \mathcal{S}_k$ for all $n \in \mathbb{N}_0^k$ and $D^n u \in \mathcal{S}'_k$ for all $u \in \mathcal{S}_k$, (6.1.3) in particular implies:

$$\langle H_n \varphi_k, u \rangle = \frac{1}{\sqrt{n!}} \langle \varphi_k, D^n u \rangle \text{ for all } u \in \mathcal{S}'_k \text{ and } n \in \mathbb{N}_0^k.$$

Given an arbitrary compact subset $K \subseteq \mathbb{R}^k$ and $f \in L^2(\gamma_k)$, it follows that

$$\int_K |f| \, d\lambda_k = \sqrt{2\pi} \int_{\mathbb{R}^k} \mathbb{1}_K(x) \exp(\frac{1}{2}\|x\|^2) |f(x)| \, \gamma_k(dx) \leq c_K \|f\|_{\gamma_k},$$

where $c_K := \sqrt{2\pi} \int_K e^{\|x\|^2} \gamma_k(dx) = \sqrt{2\pi} \int_K \exp(\frac{1}{2}\|x\|^2) \lambda_k(dx) < \infty$. Consequently, $L^2(\gamma_k) \subseteq L^1_{\text{loc}}(\mathbb{R}^k)$, implying that for any $\mathbf{X} \sim N_k(0, I_k)$

$$\mathbb{E}[H_n(\mathbf{X}) f(\mathbf{X})] = \langle H_n, f \rangle_{\gamma_k} = \langle H_n \varphi_k, \Lambda_f \rangle$$

$$= \frac{1}{\sqrt{n!}} \langle \varphi_k, D^n \Lambda_f \rangle \text{ for all } f \in L^2(\gamma_k) \text{ and } n \in \mathbb{N}_0^k. \qquad (6.3.11)$$

Example 6.6 (*n*'th Weak Derivative of $\Lambda_{\mathbb{1}_{[0,\infty)}}$) Let $k = 1$ and $n \in \mathbb{N}_0$. A very easy proof by induction on n, based on (6.3.10), together with (6.1.2) firstly reveals that

$$\langle \varphi, D^n \delta \rangle = (-1)^n \varphi^{(n)}(0) = \frac{\sqrt{n!}}{\sqrt{2\pi}} H_n(0).$$

(6.1.4) therefore implies that

$$\langle \varphi, D^{2l} \delta \rangle = (-1)^l \frac{1}{\sqrt{2\pi}} (2l-1)!! \text{ for all } l \in \mathbb{N}_0.$$

Similarly, (6.3.10) implies that $D\Lambda_{\mathbb{1}_{[0,\infty)}} = \delta$. Since $\text{sign} = 2\mathbb{1}_{[0,\infty)} - 1$ (γ_1-almost surely) and $\langle H_{2l+1}, 1\rangle_{\gamma_1} = \langle H_{2l+1}, H_0\rangle_{\gamma_1} = \delta_{2l+1,0} = 0$, it follows that

$$\langle H_{2l+1}, \text{sign}\rangle_{\gamma_1} = 2\langle H_{2l+1}, \mathbb{1}_{[0,\infty)}\rangle_{\gamma_1} \stackrel{(6.3.11)}{=} \frac{2}{\sqrt{(2l+1)!}} \langle \varphi, D^{2l+1}\Lambda_{\mathbb{1}_{[0,\infty)}}\rangle$$

$$= \frac{2}{\sqrt{(2l+1)!}} \langle \varphi, D^{2l}\delta\rangle = (-1)^l \sqrt{\frac{2}{\pi}} \frac{(2l-1)!!}{\sqrt{(2l+1)!}}$$

for all $l \in \mathbb{N}_0$. This outcome—which is solely based on a multiple weak differentiation of $\Lambda_{\mathbb{1}_{[0,\infty)}}$—should be compared now with (the derivation of) (6.3.7)!

Equation (6.3.11) can be strongly simplified if f is smooth (cf. also Proposition 6.1). More precisely, if $N \in \mathbb{N}_0$, $D^n f \in L^2(\gamma_k) \cap C(\mathbb{R}^k)$ for all $n \in \mathbb{N}_0^k$, satisfying $0 \leq |n| \leq N$, and if $\mathbf{X} \sim N_k(0, I_k)$, then $D^n \Lambda_f = \Lambda_{D^n f}$ (cf. [131, Chapter 6.13]), and

$$\mathbb{E}[H_n(\mathbf{X})f(\mathbf{X})] = \langle H_n, f\rangle_{\gamma_k} = \frac{1}{\sqrt{n!}}\mathbb{E}[D^n f(\mathbf{X})] \text{ for all } n \in \mathbb{N}_0^k, \text{ with } |n| \leq N. \tag{6.3.12}$$

In fact, even more can be said. Recall that (by construction of Sobolev spaces) the latter equality also holds without the smoothness assumption, if we just assume that $f \in L^2(\gamma_k) \cap W_{\text{loc}}^{N,1}(\mathbb{R}^k)$, so that in this case $D^n f$ denotes the distributional n'th derivative in $L_{\text{loc}}^1(\mathbb{R}^k)$ instead (cf. e. g. [63, Chapter 3.1]). Independent of an application of Theorem 6.2 to the primary topic of our work, it implies further nontrivial consequences including a generalisation of Stein's Lemma (cf. [94, Theorem 1 and Example 1]) and a certain "decorrelation property" of harmonic functions. We only have to combine (6.3.12) and Theorem 6.2, resulting at once in

Proposition 6.11 Let $k \in \mathbb{N}$, $\rho \in [-1, 1]$ and $\text{vec}(\mathbf{X}, \mathbf{Y}) \sim N_{2k}(0, \Sigma_{2k}(\rho))$. Let $f, g \in L^2(\gamma_k)$. Then

$$\text{cov}(f(\mathbf{X}), g(\mathbf{Y})) = \sum_{\nu=1}^{\infty} \Big(\sum_{n \in C(\nu,k)} \frac{1}{n!} \langle \varphi_k, D^n \Lambda_f\rangle \langle \varphi_k, D^n \Lambda_g\rangle \Big) \rho^\nu.$$

If in addition $(f, g) \in W_{\text{loc}}^{N,1}(\mathbb{R}^k) \times W_{\text{loc}}^{N,1}(\mathbb{R}^k)$, or if $(D^n f, D^n g) \in (L^2(\gamma_k) \cap C(\mathbb{R}^k)) \times (L^2(\gamma_k) \cap C(\mathbb{R}^k))$ for all $n \in \mathbb{N}_0^k$, satisfying $0 \leq |n| \leq N$ for some $N \in \mathbb{N}$, then

$$\text{cov}(f(\mathbf{X}), g(\mathbf{Y})) = \sum_{\nu=1}^{N} \Big(\sum_{n \in C(\nu,k)} \frac{1}{n!} \mathbb{E}[D^n f(\mathbf{X})]\mathbb{E}[D^n g(\mathbf{Y})] \Big) \rho^\nu$$

$$+ \sum_{\nu=N+1}^{\infty} \Big(\sum_{n \in C(\nu,k)} \frac{1}{n!} \langle \varphi_k, D^n \Lambda_f\rangle \langle \varphi_k, D^n \Lambda_g\rangle \Big) \rho^\nu.$$

6.3 Examples of Real CCP Functions, Gaussian Copulas and an Extension of...

In particular, if $N = 1$, then

$$\text{cov}(f(\mathbf{X}), Y_i) = \rho \, \mathbb{E}[\frac{\partial f}{\partial x_i}(\mathbf{X})]$$

for all $i \in [k]$, respectively

$$\mathbb{E}[f(\mathbf{X})\mathbf{Y}] = \rho \, \mathbb{E}[\nabla f(\mathbf{X})].$$

If $N = 2$, then

$$\text{cov}(f(\mathbf{X}), \|\mathbf{Y}\|^2) = \rho^2 \, \mathbb{E}[\Delta f(\mathbf{X})].$$

In the one-dimensional case (i.e., $k = 1$), a direct application of Proposition 6.11, respectively (6.3.12), implies two further results which are of their own interest. In particular, our approach enables the provision of a quick and short proof of the following generalisation of Stein's Lemma to standard Gaussian random powers (cf. [100, Theorem 1]):

Corollary 6.6 *Let $m \in \mathbb{N}_0$, $f \in L^2(\gamma_1)$ and $X \sim N_1(0, 1)$ be given. If $D^l f \equiv f^{(l)} \in L^2(\gamma_1) \cap C(\mathbb{R})$ for all $0 \leq l \leq m$, then*

$$\mathbb{E}[f(X)X^m] = \mathbb{E}[(-i)^m \, H_m(iD)(f(X))] = \sum_{\nu=0}^{\lfloor m/2 \rfloor} \binom{m}{2\nu}(2\nu-1)!! \, \mathbb{E}[D^{m-2\nu} f(X)].$$

Proof Fix $m \in \mathbb{N}_0$. Put $f_m(x) := x^m$ ($x \in \mathbb{R}$). Firstly, recall the well-known result that

$$\mathbb{E}[f_\nu(X)] = \mathbb{E}[X^\nu] = \frac{1+(-1)^\nu}{2}(\nu-1)!! \text{ for all } \nu \in \mathbb{N}_0,$$

implying that

$$\langle f_m, H_l \rangle_{\gamma_1} \stackrel{(6.3.12)}{=} \frac{1}{\sqrt{l!}} \mathbb{E}[D^l f_m(X)] = \begin{cases} \frac{1+(-1)^{m-l}}{2} \sqrt{l!} \binom{m}{l}(m-l-1)!! & \text{if } l \leq m \\ 0 & \text{if } l > m \end{cases}$$

Consequently,

$$\mathbb{E}[f(X)f_m(X)] = h_{f,f_m}(1) = \sum_{l=0}^{m} \langle f, H_l \rangle_{\gamma_1} \langle f_m, H_l \rangle_{\gamma_1}$$

$$= \sum_{l=0}^{m} \frac{1+(-1)^{m-l}}{2} \mathbb{E}[D^l f(X)] \binom{m}{l}(m-l-1)!!.$$

If m is even, it therefore follows that

$$\mathbb{E}[f(X)f_m(X)] = \sum_{k=0}^{m/2} \mathbb{E}[D^{2k}f(X)]\binom{m}{2k}(m-2k-1)!!$$

$$= \sum_{\nu=0}^{m/2} \mathbb{E}[D^{m-2\nu}f(X)]\binom{m}{2\nu}(2\nu-1)!!.$$

Similarly, if m is odd, we obtain

$$\mathbb{E}[f(X)f_m(X)] = \sum_{\nu=0}^{(m-1)/2} \mathbb{E}[D^{m-2\nu}f(X)]\binom{m}{2\nu}(2\nu-1)!!.$$

However, since $\binom{m}{2\nu}(2\nu-1)!! = H_{m,\nu}$ (due to (6.1.2)), the claim clearly follows. □

Corollary 6.7 *Let* $\nu \in \mathbb{N}_0$, $f, g \in L^2(\gamma_1)$ *and* $\rho \in (-1, 1)$. *Then*

$$h_{f,g}^{(\nu)}(\rho) = \sum_{n=0}^{\infty} \frac{1}{n!} \langle \varphi, D^n(D^\nu \Lambda_f)\rangle \langle \varphi, D^n(D^\nu \Lambda_g)\rangle \rho^n.$$

In particular, if in addition $(f, g) \in C^\infty(\mathbb{R}^k) \times C^\infty(\mathbb{R}^k)$, *then*

$$h_{f,g}^{(\nu)}(\rho) = h_{f^{(\nu)},g^{(\nu)}}(\rho) \text{ for all } \rho \in (-1, 1).$$

Proof Since $h_{f,g}$ is real analytic, $h_{f,g}^{(\nu)}$ is well-defined (where $h_{f,g}^{(0)} := h_{f,g}$, of course). Due to (6.3.11) and Theorem 6.2, applied to $k = 1$, the remaining part of the proof is a straightforward proof by induction on $\nu \in \mathbb{N}_0$. □

Remark 6.4 Firstly, observe that the strong impact of the standard (multivariate) *Gaussian* law, is reflected in (6.3.11), primarily implied by the fact that the Gaussian density function φ_k and hence each $H_n \varphi_k$, is rapidly decreasing: $H_n \varphi_k \in \mathcal{S}_k$ for all $n \in \mathbb{N}_0^k$ (since each H_n is a polynomial). This fact and the structure of the (multivariate) Hermite polynomials (which itself is also induced by the structure of φ_k) namely enables a reiterated use of the integration by parts formula in the smooth case, respectively a use of the Sobolev space $W_{\text{loc}}^{N,1}(\mathbb{R}^k)$ in the non-smooth case, implying the transition of each inner product $\langle H_n, f\rangle_{\gamma_k}$ on the Hilbert space $L^2(\gamma_k)$ into the duality bracket $\langle \varphi_k, D^n \Lambda_f\rangle$ on $\mathcal{S}_k \times \mathcal{S}_k'$. The latter, however, seems to be more convenient for performing specific computations.

6.4 Upper Bounds of $K_G^{\mathbb{R}}$ and Inversion of Real CCP Functions

Our next step is to embed Grothendieck's original approach as well as Krivine's improvement into a general framework. We are going to show that we may *substitute* the Grothendieck function $h_{\text{sign},\text{sign}} = \frac{2}{\pi}\arcsin$ through *invertible* CCP functions $h_{f,f}$, generated by *bounded* functions $f : \mathbb{R}^k \longrightarrow \mathbb{R}$ (see Theorem 6.8 and Theorem 6.9). Various proofs of Krivine's main result ($K_G^{\mathbb{R}} \leq \frac{\pi}{2\ln(1+\sqrt{2})} \approx 1.782$) are set out in detail in [72, Section 5], [79, proof of Lemma 10.5] and [158]. We will recognise soon that inverses of odd CCP functions also lead to a further crucial construction of correlation matrices, lurking in the following important implication of Lemma 5.3:

Theorem 6.4 *Let $r > 0$, $0 < c \leq r$ and $0 \neq \psi \in W_+^{\omega}((-r,r))$. Then $\psi_{abs}|_{[0,r]}$ is strictly increasing and $\psi_{abs}(c) > 0$. For any $k \in \mathbb{N}$ there exist $\alpha_k \equiv \alpha_k^{\psi,c} \in S_{L^2(\gamma_k)}$ and $\beta_k \equiv \beta_k^{\psi,c} \in S_{L^2(\gamma_k)}$, such that the following properties hold:*

(i) *If $\mathbf{X} \sim N_k(0, I_k)$, then $\sqrt{\psi_{abs}(c)}\,\mathbb{E}[\alpha_k(\mathbf{X})] = \text{sign}(\psi(0))\sqrt{|\psi(0)|}$ and $\sqrt{\psi_{abs}(c)}\,\mathbb{E}[\beta_k(\mathbf{X})] = \sqrt{|\psi(0)|}$. In particular, $\mathbb{E}[\alpha_k(\mathbf{X})] = \text{sign}(\psi(0))\,\mathbb{E}[\beta_k(\mathbf{X})]$.*

(ii) *If $c < r$, then $[-c,c] \subseteq (-r,r)$ and*
$$\psi(c\rho) = \psi_{abs}(c) h_{\alpha_k,\beta_k}(\rho) \text{ for all } \rho \in [-1,1]. \tag{6.4.1}$$

In particular, $\psi(c) = \psi_{abs}(c)\langle \alpha_k, \beta_k\rangle_{\gamma_k}$.

(iii) *If $c \leq r$, then*
$$\psi_{abs}(c\rho) = \psi_{abs}(c)\, h_{\alpha_k,\alpha_k}(\rho) = \psi_{abs}(c)\, h_{\beta_k,\beta_k}(\rho) \text{ for all } \rho \in [-1,1].$$

(iv) *If $c < r$, H is an arbitrary $\mathbb{R}$-Hilbert space, then there exists a $\mathbb{R}$-Hilbert space $\mathbb{H}$ such that for any $u, v \in S_H$*
$$\psi(c\langle u, v\rangle_H) = \psi_{abs}(c)\langle \psi_u(\alpha_k), \psi_v(\beta_k)\rangle_{\mathbb{H}} = \langle a_u, b_v\rangle_{\mathbb{H}}, \tag{6.4.2}$$

where for any $w \in S_H$, $\|a_w\|_{\mathbb{H}}^2 = \|b_w\|_{\mathbb{H}}^2 = \psi_{abs}(c)$ and $\psi_w : L^2(\gamma_k) \longrightarrow \mathbb{H}$ is a mapping which satisfies $\psi_w(S_{L^2(\gamma_k)}) \subseteq S_{\mathbb{H}}$. In particular, $\psi(c) = \langle a_w, b_w\rangle_{\mathbb{H}}$ for all $w \in S_H$.
If $c \leq r$, then
$$\psi_{abs}(c\langle u,v\rangle_H) = \psi_{abs}(c)\langle \psi_u(\alpha_k), \psi_v(\alpha_k)\rangle_{\mathbb{H}} = \psi_{abs}(c)\langle \psi_u(\beta_k), \psi_v(\beta_k)\rangle_{\mathbb{H}}$$
$$= \langle a_u, a_v\rangle_{\mathbb{H}} = \langle b_u, b_v\rangle_{\mathbb{H}},$$

Proof Let $0 < c \leq r$. Put $b_\nu := \frac{\psi^{(\nu)}(0)}{\nu!}$ ($\nu \in \mathbb{N}_0$). Consider the family $\{q_n : n \in \mathbb{N}_0^k\}$, defined as $q_0 := \sqrt{|\psi(0)|} = \sqrt{|b_0|}$ and

$$q_n \equiv q_n(k) := \sqrt{\frac{|\psi^{(|n|)}(0)|}{n! k^{|n|}}} = \sqrt{\frac{|n|! |b_{|n|}|}{n! k^{|n|}}} = \sqrt{\frac{|n|!}{n!} a_{|n|}^{|n|}} = \sqrt{\frac{|n|!}{n!} \prod_{i=1}^k a_{|n|}^{n_i}}$$

for $n \in \mathbb{N}_0^k \setminus \{0\}$, where $a_\nu \equiv a_\nu(k) := \left(\frac{|b_\nu|}{k^\nu}\right)^{1/\nu}$ ($\nu \neq 0$). As before, let $C(\nu; k) := \{n \in \mathbb{N}_0^k : |n| = \nu\}$, where $\nu \in \mathbb{N}_0$. A straightforward application of the multinomial theorem implies that for all $\nu \in \mathbb{N}$ the following equality holds:

$$\sum_{n \in C(\nu;k)} q_n^2(k) = \sum_{n \in C(\nu;k)} \frac{\nu!}{n!} \prod_{i=1}^k a_\nu^{n_i} = \sum_{n \in C(\nu;k)} \binom{\nu}{n_1, n_2, \ldots, n_k} \prod_{i=1}^k a_\nu^{n_i}$$

$$= \left(\sum_{i=1}^k a_\nu\right)^\nu = k^\nu a_\nu^\nu = |b_\nu|. \quad (6.4.3)$$

It should be noted here, that by construction also $\sum_{n \in C(0;k)} q_n^2 = q_0^2 = |b_0|$ holds. Consequently, since $\{H_\alpha : \alpha \in \mathbb{N}_0^k\}$ is an orthonormal basis in $L^2(\gamma_k)$, it follows that

$$\alpha_k^{\psi,c} := \frac{1}{\sqrt{\psi_{\text{abs}}(c)}} \sum_{n \in \mathbb{N}_0^k} \text{sign}(b_{|n|}) q_n(k) (\sqrt{c})^{|n|} H_n \in S_{L^2(\gamma_k)}$$

and

$$\beta_k^{\psi,c} := \frac{1}{\sqrt{\psi_{\text{abs}}(c)}} \sum_{n \in \mathbb{N}_0^k} q_n(k) (\sqrt{c})^{|n|} H_n \in S_{L^2(\gamma_k)}$$

are well-defined (including the case $c = r$), from which the assertions (i), (ii) and (iii) follow (due to (6.4.3)). (iv) follows from (ii), (iii) and Proposition 6.2, with $\rho := \langle u, v \rangle_H$, and (v) is an implication of Lemma 5.3 and the construction of α_k and β_k. □

If we combine Theorem 5.3, Theorem 6.2, Proposition 6.10 and Theorem 6.4, (iii), a further significant characterisation of continuous odd CCP functions follows at once:

Theorem 6.5 (CCP Representation Theorem) *Let $\psi : [-1, 1] \longrightarrow \mathbb{R}$ be a continuous odd function and $k \in \mathbb{N}$. Then the following statements are equivalent:*

(i) *ψ is a CCP function.*
(ii) *$\psi = h_{f,f}$ for some odd $f \in S_{L^2(\gamma_k)}$.*

6.4 Upper Bounds of $K_G^{\mathbb{R}}$ and Inversion of Real CCP Functions

Theorem 6.4, together with Lemma 6.1 directly enriches us with another crucial result, related to a construction of real quantum correlation matrices which actually include lurking upper bounds of $K_G^{\mathbb{R}}$. At this point we should especially recall the representation (6.2.26) if the sign condition is present.

Corollary 6.8 *Let $m, n, k \in \mathbb{N}$ and $f \in S_{L^2(\gamma_k)}$ be odd. Assume that $h_{f,f}^{-1}|_{(-1,1)} \in W_+^\omega((-1,1))$. Put*

$$c(f) := h_{f,f}^{hyp}(1).$$

Then $c(f) \in (0, 1)$ and

$$h_{f,f}^{-1}[c(f)S] \in \mathcal{Q}_{m,n} \text{ for all } S \in \mathcal{Q}_{m,n}.$$

If $\Sigma = \begin{pmatrix} M & S \\ S^\top & N \end{pmatrix} \in C(m+n; \mathbb{R})$ is an arbitrary real $(m+n) \times (m+n)$ correlation matrix (with block elements $M \in C(m; \mathbb{R})$, $N \in C(n; \mathbb{R})$ and $S \in \mathcal{Q}_{m,n}$), then

$$\begin{pmatrix} (h_{f,f}^{-1})_{abs}[c(f)M] & h_{f,f}^{-1}[c(f)S] \\ h_{f,f}^{-1}[c(f)S^\top] & (h_{f,f}^{-1})_{abs}[c(f)N] \end{pmatrix} \in C(m+n; \mathbb{R}) \quad (6.4.4)$$

again is an $(m+n) \times (m+n)$ correlation matrix with real entries.

Proof We just have to apply Theorem 6.4, (iv), including (6.4.2), to the function $\psi := h_{f,f}^{-1}|_{(-1,1)}$ and the constant $c := h_{f,f}^{hyp}(1)$ ($0 < c < 1$, due to Lemma 6.1). □

If $\psi : D_\mathbb{F} \longrightarrow D_\mathbb{F}$ is an arbitrary CCP function and $\Sigma \in C(m+n; \mathbb{F})$ is an arbitrary $m+n$-correlation matrix, then

$$\psi[\Sigma] = \begin{pmatrix} \psi[\Gamma_H(u,u)] & \psi[\Gamma_H(u,v)] \\ \psi[\Gamma_H(u,v)^*] & \psi[\Gamma_H(v,v)] \end{pmatrix} \in C(m+n; \mathbb{F})$$

again is a correlation matrix. Consequently, it follows that $\psi[\Gamma_H(u,u)] = \Gamma_K(x,x) \in C(m; \mathbb{F})$, $\psi[\Gamma_H(v,v)] = \Gamma_K(y,y) \in C(n; \mathbb{F})$ and $\psi[\Gamma_H(u,v)] = \Gamma_K(x,y) \in \mathcal{Q}_{m,n}(\mathbb{F})$ for some Hilbert space K over $\mathbb{F}$ and some $(x,y) \in S_K^m \times S_K^n$ (due to Corollary 3.1). However, a considerably stronger result holds (cf. also (6.4.4)):

Theorem 6.6 *Let $m, n \in \mathbb{N}$, $A \in \mathbb{M}_m(\mathbb{R})$, $S \in \mathbb{M}_{m,n}(\mathbb{R})$ and $B \in \mathbb{M}_n(\mathbb{R})$. Consider the block matrix*

$$M := \begin{pmatrix} A & S \\ S^\top & B \end{pmatrix} \in \mathbb{M}_{m+n}(\mathbb{R}).$$

Let $0 < r < \infty$ and $f, g : (-r, r) \longrightarrow \mathbb{R}$ be two functions, such that $(-r, r) \ni x \mapsto f(x) = \sum_{\nu=0}^{\infty} a_\nu x^\nu \in W_+^\omega((-r, r))$ and $(-r, r) \ni x \mapsto g(x) = \sum_{\nu=0}^{\infty} b_\nu x^\nu \in W_+^\omega((-r, r))$. Let $q \in C^\omega((-r, r))$, such that

$$|c_\nu| \leq \sqrt{|a_\nu| |b_\nu|} \text{ for all } \nu \in \mathbb{N}_0, \tag{6.4.5}$$

where $c_\nu := \frac{q^{(\nu)}(0)}{\nu!}$. Then $q \in W_+^\omega((-r, r))$, and the following properties hold:

(i) If $M \in \mathbb{M}_{m+n}([-r, r])^+$ is positive semidefinite, and if any entry of the matrices A, S and B is an element of $[-r, r]$, then also

$$\begin{pmatrix} f_{abs}[A] & \widetilde{q}[S] \\ \widetilde{q}[S]^\top & g_{abs}[B] \end{pmatrix} \in \mathbb{M}_{m+n}([-r, r])^+$$

is positive semidefinite, where $\widetilde{q}$ is defined as in Lemma 5.3. In particular, if $0 < c^* \leq r$ is a root of $f_{abs} - 1$, then

$$\widetilde{q}[c^* \Gamma] \in \mathcal{Q}_{m,n} \text{ for all } \Gamma \in \mathcal{Q}_{m,n}.$$

(ii) If $M \in C(m + n; \mathbb{R})$ is a real correlation matrix, then also

$$\begin{pmatrix} h_{\alpha, \alpha}[A] & h_{\alpha, \beta}[S] \\ h_{\alpha, \beta}[S^\top] & h_{\beta, \beta}[B] \end{pmatrix} \in C(m + n; \mathbb{R})$$

is a real correlation matrix for all $\alpha, \beta \in S_{L^2(\gamma_k)}$. In particular,

$$h_{\alpha, \beta}[\cdot] : \mathcal{Q}_{m,n} \longrightarrow \mathcal{Q}_{m,n} \text{ for all } \alpha, \beta \in S_{L^2(\gamma_k)}. \tag{6.4.6}$$

(iii) For all $\alpha, \beta \in S_{L^2(\gamma_k)}$, for all Hilbert spaces H and $u, v \in S_H$ there exist $d \in \mathbb{N}$ and $x, y \in \mathbb{S}^{d-1}$, such that

$$h_{\alpha, \beta}(\langle u, v \rangle_H) = \langle x, y \rangle_{\mathbb{R}_2^d} = x^\top y.$$

Proof

(i) Our proof is built on a sufficient and necessary characterisation of positive semidefinite block matrices, Schur multiplication of matrices, and a straightforward application of Hölder's inequality in l_2. Due to [59, Lemma 1.1.13, statement 2], it suffices to show that the following inequality is satisfied and well-defined:

$$|h^\top q[S]k| \leq \sqrt{(h^\top f_{abs}[A]h)} \sqrt{(k^\top g_{abs}[B]k)} \text{ for all } (h, k) \in \mathbb{R}^m \times \mathbb{R}^n. \tag{6.4.7}$$

6.4 Upper Bounds of $K_G^{\mathbb{R}}$ and Inversion of Real CCP Functions

So, fix $h \in \mathbb{R}^m$ and $k \in \mathbb{R}^n$. Given the assumption that M is positive semidefinite, it follows that also A and B are positive semidefinite matrices. Since $M^{*0} = \mathbf{1}_{m+n}\mathbf{1}_{m+n}^\top$, $A^{*0} = \mathbf{1}_m \mathbf{1}_m^\top$, $B^{*0} = \mathbf{1}_n \mathbf{1}_n^\top$ and $S^{*0} = \mathbf{1}_m \mathbf{1}_n^\top$, the Schur multiplication theorem therefore implies that in fact for any $\nu \in \mathbb{N}_0$

$$M^{*\nu} = \begin{pmatrix} A^{*\nu} & S^{*\nu} \\ (S^{*\nu})^\top & B^{*\nu} \end{pmatrix}$$

is positive semidefinite, such as $A^{*\nu}$ and $B^{*\nu}$. Hence,

$$|h^\top S^{*\nu} k| \leq \sqrt{h^\top A^{*\nu} h} \sqrt{k^\top B^{*\nu} k} \text{ for all } \nu \in \mathbb{N}_0$$

(due to [59, Lemma 1.1.13.2]). Put $\alpha_\nu := \sqrt{|a_\nu|} \sqrt{h^\top (A^{*\nu}) h} = \sqrt{h^\top (|a_\nu| A^{*\nu}) h}$ and $\beta_\nu := \sqrt{|b_\nu|} \sqrt{h^\top (B^{*\nu}) h} = \sqrt{k^\top (|b_\nu| B^{*\nu}) k}$. An application of Hölder's inequality in l_2 implies that

$$\sum_{\nu=0}^{\infty} \alpha_\nu \beta_\nu \leq \Big(\sum_{\nu=0}^{\infty} \alpha_\nu^2\Big)^{\frac{1}{2}} \Big(\sum_{\nu=0}^{\infty} \beta_\nu^2\Big)^{\frac{1}{2}} = \sqrt{(h^\top f_{\text{abs}}[A] h)} \sqrt{(k^\top g_{\text{abs}}[B] k)}. \tag{6.4.8}$$

Due to (6.4.5), a further application of Hölder's inequality in l_2 implies that

$$\sum_{\nu=0}^{\infty} |c_\nu| r^\nu \overset{(6.4.5)}{\leq} \sum_{\nu=0}^{\infty} \sqrt{|a_\nu| r^\nu} \sqrt{|b_\nu| r^\nu} \leq \Big(\sum_{\nu=0}^{\infty} |a_\nu| r^\nu\Big)^{\frac{1}{2}} \Big(\sum_{\nu=0}^{\infty} |b_\nu| r^\nu\Big)^{\frac{1}{2}}$$

$$= \sqrt{f_{\text{abs}}(r) g_{\text{abs}}(r)} < \infty.$$

Thus, $q \in W_+^\omega((-r, r))$. Hence,

$$|h^\top \widetilde{q}[S] k| = \Big| h^\top \Big(\sum_{\nu=0}^{\infty} c_\nu S^{*\nu}\Big) k \Big| \leq \sum_{\nu=0}^{\infty} |c_\nu| |h^\top S^{*\nu} k|$$

$$\leq \sum_{\nu=0}^{\infty} \sqrt{|a_\nu| |b_\nu|} \sqrt{h^\top A^{*\nu} h} \sqrt{k^\top B^{*\nu} k} = \sum_{\nu=0}^{\infty} \alpha_\nu \beta_\nu.$$

Thus, (6.4.8) implies (6.4.7). Suppose that $f_{\text{abs}}(c^*) = 1$ for some $0 < c^* \leq r$. Let $S \in \mathcal{Q}_{m,n}$. Due to Corollary 3.1, we may assume without loss of generality that $A \in C(m; \mathbb{R})$ and $B \in C(n; \mathbb{R})$, implying that $M \in C(m+n; \mathbb{R})$.

In particular, $c^*M = \begin{pmatrix} c^*A & c^*S \\ (c^*S)^\top & c^*B \end{pmatrix}$ is positive semidefinite. Lemma 3.1 therefore implies that

$$_{f,g}M_q := \begin{pmatrix} f_{\text{abs}}[c^*A] & \widetilde{q}[c^*S] \\ \widetilde{q}[c^*S]^\top & g_{\text{abs}}[c^*B] \end{pmatrix}$$

is already a correlation matrix (since $f_{\text{abs}}(c^*a_{ii}) = f_{\text{abs}}(c^*) = 1 = f_{\text{abs}}(c^*b_{ii})$). Corollary 3.1 therefore concludes the proof of (i).

(ii) Put $f := h_{\alpha,\alpha}$, $g := h_{\beta,\beta}$ and $q := h_{\alpha,\beta}$. Observe that $h_{\alpha,\beta} = q = \widetilde{q}|_{[-1,1]}$ (since $h_{\alpha,\beta}$ is continuous). Firstly, since $\alpha, \beta \in S_{L^2(\gamma_k)}$, Theorem 6.2 implies that $f(1) = 1 = g(1)$, so that the diagonal of the matrix $_{f,g}M_q$ is occupied with $m + n$ 1's. It remains to prove that $_{f,g}M_q$ is positive semidefinite. Recall that $|p_\nu(\alpha, \beta)| \leq \sqrt{p_\nu(\alpha, \alpha)}\sqrt{p_\nu(\beta, \beta)}$ for all $\nu \in \mathbb{N}_0$, where $p_\nu(\psi, \kappa) := \sum_{n \in C(\nu,k)} \langle \psi, H_n \rangle_{\gamma_k} \langle \kappa, H_n \rangle_{\gamma_k} = \frac{h_{\psi,\kappa}^{(\nu)}(0)}{\nu!}$, $\psi, \kappa \in \{\alpha, \beta\}$ (cf. (6.2.2)). Theorem 6.2 and Theorem 6.5 therefore imply that all assumptions of (i) are satisfied for the so defined functions f, g and q (including (6.4.5)), so that we may apply statement (i) also to these functions. Finally, (6.4.6) and hence (iii) follows from Corollary 3.1 and Proposition 3.3. □

However, we will recognise that Theorem 6.6 cannot be fully transferred to the complex field, so that we have to distinguish carefully between the real case and the complex case here (cf. Proposition 7.1). If we link (5.1.4) and Theorem 6.6, we obtain

Corollary 6.9 *Let $m, n \in \mathbb{N}$, $A \in \mathbb{M}_{m,n}(\mathbb{R})$ and*

$$\Sigma = \begin{pmatrix} M & S \\ S^\top & N \end{pmatrix} \in C(m+n; \mathbb{R})$$

be an arbitrary real $(m+n) \times (m+n)$ correlation matrix (with block elements $M \in C(m; \mathbb{R})$, $N \in C(n; \mathbb{R})$ and $S \in \mathcal{Q}_{m,n}$). Let $r > 0$ and $0 \neq \psi \in W_+^\omega((-r,r))$. If $0 < c \leq r$, then

$$\frac{1}{\psi_{\text{abs}}(c)} \begin{pmatrix} \psi_{\text{abs}}[c M] & \widetilde{\psi}[c S] \\ \widetilde{\psi}[c S]^\top & \psi_{\text{abs}}[c N] \end{pmatrix} \in C(m+n; \mathbb{R}) \qquad (6.4.9)$$

again is a correlation matrix with real entries. In particular,

$$\frac{1}{\psi_{\text{abs}}(c)} \widetilde{\psi}[cS] \in \mathcal{Q}_{m,n} \text{ for all } S \in \mathcal{Q}_{m,n}. \qquad (6.4.10)$$

6.4 Upper Bounds of $K_G^{\mathbb{R}}$ and Inversion of Real CCP Functions

If we apply Theorem 6.4, (iv) and (6.4.10) to the *inverses* of invertible functions in $W_+^\omega((-1, 1))$, Bolzano's intermediate value theorem from calculus immediately implies a further crucial result. To this end, recall also (5.1.3) and (5.1.4).

Theorem 6.7 (Real Inner Product Rounding) *Let* $\psi : [-1, 1] \longrightarrow [-1, 1]$ *be a bijective real function. Assume that*

$$\psi|_{(-1,1)} \in W_+^\omega((-1, 1)) \text{ and } \psi^{-1}|_{(-1,1)} \in W_+^\omega((-1, 1)).$$

Then $|\psi^{-1}(0)| = (\psi^{-1}|_{(-1,1)})_{abs}(0) \leq (\psi^{-1}|_{(-1,1)})_{abs}(1)$. *Assume that*

$$|\psi^{-1}(0)| < 1 < (\psi^{-1}|_{(-1,1)})_{abs}(1).$$

Then there is a unique number $c^* \in (0, 1)$, *such that* $(\psi^{-1})_{abs}(c^*) = 1$. *Let* $k \in \mathbb{N}$. *There is* $(\alpha, \beta) \in S_{L^2(\gamma_k)} \times S_{L^2(\gamma_k)}$ *(dependent on* c^* *and k) such that for any separable* $\mathbb{R}$*-Hilbert space* H *and any* $u, v \in S_H$, *the following statements apply:*

(i)
$$\langle u, v \rangle_H = \frac{1}{c^*} \psi(\rho_{u,v}) \tag{6.4.11}$$

where $\rho_{u,v} := h_{\alpha,\beta}(\langle u, v \rangle_H)$.

(ii)
$$c^* = \psi(\langle \alpha, \beta \rangle_{\gamma_1}).$$

(iii) *Suppose that* $\psi = h_{f,g}$ *for some* $v \in \mathbb{N}$ *and* $f, g \in L^2(\gamma_v)$. *If* $\rho_{u,v} \in (-1, 1)$, *then*

$$c^* \langle u, v \rangle_H = \frac{1}{(2\pi)^v (1 - \rho_{u,v}^2)^{v/2}}$$

$$\int_{\mathbb{R}^v} \int_{\mathbb{R}^v} f(x) g(y) \exp\left(-\frac{\|x\|^2 + \|y\|^2 - 2\rho_{u,v}\langle x, y \rangle_2}{2(1 - \rho_{u,v}^2)}\right) d^v x \, d^v y.$$

If $m, n \in \mathbb{N}$ *and* $(x, y) \in S_H^m \times S_H^n$, *then there exist* $m + n$ $\mathbb{R}^k$*-valued random vectors* $X_1, \ldots, X_m, Y_1, \ldots, Y_n$, *such that* $vec(X_i, Y_j) \sim N_{2k}(0, \Sigma_{2k}(\rho_{ij}))$ *for all* $(i, j) \in [m] \times [n]$, *and*

$$\Gamma_H(x, y) = \frac{1}{c^*} h_{f,g}[R] = \frac{1}{c^*} \mathbb{E}[P_f Q_g^\top] = \frac{1}{c^*} \mathbb{E}[\Gamma_\mathbb{R}(P_f, Q_g)], \tag{6.4.12}$$

where $(P_f)_i := f(X_i)$, $(Q_g)_j := g(Y_j)$, $\rho_{ij} := h_{\alpha,\beta}(\langle x_i, y_j \rangle_H)$ *and* $R := h_{\alpha,\beta}[\Gamma_H(x, y)] = (\rho_{ij})_{(i,j) \in [m] \times [n]} \in \mathcal{Q}_{m,n}$.

(iv) *Moreover,* $c^* \langle u, v \rangle_H \in (-1, 1)$ *for all* $u, v \in S_H$ *and*

$$tr(A^\top S) = \frac{1}{c^*} tr(A^\top \psi[Q_{c^*,\psi}]) \tag{6.4.13}$$

for all $m, n \in \mathbb{N}$, for all $A \in \mathbb{M}_{m,n}(\mathbb{R})$, for all $S \in \mathcal{Q}_{m,n}$, where $Q_{c^*, \psi} := \psi^{-1}[c^* S] \in \mathcal{Q}_{m,n}$.

It is far from being trivial that it is possible to transfer Theorem 6.7 from the real field $\mathbb{R}$ to the complex field $\mathbb{C}$; at least if $f = g$ is odd (cf. Theorem 7.5). In order to achieve this, we have to develop and implement certain non-trivial structural properties of the class of complex Hermite polynomials (cf. Theorem 7.3 and Theorem 7.4 below).

If $\nu \in \mathbb{N}_0$, $m, n \in \mathbb{N}$ and $A \in \mathbb{M}_{m,n}(\mathbb{F})$, then $A^{*\nu}$ denotes the ν-th entrywise power of the matrix A (in terms of Schur multiplication), where $A^{*0} := \mathbf{1}_m \mathbf{1}_n^*$ is the (rank 1) $m \times n$ matrix of all ones, where $\mathbf{1}_l := (1, 1, \ldots, 1)^\top \in \mathbb{F}^l, l \in \mathbb{N}$ (by adapting the convention that $0^0 := 1$—cf. [83, Remark 9.2]).

Now, we are fully prepared to embed both, Grothendieck's original estimation $K_G^\mathbb{R} \leq \sinh(\frac{\pi}{2}) \approx 2.301$, and Krivine's original estimation $K_G^\mathbb{R} \leq \frac{\pi}{2\ln(1+\sqrt{2})} \approx 1.782$ into a general framework. In particular, we will provide a further proof of the Grothendieck inequality itself (see Theorem 6.8 below). Moreover, we will make a cute use of the little Grothendieck inequality, yielding a quite surprising outcome. To this end, we have to work with functions $h_{f,g}$, which are "generated" by *bounded functions* $f, g \in L^\infty(\mathbb{R}^k, \gamma_k) = L^\infty(\mathbb{R}^k, \lambda_k) \equiv L^\infty(\mathbb{R}^k), k \in \mathbb{N}$. (6.2.8) obviously implies that

$$|h_{f,g}(\rho)| \leq \|f\|_\infty \|g\|_\infty \text{ for all } f, g \in L^\infty(\mathbb{R}^k) \text{ and } \rho \in [-1, 1].$$

Remark 6.5 In statistical machine learning a function $f : \mathbb{R}^k \longrightarrow \{-1, 1\}, k \in \mathbb{N}$ is a particular example of a mapping from some general domain of definition to the set $\{-1, 1\}$, where the latter is known as "concept" (cf. e.g. [145]). The concept f obviously satisfies the condition $\|f\|_{\gamma_k} = 1 = \|f\|_\infty$.

We also need the following important result which holds for both fields interchangeably.

Lemma 6.2 *Let* $\mathbb{F} \in \{\mathbb{R}, \mathbb{C}\}$ *and* $m, n \in \mathbb{N}$. *Let* $\boldsymbol{R}$ *be a m-dimensional random vector in* $(\mathbb{F} \cap \overline{\mathbb{D}})^m$ *and* $\boldsymbol{S}$ *be a n-dimensional random vector in* $(\mathbb{F} \cap \overline{\mathbb{D}})^n$, *such that* $\mathbb{E}[\boldsymbol{R}\boldsymbol{S}^*]$ *exists. Then*

$$\|B * \mathbb{E}[\boldsymbol{R}\boldsymbol{S}^*]^{*\nu}\|_{\infty,1}^\mathbb{F} \leq \|B\|_{\infty,1}^\mathbb{F} \text{ for all } B \in \mathbb{M}_{m,n}(\mathbb{F}) \text{ and } \nu \in \mathbb{N}.$$

Proof Clearly, it is sufficient to verify the statement for $\nu = 1$. Without loss of generality, we may assume that $m = n$ (just to simplify the spelling). So, fix $B \in \mathbb{M}_n(\mathbb{F})$. Put $\boldsymbol{\Lambda} := \boldsymbol{R}\boldsymbol{S}^* = \Gamma_\mathbb{F}(\overline{\boldsymbol{R}}, \overline{\boldsymbol{S}})$. Because of the structure of the operator norm $\| \cdot \|_{\infty,1}^\mathbb{F}$ and Lemma 5.2, it follows that

$$\|B * \mathbb{E}[\boldsymbol{\Lambda}]\|_{\infty,1}^\mathbb{F} = |\text{tr}((B * \mathbb{E}[\boldsymbol{\Lambda}])^* p q^*)| = |\text{tr}(B^* \mathbb{E}[(p q^*) * \overline{\boldsymbol{\Lambda}}])|$$

6.4 Upper Bounds of $K_G^{\mathbb{R}}$ and Inversion of Real CCP Functions

for some non-random vectors $p, q \in S_{\mathbb{F}}^n$. Given the assumed structure of the random matrix Λ, it follows that $\overline{\Lambda} = \overline{\mathbf{R}}\,\overline{\mathbf{S}}^*$. Consequently,

$$(p\,q^*) * \overline{\Lambda} = (p\,q^*) * (\overline{\mathbf{R}}\,\overline{\mathbf{S}}^*) = (p * \overline{\mathbf{R}})(q * \overline{\mathbf{S}})^* = \Gamma_{\mathbb{F}}(\overline{p} * \mathbf{R}, \overline{q} * \mathbf{S})$$

is a random Gram matrix, such that both random vectors $\overline{p} * \mathbf{R}$, and $\overline{q} * \mathbf{S}$ map into $(\mathbb{F} \cap \overline{\mathbb{D}})^n$. Hence,

$$\|B * \mathbb{E}[\Lambda]\|_{\infty,1}^{\mathbb{F}} = \left|\mathrm{tr}(\mathbb{E}[B^* \, \Gamma_{\mathbb{F}}(\overline{p} * \mathbf{R}, \overline{q} * \mathbf{S})])\right|$$

$$\leq \mathbb{E}\left[\left|\mathrm{tr}(B^* \, \Gamma_{\mathbb{F}}(\overline{p} * \mathbf{R}, \overline{q} * \mathbf{S}))\right|\right] \leq \|B\|_{\infty,1}.$$

$\square$

Theorem 6.8 *Let $k, m, n \in \mathbb{N}$ and $f, g \in L^\infty(\mathbb{R}^k)$. Put $r_\infty \equiv r_\infty(f, g) := \|f\|_\infty \|g\|_\infty$. Then:*

(i)
$$|tr(A^\top h_{f,g}[S])| \leq r_\infty \|A\|_{\infty,1} \text{ for all } A \in \mathbb{M}_{m,n}(\mathbb{R}) \text{ and } S \in \mathcal{Q}_{m,n} \tag{6.4.14}$$

and

$$|tr(A^\top h_{f,f}[\Sigma])| \leq \|f\|_\infty^2 \max_{x \in [-1,1]^n} |x^\top A x|$$

for all $A \in \mathbb{M}_n(\mathbb{R})$ and $\Sigma \in C(n; \mathbb{R})$. (6.4.15)

(ii) *Assume that $r_\infty > 0$ and f or g is odd. Let $h_{f,g} : [-1, 1] \longrightarrow \mathbb{R}$ be continuous and injective. Then $h_{f,g}$ is a homeomorphism which is either strictly increasing or strictly decreasing and satisfies $h_{f,g}([-1, 1]) = [-r, r]$, where $r \equiv r_k(f, g) := \max\{-h_{f,g}(1), h_{f,g}(1)\} = \max\{-\langle f, g\rangle_{\gamma_k}, \langle f, g\rangle_{\gamma_k}\}$. Moreover, $0 < r \leq r_\infty$. Assume that $h_{f,g}^{-1}\big|_{(-r,r)} \in W_+^\omega((-r, r))$. Then the following two statements hold:*

(ii-1) *If $\left(h_{f,g}^{-1}\big|_{(-r,r)}\right)_{abs}(r) > 1$, then there is exactly one number $0 < c_k^* \equiv c_k^*(f, g) < r$, such that $\left(h_{f,g}^{-1}\big|_{(-r,r)}\right)_{abs}(c_k^*) = 1$ and*

$$K_G^{\mathbb{R}} \leq \frac{r_\infty}{c_k^*}. \tag{6.4.16}$$

(ii-2) *If* $r = r_\infty$, *then there exists a unique number* $0 < \gamma_k^* \equiv \gamma_k^*(f,g) \in (0,r]$, *such that*

$$K_G^{\mathbb{R}} = \left(h_{f,g}^{-1}\big|_{(-r,r)}\right)_{abs}(\gamma_k^*) \le \min\left\{\frac{r}{c_k^*},\, \left(h_{f,g}^{-1}\big|_{(-r,r)}\right)_{abs}(r)\right\}. \tag{6.4.17}$$

Proof If $r_\infty > 0$, we can assume, throughout the proof, that $\|f\|_\infty = 1$ and $\|g\|_\infty = 1$, implying that $r_\infty = 1$. Fix $A \in \mathbb{M}_{m,n}(\mathbb{R})$ and $S \in \mathcal{Q}_{m,n}$. Then $S = \Gamma_H(u,v)$ for some real finite-dimensional Hilbert space H and some $(u,v) \in S_H^m \times S_H^n$ (due to Proposition 3.3).

(i) Because of (6.2.17), it follows that

$$h_{f,g}[S] = h_{f,g}[\Gamma_H(u,v)] = \mathbb{E}[\mathbf{R}_f\, \mathbf{T}_g^\top] \tag{6.4.18}$$

where $(\mathbf{R}_f)_i := f(\mathbf{X}_{u_i})$ and $(\mathbf{T}_g)_j := g(\mathbf{X}_{v_j})$, for some $\mathrm{vec}(\mathbf{X}_{u_i}, \mathbf{X}_{v_j}) \sim N_{2k}(0, \Sigma_{2k}(\langle u_i, v_j\rangle_H))$ for all $(i,j) \in [m] \times [n]$. Since $\|f\|_\infty = 1 = \|g\|_\infty$, the stochastic inequality

$$|\mathrm{tr}(A^\top \mathbf{R}_f\, \mathbf{T}_g^\top)| \le \|A\|_{\infty,1} \tag{6.4.19}$$

is satisfied (due to (3.2.2) and (3.2.3)). Hence,

$$|\mathrm{tr}(A^\top h_{f,g}[S])| = |\mathrm{tr}(A^\top \mathbb{E}[\mathbf{R}_f\, \mathbf{T}_g^\top])| = |\mathbb{E}[\mathrm{tr}(A^\top \mathbf{R}_f\, \mathbf{T}_g^\top)]|$$

$$\le \mathbb{E}[|\mathrm{tr}(A^\top \mathbf{R}_f\, \mathbf{T}_g^\top)|] \overset{(6.4.19)}{\le} \|A\|_{\infty,1}.$$

Similarly, by making use of (6.2.18), the proof of (6.4.15) can be performed straightforwardly, so that (i) follows.

(ii) Firstly, since $h_{f,g}$ is odd, it follows that $-h_{f,g}(1) = h_{f,g}(-1)$. If $h_{f,g}$ is strictly decreasing, then $h_{f,g}(1) < h_{f,g}(-1) = -h_{f,g}(1)$, implying that $r = -h_{f,g}(1) = -\langle f,g\rangle_{\gamma_k} > -h_{f,g}(0) = 0$. Consequently, $0 < r = |h_{f,g}(1)| \le \|f\|_{\gamma_k} \|g\|_{\gamma_k} \le r_\infty$ (due to (6.2.7), respectively (6.2.8)). A similar proof obviously holds if $h_{f,g}$ is strictly increasing. All described topological properties of $h_{f,g}$ are an immediate application of well-known facts from classical real analysis.

(ii-1) Assume that $\left(h_{f,g}^{-1}\big|_{(-r,r)}\right)_{abs}(r) > 1$. Since $h_{f,g}^{-1}(0) = 0 = \left(h_{f,g}^{-1}\big|_{(-r,r)}\right)_{abs}(0)$, we may apply Theorem 6.7, (iv) to $\psi := \frac{1}{r} h_{f,g}$ (since $\left(\psi^{-1}\big|_{(-1,1)}\right)_{abs} = \left(h_{f,g}^{-1}\big|_{(-r,r)}\right)_{abs}(r\,\cdot)$). It follows that there exists a uniquely determined number $0 < \gamma_k^* \equiv \gamma_k^*(f,g) < 1$ such that $\left(\psi^{-1}\big|_{(-1,1)}\right)_{abs}(\gamma_k^*) = 1$ and $\mathrm{tr}(A^\top S) = \frac{1}{\gamma_k^*} \mathrm{tr}(A^\top \psi[S_*]) = \frac{1}{r\gamma_k^*} \mathrm{tr}(A^\top h_{f,g}[S_*])$ for some $S_* \in \mathcal{Q}_{m,n}$. Put $0 < c_k^* := r\gamma_k^* < r$. Consequently, we may apply statement (i), and it

6.4 Upper Bounds of $K_G^{\mathbb{R}}$ and Inversion of Real CCP Functions

follows that

$$|\mathrm{tr}(A^\top S)| = \frac{1}{c_k^*}|\mathrm{tr}(A^\top h_{f,g}[S_*])| \overset{(6.4.14)}{\leq} \frac{1}{c_k^*}\|A\|_{\infty,1}. \qquad (6.4.20)$$

Corollary 3.3 therefore implies (6.4.16).

(ii-2) Assume that $r \geq r_\infty = 1$. Due to the existence of the inverse function $h_{f,g}^{-1} : [-r, r] \longrightarrow [-1, 1]$, (6.4.18) implies that (algebraically)

$$S = h_{f,g}^{-1}[h_{f,g}[S]] = h_{f,g}^{-1}[\mathbb{E}[\mathbf{\Lambda}_{f,g}]],$$

where $\mathbf{\Lambda}_{f,g} := \mathbf{R}_f \mathbf{T}_g^\top$. Put $a_\nu := \frac{(h_{f,g}^{-1})^{(\nu)}(0)}{\nu!}$, $\nu \in \mathbb{N}$. Since by assumption $r \geq 1$, it follows that $\sum_{\nu=0}^\infty |a_\nu| \leq \sum_{\nu=0}^\infty |a_\nu| r^\nu = \left(h_{f,g}^{-1}\big|_{(-r,r)}\right)_{\mathrm{abs}}(r) < \infty$. Lemma 5.3, applied to the real analytic function $h_{f,g}^{-1}\big|_{(-r,r)}$ and the continuity of the function $h_{f,g}^{-1} : [-r, r] \longrightarrow [-1, 1]$ therefore imply that

$$S = \sum_{\nu=0}^\infty a_\nu \, \mathbb{E}[\mathbf{\Lambda}_{f,g}]^{*\nu},$$

where $*$ denotes entrywise multiplication on $M_{m,n}(\mathbb{R})$ (i.e., the Hadamard product of matrices). Since $\|f\|_\infty = \|g\|_\infty = 1$, it follows that

$$\|A_m\|_{\infty,1} \leq \|A\|_{\infty,1} \text{ for all } m \in \mathbb{N}_0, \qquad (6.4.21)$$

where $A_m := A * \mathbb{E}[\mathbf{\Lambda}_{f,g}]^{*m}$ ($m \in \mathbb{N}_0$) (due to Lemma 6.2). Therefore, Lemma 5.2 leads to

$$|\mathrm{tr}(A^\top S)| \leq \sum_{\nu=0}^\infty |a_\nu| \, |\mathrm{tr}(A^\top \mathbb{E}[\mathbf{\Lambda}_{f,g}]^{*\nu})|$$

$$= |a_0||\mathrm{tr}(A^\top \mathbf{1}_m \mathbf{1}_n^*)| + \sum_{\nu=1}^\infty |a_\nu| \, |\mathrm{tr}(A_{\nu-1}^\top \mathbb{E}[\mathbf{\Lambda}_{f,g}])|$$

$$\leq |a_0||\mathrm{tr}(A^\top \mathbf{1}_m \mathbf{1}_n^*)| + \sum_{\nu=1}^\infty |a_\nu| \, \mathbb{E}[|\mathrm{tr}(A_{\nu-1}^\top \mathbf{\Lambda}_{f,g})|]$$

$$\overset{(6.4.19)}{\leq} |a_0|\|A\|_{\infty,1} + \sum_{\nu=1}^\infty |a_\nu| \, \|A_{\nu-1}\|_{\infty,1} \overset{(6.4.21)}{\leq} \left(\sum_{\nu=0}^\infty |a_\nu|\right)\|A\|_{\infty,1}$$

$$\leq \left(h_{f,g}^{-1}\big|_{(-r,r)}\right)_{\mathrm{abs}}(r)\|A\|_{\infty,1}.$$

Consequently, Corollary 3.3 implies that $1 < K_G^{\mathbb{R}} \leq \left(h_{f,g}^{-1}|_{(-r,r)}\right)_{\text{abs}}(r)$. Equation (6.4.17) now follows from (6.4.16) and Bolzano's intermediate value theorem, where the latter is applied to the strictly increasing function $\left(h_{f,g}^{-1}|_{(-r,r)}\right)_{\text{abs}} - K_G^{\mathbb{R}}$ (cf. proof of Theorem 6.3). □

Inequality (6.4.15), together with the equality (3.2.8) allows us to recover as special case (applied to $f = \operatorname{sign}$) an interesting result of Y. Nesterov, yet without having to make use of their random hyperplane rounding technique (cf. [107]):

Corollary 6.10 *Let* $k, n \in \mathbb{N}$, $A \in \mathbb{M}_n(\mathbb{R})^+$ *be positive semidefinite and* $f \in L^\infty(\mathbb{R}^k)$ *be odd. Then*

$$h'_{f,f}(0) \sup_{\Sigma \in C(n;\mathbb{R})} \operatorname{tr}(A\Sigma) \leq \sup_{\Sigma \in C(n;\mathbb{R})} \operatorname{tr}(A\, h_{f,f}[\Sigma])$$

$$\leq \|f\|_\infty^2 \sup_{x \in \{-1,1\}^n} x^\top A x \leq \|f\|_\infty^2 \|A\|_{\infty,1}.$$

In particular,

$$0 \leq h'_{f,f}(0) \leq \frac{2}{\pi} \|f\|_\infty^2. \tag{6.4.22}$$

Proof Let $\Sigma \in C(n;\mathbb{R})$. Put $h \equiv h_{f,f}$. Since f is odd, Theorem 6.2 implies that

$$h(\rho) = \sum_{\nu=0}^\infty b_{2\nu+1} \rho^{2\nu+1} = h'(0)\rho + \psi(\rho),$$

for all $\rho \in [-1, 1]$, where $\psi(\rho) := \sum_{\nu=1}^\infty b_{2\nu+1} \rho^{2\nu+1}$ and $b_{2\nu+1} \geq 0$ for all $\nu \in \mathbb{N}_0$. Hence, $\psi[\Sigma] = h[\Sigma] - h'(0)\Sigma \in \mathbb{M}_n(\mathbb{R})^+$ (due to Theorem 5.2), whence $\operatorname{tr}(\psi[\Sigma]) = \operatorname{tr}(A(h[\Sigma]) - h'(0)\operatorname{tr}(\Sigma)) \geq 0$ (since $\operatorname{tr}(AB) = \|B^{1/2}A^{1/2}\|_F^2 \geq 0$ for all $B \in \mathbb{M}_n(\mathbb{R})^+$). Thus,

$$h'(0)\operatorname{tr}(A\Sigma) \leq \operatorname{tr}(A\,h[\Sigma]) \stackrel{(6.4.15)}{\leq} \|f\|_\infty^2 \sup_{x \in \{-1,1\}^n} x^\top A x$$

$$= \|f\|_\infty^2 \sup_{x \in \{-1,1\}^n} \operatorname{tr}(Axx^\top) \leq \|f\|_\infty^2 \|A\|_{\infty,1}.$$

In order to verify (6.4.22), we are going to make use of the little Grothendieck inequality. Obviously, we may assume that $h'(0) > 0$, implying that

$$\operatorname{tr}(B\Sigma) \leq \frac{1}{h'(0)} \|f\|_\infty^2 \|B\|_{\infty,1} \text{ for all } B \in \mathbb{M}_n(\mathbb{R})^+ \text{ and } \Sigma \in C(n;\mathbb{R}).$$

6.4 Upper Bounds of $K_G^{\mathbb{R}}$ and Inversion of Real CCP Functions

Consequently, the "correlation matrix version" of the little Grothendieck inequality (see (3.2.10)) implies

$$\frac{\pi}{2} = k_G^{\mathbb{R}} \leq \frac{1}{h'(0)} \|f\|_\infty^2,$$

and (6.4.22) follows. □

In a similar vain, we directly obtain a further (and very short) proof for the value of the real little Grothendieck constant $k_G^{\mathbb{R}}$. To this end, we only have to combine (6.4.14) and Proposition 6.7:

Corollary 6.11 (Grothendieck, 1953)

$$k_G^{\mathbb{R}} = \frac{\pi}{2}.$$

Proof Let $A \in \mathbb{M}_n(\mathbb{R})^+$ and $\Sigma \in C(n; \mathbb{R})$ be arbitrary. We just have to work with the CCP function $[-1, 1] \ni \rho \mapsto h_{f_1, f_1}(\rho) - c_1^2 \rho = \frac{2}{\pi} \arcsin(\rho) - \frac{2}{\pi} \rho$. Because then

$$0 \leq \tfrac{2}{\pi} tr(A\Sigma) = tr(A(\tfrac{2}{\pi}\Sigma)) \leq tr(Ah_{f_1,f_1}[\Sigma]) \overset{(6.4.14)}{\leq} \|A\|_{\infty,1}.$$

□

Remark 6.6 A direct estimation of (6.2.19) leads to the upper bound $k\|f\|_\infty^2 (\sqrt{\frac{2}{\pi}})^2 = \frac{2k}{\pi}\|f\|_\infty^2$, which, however, strongly depends on the dimension $k \in \mathbb{N}$. That upper bound, viewed as a function of k is even strictly increasing. Our application of the little Grothendieck inequality implies the non-trivial result that *for all $k \in \mathbb{N}$, $h'_{f,f}(0)$ actually is bounded above by $\frac{2}{\pi}\|f\|_\infty^2$ "uniformly"*. Moreover, if f were an even function, then (6.4.22) would be trivial, since $h'_{f,f}(0) = 0$.

Theorem 6.8 also implies a remarkable property of CCP functions. To this end, let $S \in \mathcal{Q}_{m,n}$ and $h = h_{f,f}$ be an arbitrary CCP function (not necessarily odd!), generated by some $f \in L^\infty(\mathbb{R}^k)$. Firstly, note that in any case,

$$\frac{\pi}{2}\frac{1}{K_G^{\mathbb{R}}} \overset{(1.1.1)}{<} 1 = \|f\|_{\gamma_k} \leq \|f\|_\infty \leq \|f\|_\infty^2.$$

If also $h_{f,f}^{-1}$ were a CCP function, then $\widetilde{S} := h_{f,f}^{-1}[S] \in \mathcal{Q}_{m,n}$ (due to (6.4.6)). Hence,

$$|tr(A^\top S)| = |tr(A^\top h_{f,f}[\widetilde{S}])| \leq \|f\|_\infty^2 \|A\|_{\infty,1} \text{ for all } A \in \mathbb{M}_{m,n}(\mathbb{R}),$$

implying that $K_G^{\mathbb{R}} \leq \|f\|_\infty^2$. If we join the latter observation and Theorem 5.3, we obtain another interesting fact (which should be carefully compared with Corollary 6.4):

Remark 6.7 Let $k, m, n \in \mathbb{R}$ and $h = h_{f,f}$ be CCP for some $f \in S_{L^2(\gamma_k)} \cap L^\infty(\mathbb{R}^k)$. If $\|f\|_\infty < \sqrt{K_G^\mathbb{R}}$, then the inverse function $h_{f,f}^{-1}$ is not CCP.

Next, we are going to summarise the remarkable properties of *odd CCP functions* which we found so far and shed some light on an additional, quite surprising estimate for odd CCP functions $h_{f,f}$ which emerges if we assume in addition that $f \in S_{L^2(\gamma_k)}$ is *bounded* (a.s.); i.e., if $f \in S_{L^2(\gamma_k)} \cap L^\infty(\mathbb{R}^k)$. In particular, if we combine Theorems 6.3 and 6.8 in this case, we recover Grothendieck's upper bound as well as Krivine's upper bound at once. This follows from Example 6.7, which is a special case of our following key result for the real odd CCP case:

Theorem 6.9 Let $k \in \mathbb{N}$ and $f \in S_{L^2(\gamma_k)} \cap L^\infty(\mathbb{R}^k)$ such that $\|f\|_{\gamma_k} = 1$. Then $\|f\|_\infty \geq 1$. Assume that f is odd and $h_{f,f}^{-1}|_{(-1,1)} \in W_+^\omega((-1,1))$. Then the following statements hold:

(i)
$$\left(h_{f,f}^{-1}|_{(-1,1)}\right)_{abs}(y) \geq \frac{\pi}{2} \frac{1}{\|f\|_\infty^2} y \text{ for all } y \in [0,1].$$

(ii)
$$K_G^\mathbb{R} \leq \frac{\|f\|_\infty^2}{h_{f,f}^{hyp}(1)}.$$

In particular, if $\text{sign}\left((h_{f,f}^{-1}|_{(-1,1)})^{(2n+1)}(0)\right) = (-1)^n$ for all $n \in \mathbb{N}_0$, then

$$K_G^\mathbb{R} \leq i \frac{\|f\|_\infty^2}{\widetilde{\psi}_f(i)},$$

where $\psi_f := h_{f,f}|_{(-1,1)}$ and $\widetilde{\psi}_f : \overline{\mathbb{D}} \longrightarrow \overline{\mathbb{D}}$ is defined as in Lemma 5.3.

(iii) Let $1 \leq c_* < K_G^\mathbb{R}$. If $\|f\|_\infty = 1$, then $0 < h_{f,f}^{hyp}(c_*) < 1$ and there is exactly one number $\gamma^*(f) \in (h_{f,f}^{hyp}(c_*), 1]$, such that

$$K_G^\mathbb{R} = \left(h_{f,f}^{-1}|_{(-1,1)}\right)_{abs}(\gamma^*(f)) \leq \min\left\{\frac{1}{h_{f,f}^{hyp}(1)}, \left(h_{f,f}^{-1}|_{(-1,1)}\right)_{abs}(1)\right\}.$$

(6.4.23)

6.4 Upper Bounds of $K_G^{\mathbb{R}}$ and Inversion of Real CCP Functions

Proof Since $f \in L^\infty$, it follows that $f \in L^2(\gamma_k)$ and $1 = \|f\|_{\gamma_k} \leq \|f\|_\infty$.

(i) and (ii) The claimed inequalities directly follow from (5.1.3), (6.2.20), Lemma 6.1, (6.2.26), Theorem 6.8, and (6.4.22).

(iii) Since $1 \leq c_* < K_G^{\mathbb{R}} \leq \left(h_{f,f}^{-1}\big|_{(-1,1)}\right)_{abs}(1)$ (due to Theorem 6.8), it follows from Lemma 6.1 that $0 < h_{f,f}^{\text{hyp}}(1) \leq h_{f,f}^{\text{hyp}}(c_*) < \gamma^*(f) := h_{f,f}^{\text{hyp}}(K_G^{\mathbb{R}}) \leq 1$. (iii) now follows from (ii). □

If we only assume that $f \in L^\infty(\mathbb{R}^k) \setminus \{0\}$, then an application of Theorem 6.9 to $\frac{1}{\|f\|_{\gamma_k}} f \in S_{L^2(\gamma_k)}$ directly leads to

Corollary 6.12 Let $k \in \mathbb{N}$ and $f \in L^\infty(\mathbb{R}^k)\setminus\{0\}$. Then $0 < r \equiv r_k(f) := \|f\|_{\gamma_k}^2 < \infty$ and $\|f\|_\infty \geq \sqrt{r}$. Assume that f is odd and $h_{f,f}^{-1}\big|_{(-r,r)} \in W_+^\omega((-r,r))$. Then $\left(h_{f,f}^{-1}\big|_{(-r,r)}\right)_{abs}(r) > 1$, and the following statements hold:

(i)
$$\left(h_{f,f}^{-1}\big|_{(-r,r)}\right)_{abs}(y) \geq \frac{\pi}{2} \frac{1}{\|f\|_\infty^2} y \text{ for all } y \in [0,r].$$

(ii)
$$r K_G^{\mathbb{R}} \leq \frac{\|f\|_\infty^2}{h_{\frac{f}{\sqrt{r}},\frac{f}{\sqrt{r}}}^{\text{hyp}}(1)}. \tag{6.4.24}$$

In particular, if $\text{sign}\left((h_{f,f}^{-1}\big|_{(-r,r)})^{(2n+1)}(0)\right) = (-1)^n$ for all $n \in \mathbb{N}_0$, then

$$K_G^{\mathbb{R}} \leq i \frac{\|f\|_\infty^2}{\widetilde{\psi}_f(i)},$$

where $\psi_f := h_{f,f}\big|_{(-1,1)}$ and $\widetilde{\psi}_f : \overline{\mathbb{D}} \longrightarrow r\overline{\mathbb{D}}$ is defined as in Lemma 5.3.

(iii) Let $1 \leq c_* < K_G^{\mathbb{R}}$. If $\|f\|_\infty = \sqrt{r}$, then $0 < h_{\frac{f}{\sqrt{r}},\frac{f}{\sqrt{r}}}^{\text{hyp}}(c_*) < 1$ and there is exactly one number number $\gamma^*(f) \in (h_{\frac{f}{\sqrt{r}},\frac{f}{\sqrt{r}}}^{\text{hyp}}(c_*), 1]$, such that

$$K_G^{\mathbb{R}} = \left(h_{f,f}^{-1}\big|_{(-r,r)}\right)_{abs}(r\, \gamma^*(f)) \leq \min\left\{\frac{1}{h_{\frac{f}{\sqrt{r}},\frac{f}{\sqrt{r}}}^{\text{hyp}}(1)}, \left(h_{f,f}^{-1}\big|_{(-r,r)}\right)_{abs}(r)\right\}.$$

The proof of Theorem 6.9 shows us that here we may circumvent the rather strong assumption of $h_{f,f}^{-1}\big|_{(-r,r)}$ being completely real analytic on $(-r,r)$ at 0; at least in the following sense (cf. Example 6.10 as application for this):

Proposition 6.12 Let $k \in \mathbb{N}$ and $f \in L^\infty(\mathbb{R}^k) \setminus \{0\}$. Then $0 < r \equiv r_k := \|f\|_{\gamma_k}^2 < \infty$ and $\|f\|_\infty \geq \sqrt{r}$. Assume that f is odd, $h'_{f,f}(0) > 0$ and

$$\sum_{n=0}^{\infty} \frac{|(h_{f,f}^{-1})^{(2n+1)}(0)|}{(2n+1)!} (c^*)^{2n+1} = 1, \qquad (6.4.25)$$

for some $c^* \in (0, r]$. Then $h_{f,f}^{-1}|_{(-c^*, c^*)} \in W_+^\omega((-c^*, c^*))$, and the following statements hold:

(i)
$$\left(h_{f,f}^{-1}\big|_{(-c^*, c^*)}\right)_{abs}(y) \geq \frac{\pi}{2} \frac{1}{\|f\|_\infty^2} y \text{ for all } y \in [0, c^*]$$

$$\text{and } \left(h_{f,f}^{-1}\big|_{(-c^*, c^*)}\right)_{abs}(c^*) = 1.$$

(ii)
$$h_{f,f}^{-1}[c^* S] \in \mathcal{Q}_{m,n} \text{ for all } S \in \mathcal{Q}_{m,n}. \qquad (6.4.26)$$

(iii)
$$K_G^\mathbb{R} \leq \frac{\|f\|_\infty^2}{c^*}.$$

Proof

(i) The assumption clearly implies that $h_{f,f}^{-1}|_{(-c^*, c^*)} \in W_+^\omega((-c^*, c^*))$ and $\left(h_{f,f}^{-1}|_{(-c^*, c^*)}\right)_{abs}(c^*) = 1$.

(ii) and (iii) Observe that in particular the restriction $h_{f,f}^{-1}|_{[-c^*, c^*]} : [-c^*, c^*] \to [-1, 1]$ is continuous. Consequently, if $S \in \mathcal{Q}_{m,n}$, then $S_0 := h_{f,f}^{-1}[c^* S] \in \mathcal{Q}_{m,n}$ (due to (6.4.10)), and (ii) follows. Let $m, n \in \mathbb{N}$, $A \in \mathbb{M}_{m,n}(\mathbb{R})$ and $S \in \mathcal{Q}_{m,n}$ be arbitrarily given. Since $S_0 \in \mathcal{Q}_{m,n}$, we therefore obtain

$$|\text{tr}(A^\top S)| = \frac{1}{c^*} |\text{tr}(A^\top h_{f,f}[S_0])| \overset{(6.4.14)}{\leq} \frac{1}{c^*} \|A\|_{\infty,1},$$

and (iii) follows as well. $\square$

Example 6.7 (Grothendieck and Krivine) Once again, we consider the CCP function $\psi := \frac{2}{\pi} \arcsin$. Recall that $\psi = h_{f,f}$, where $f := \text{sign} \in S_{L^\infty(\mathbb{R})} \cap S_{L^2(\gamma_k)}$. Due to (6.2.24), it follows that

$$\psi^{\text{hyp}}(1) = \frac{2}{\pi} \sinh^{-1}(1) = \frac{2}{\pi} \ln(1 + \sqrt{2}).$$

6.4 Upper Bounds of $K_G^{\mathbb{R}}$ and Inversion of Real CCP Functions

Hence,

$$K_G^{\mathbb{R}} \overset{(6.4.23)}{\leq} \min\left\{\frac{1}{\psi^{\text{hyp}}(1)}, \left(\psi^{-1}\big|_{(-1,1)}\right)_{\text{abs}}(1)\right\}$$

$$\leq \frac{\pi}{2\ln(1+\sqrt{2})}(\approx 1.78221)$$

$$\leq \sinh(\frac{\pi}{2})(\approx 2.30129)$$

precisely reflects Krivine's upper bound of $K_G^{\mathbb{R}}$ as well as Grothendieck's (larger) upper bound of $K_G^{\mathbb{R}}$!

Example 6.8 Consider the CCP function $\kappa := \sqrt{3}(2\phi - 1)$ (cf. Proposition 6.9). Due to Theorem 6.9, (i), respectively (6.4.24), we obtain the following (weaker) estimation:

$$K_G^{\mathbb{R}} \leq \frac{\pi}{6\ln(\frac{1}{2}(1+\sqrt{5}))} \cdot \|\kappa\|_\infty^2 \leq \frac{\pi}{2\ln(\frac{1}{2}(1+\sqrt{5}))}(\approx 3.26425).$$

We highly recommend the readers to check whether this estimation can be improved, if more generally the function $\kappa_\alpha := \alpha(2\phi - 1)$ is considered, where $0 < \alpha < \sqrt{3}$ is given (instead of the CCP function $\kappa = \kappa_{\sqrt{3}}$)! Observe also that $\|\kappa\|_\infty = \sqrt{3} \neq 1 = \|\kappa\|_{\gamma_k}$.

Example 6.9 Fix $k \in \mathbb{N}_3$ and consider the function $\psi := h_{f_k,f_k}$, introduced in Proposition 6.7. *Assume that* $\psi^{-1} \in W_+^\omega((-1,1))$. ψ then satisfies all assumptions, listed in Theorem 6.9, and it follows that $0 < c_k^* := h_{f_k,f_k}^{\text{hyp}}(1) < 1$ satisfies

$$K_G^{\mathbb{R}} \leq \frac{k}{c_k^*} \qquad (6.4.27)$$

(since $\|f_k\|_\infty^2 = k$). However, observe that the sequence $\left(\frac{k}{c_k^*}\right)_{k\in\mathbb{N}}$ is not bounded and hence cannot converge (since $0 < c_k^* \leq 1$). Moreover, in contrast to the previous two examples, we do not know whether also $\left(h_{f_k,f_k}^{-1}\big|_{(-1,1)}\right)_{\text{abs}}$ can be represented in a closed analytical form; one of the major open problems in our search for the smallest upper bound of $K_G^{\mathbb{R}}$ (cf. Sect. 9.1 and Example 7.1, where the latter includes the approximation of the constant $c_k^* \approx 0.71200$ if $k = 2$). A straightforward, yet a bit laborious calculation with fractions, based on the table (9.1.8) (similarly to the special case $k = 2$, studied in Example 7.1), yields that the Maclaurin series of $h_{f_k,f_k}^{-1}(s) \equiv \sum_{\nu=0}^\infty \beta_{2\nu+1}(k) s^{2\nu+1}$ can e.g. be approximated by the Taylor polynomial of degree 7 as:

$$\begin{aligned}
h_{f_k,f_k}^{-1}(s) &= \beta_1(k)s + \beta_3(k)s^3 + \beta_5(k)s^5 + \beta_7(k)s^7 + o(|s|^7) \\
&= \frac{1}{c_k^2}s + \left(-\frac{1}{c_k^6}\frac{1}{2(k+2)}s^2\right)\left(s + \frac{3}{4c_k^4}\frac{k-2}{(k+2)(k+4)}s^3 \right.\\
&\quad \left. + \frac{3}{8c_k^8}\frac{9(k+2)^2 - 48(k+2) + 64}{(k+2)^2(k+4)(k+6)}s^5\right) + o(|s|^7).
\end{aligned} \tag{6.4.28}$$

Observe also that $\beta_5(k) := -\frac{3}{8c_k^{10}}\frac{k-2}{(k+2)^2(k+4)} = 0$ if and only if $k = 2$ and that $\beta_5(k) \leq 0$ if and only if $k \geq 2$. Similarly, since the single (local) minimum of the function $\mathbb{R} \ni x \mapsto g(x) := 9x^2 - 48x + 64$ is attained at $x_* := \frac{8}{3}$ and $g(x_*) = 0$, it follows that $\beta_7(k) \leq 0$.

Example 6.10 Our approach also can be applied to slightly modify the proof of the strongest result to date, namely that $K_G^{\mathbb{R}} < \frac{\pi}{2\ln(1+\sqrt{2})}$. To this end, firstly observe that a simple change of variables reveals that *for any $f, g \in L^2(\gamma_k)$*, the corresponding generalised function $H_{f \circ \sqrt{2}, g \circ \sqrt{2}}$, listed in [22, Definition 2.1], satisfies

$$H_{f \circ \sqrt{2}, g \circ \sqrt{2}} = h_{f,g}\big|_{(-1,1)} \tag{6.4.29}$$

(due to (6.2.9)). So, $H_{f \circ \sqrt{2}, g \circ \sqrt{2}}$ is well-defined on $(-1, 1)$. The Grothendieck function $\frac{2}{\pi}\arcsin$ is then generalised in [22] to the complex-valued function

$$F_{p,\eta} := (1-p)H_0 + pH_\eta, \tag{6.4.30}$$

where $0 \leq p \leq 1$, $0 \leq \eta < 1$ and $H_\eta : \mathbb{S} \longrightarrow \mathbb{C}$ is defined as in [22, (41)], where $\mathbb{S} := \{z \in \mathbb{C} : |\mathrm{Re}(z)| < 1\}$. Independent of any complex analysis a priori, the construction of H_η, together with (6.4.29) implies that

$$H_\eta\big|_{\mathbb{S} \cap \mathbb{R}} = H_\eta\big|_{(-1,1)} = H_{g_\eta \circ \sqrt{2}, g_\eta \circ \sqrt{2}} \stackrel{(6.4.29)}{=} h_{g_\eta, g_\eta}\big|_{(-1,1)}.$$

Here, the odd function $g_\eta : \mathbb{R}^2 \longrightarrow \mathbb{R}$ is defined as $g_\eta(x) := \sqrt{\frac{\pi}{2}} f_\eta(\frac{1}{\sqrt{2}}x)$, where the 2-dimensional sign concept $f_\eta : \mathbb{R}^2 \longrightarrow \{-1, 1\}$ satisfies [22, (40)]. Consequently, the function $\frac{2}{\pi} H_\eta\big|_{(-1,1)}$ actually is a restriction of the well-defined odd—and hence invertible—CCP function $\frac{2}{\pi} h_{g_\eta, g_\eta} = h_{\sqrt{2/\pi} g_\eta, \sqrt{2/\pi} g_\eta}$ to the open interval $(-1, 1)$ (due to Theorem 6.5).

Let $\rho \in (-1, 1)$ and $\mathrm{vec}(\mathbf{X}, \mathbf{Y}) \sim N_4(0, \Sigma_4(\rho))$. Since $g_0(x) = \sqrt{\frac{\pi}{2}} \mathrm{sign}(x_2)$ for all $x = (x_1, x_2)^\top \in \mathbb{R}^2$, we may apply Proposition 2.5 to the partitioned Gaussian

6.4 Upper Bounds of $K_G^{\mathbb{R}}$ and Inversion of Real CCP Functions

random vector $\text{vec}(\mathbf{X}, \mathbf{Y})$, and it follows that

$$H_0(\rho) = h_{g_0,g_0}(\rho) = \frac{\pi}{2} \mathbb{E}_{\mathbb{P}_{\text{vec}(\mathbf{X},\mathbf{Y})}}[\text{sign}(X_2)\text{sign}(Y_2)] = \arcsin(\rho),$$

which slightly shortens the proof of [22, Lemma 4.3]. In particular, for any $p \in [0, 1]$,

$$\frac{2}{\pi} F_{p,\eta}\Big|_{(-1,1)} = \psi_{p,\eta}\Big|_{(-1,1)} \tag{6.4.31}$$

emerges as a restriction of the odd and hence *strictly increasing, homeomorphic* CCP function $\psi_{p,\eta} := (1-p)\frac{2}{\pi}\arcsin + p\frac{2}{\pi}h_{g_\eta,g_\eta}$ on $(-1, 1)$ (due to Theorems 5.3 and 6.3). Observe that $\psi'_{p,\eta}(0) \geq (1-p)\frac{2}{\pi}$, implying that $\psi_{p,\eta}^{-1}\big|_{(-1,1)} \in C^\omega((-1, 1))$ if $p < 1$ (due to Theorem 6.3). Because of the non-trivial result [22, Theorem 5.1] (including its technically demanding proof) it follows the existence of $(p_0, \eta_0) \in (0, 1) \times (0, 1)$ and $c^* \in (\frac{2}{\pi}\ln(1+\sqrt{2}), \frac{9}{5\pi})$, such that

$$\sum_{n=0}^{\infty} \frac{|(\psi_{p_0,\eta_0}^{-1})^{(2n+1)}(0)|}{(2n+1)!} (c^*)^{2n+1} = 1.$$

(Given the outcome of [22, Theorem 5.1], we only have to set $c^* := \frac{2\gamma}{\pi}$, where $\ln(1+\sqrt{2}) < \gamma < \frac{9}{10}$ satisfies [22, (63)].) Hence, $\psi_{p_0,\eta_0}^{-1}\big|_{(-c^*,c^*)} \in W_+^\omega((-c^*, c^*))$. Since in particular $\psi_{p_0,\eta_0}^{-1}\big|_{[-c^*,c^*]}$ is continuous, we may apply Proposition 6.12. Consequently, if $S \in \mathcal{Q}_{m,n}$, then $S_0 := \psi_{p_0,\eta_0}^{-1}[c^* S] \in \mathcal{Q}_{m,n}$ (due to (6.4.26)). Observe that for any $0 \leq \eta < 1$ the CCP function $\frac{2}{\pi}h_{g_\eta,g_\eta} = h_{\sqrt{2/\pi}g_\eta,\sqrt{2/\pi}g_\eta}$ actually originates from the *bounded* function $\sqrt{\frac{2}{\pi}}g_\eta \in S_{L^\infty} \cap S_{L^2(\gamma_2)}$, such as $\frac{2}{\pi}\arcsin = h_{\text{sign},\text{sign}}$. Let $S \in \mathcal{Q}_{m,n}$ be arbitrarily given. Since $S_0 \in \mathcal{Q}_{m,n}$, we therefore obtain

$$|\text{tr}(A^\top S)| = \frac{1}{c^*}|\text{tr}(A^\top \psi_{p_0,\eta_0}[S_0])|$$

$$= \frac{1}{c^*}\left|\text{tr}\left(A^\top\left((1-p)\frac{2}{\pi}\arcsin[S_0] + p\frac{2}{\pi}h_{g_\eta,g_\eta}[S_0]\right)\right)\right|$$

$$\stackrel{(6.4.14)}{\leq} \frac{1}{c^*}((1-p)\|A\|_{\infty,1} + p\|A\|_{\infty,1}) = \frac{1}{c^*}\|A\|_{\infty,1},$$

whence

$$K_G^{\mathbb{R}} \leq \frac{1}{c^*} < \frac{\pi}{2\ln(1+\sqrt{2})}.$$

In summary, given the—crucial—result [22, Theorem 5.1], our general framework can be applied here as well, leading to a slightly modified proof of [22, Theorem 1.1]. At this point, we would like to highlight another, very recent and partially altered proof of [22, Theorem 1.1]—provided by Krivine again (cf. [93, Theorem 1]).

Despite the quite remarkable outcome of Theorem 6.9, we should observe that its practical implementation seems to be quite difficult (at least without sufficiently large computer power). Primarily, as we already have seen, this is due to the following facts:

(i) Either we have to know a closed form representation (or at least a "close" approximation of the Maclaurin series) of $h_{f,f}$, $h_{f,f}^{-1}$ and $\left(h_{f,f}^{-1}\right)_{\text{abs}}$ if $f \in L^\infty(\mathbb{R}^k)$ is given (such as is the case for $k = 1$ and $f := \text{sign} \in L^\infty(\mathbb{R}^1) \cap S_{L^2(\gamma_1)}$), or we have to check whether $h = h_{f,f}$ "originates" from some $f \in L^\infty(\mathbb{R}^k) \cap S_{L^2(\gamma_k)}$, if the functions h, h^{-1} and $\left(h^{-1}\right)_{\text{abs}}$ are known to us.

(ii) However, already in the one-dimensional case (i.e., for $k = 1$) that search requires rather complex calculation techniques, respectively some very helpful knowledge about Hermite polynomials. Moreover, if k increases, we are confronted with a "curse of combinatorial dimensionality", since for any $\nu \in \mathbb{N}_0$ and $k \in \mathbb{N}$ it can be easily shown by induction on k that the set $C(\nu, k) := \{n \in \mathbb{N}_0^k : |n| = \nu\}$ which determines the structure of $h_{f,f}$ (cf. Theorem 6.2) actually consists of $\binom{\nu+k-1}{k-1} = \frac{(\nu+k-1)!}{\nu!\,(k-1)!}$ elements. In particular, already $C(\nu, 2)$ consists of $\nu + 1$ elements ($\nu \in \mathbb{N}_0$). For example, to determine $C(2, 11)$ explicitly, we would have to know all of its 66 elements!

We will recognise how deep actually we are confronted with a "curse of combinatorial dimensionality" if only a Maclaurin series representation of the function $h_{f,f}$ is given to us (cf. Sect. 9.1).

Chapter 7
The Complex Case: Towards Extending Haagerup's Approach

7.1 Multivariate Complex CCP Functions and Their Relation to the Real Case

Next, we are going to transfer the main results in the previous chapter from the real field $\mathbb{R}$ to the complex field $\mathbb{C}$. In order to achieve this, we have to implement non-trivial structural properties of the class of *complex* Hermite polynomials (cf. Theorems 7.3 and 7.4). The complex versions of Theorems 6.2 and 6.3 also allow a generalisation of the Haagerup equality by transition from the $\Sigma_2(\zeta)$-correlated couple of two complex one-dimensional signum functions sign : $\mathbb{C} \longrightarrow \mathbb{T}$ to a $\Sigma_{2k}(\zeta)$-correlated couple of two (possibly different) arbitrary square-integrable functions $b, c : \mathbb{C}^k \longrightarrow \mathbb{C}$, where $k \in \mathbb{N}$ can be arbitrarily large (cf. Corollary 7.1 and Example 7.1). By definition (cf. [60, Lemma 3.2. and Proof of Theorem 3.1]), the complex sign-function is given as

$$\operatorname{sign}(z) := \begin{cases} \frac{z}{|z|} & \text{if } z \in \mathbb{C}^* \\ 0 & \text{if } z = 0 \end{cases}.$$

We introduce the following helpful symbolic constructions and shortcuts. Fix $k, l \in \mathbb{N}$. Let $b : \mathbb{C}^k \longrightarrow \mathbb{C}$ and $c : \mathbb{C}^l \longrightarrow \mathbb{C}$ be two functions. Put

$$b_k \otimes_l c := (b \circ P_k)(c \circ Q_l),$$

where

$$P_k := \begin{pmatrix} 1 & 0 & \cdots & 0 & 0 & 0 & \cdots & 0 \\ 0 & 1 & \cdots & 0 & 0 & 0 & \cdots & 0 \\ \vdots & \vdots & \ddots & \vdots & \vdots & \vdots & \ddots & \vdots \\ 0 & 0 & \cdots & 1 & 0 & 0 & \cdots & 0 \end{pmatrix} = (I_k \mid 0) \in \mathbb{M}_{k,k+l}(\mathbb{C})$$

and

$$Q_l := \begin{pmatrix} 0 & 0 & \cdots & 0 & 1 & 0 & \cdots & 0 \\ 0 & 0 & \cdots & 0 & 0 & 1 & \cdots & 0 \\ \vdots & \vdots & \ddots & \vdots & \vdots & \vdots & \ddots & \vdots \\ 0 & 0 & \cdots & 0 & 0 & 0 & \cdots & 1 \end{pmatrix} = (0 \mid I_l) \in \mathbb{M}_{l,k+l}(\mathbb{C}),$$

implying that

$$b_{\,k} \otimes_l c(\text{vec}(z, w)) = b(z)c(w) \text{ for all } (z, w) \in \mathbb{C}^k \times \mathbb{C}^l. \tag{7.1.1}$$

Given the construction of $b_{\,k} \otimes_l c$ we may unambiguously shorten it simply to $b \otimes c$ (and suppress the listing of the dimensions of the domains of definition of b, respectively c). Let $d : \mathbb{C}^k \longrightarrow \mathbb{C}$ be a given function. Recall the induced functions $r(d) : \mathbb{R}^{2k} \longrightarrow \mathbb{R}$ and $s(d) : \mathbb{R}^{2k} \longrightarrow \mathbb{R}$, defined as $r(d) := \text{Re}(d) \circ \frac{1}{\sqrt{2}} J_2^{-1}$ and $s(d) := \text{Im}(d) \circ \frac{1}{\sqrt{2}} J_2^{-1}$ (cf. (2.1.8)). In order to facilitate reading, we put $d_\alpha(z) := d(\text{sign}(\alpha)z)$, where $\alpha \in \mathbb{C}$ and $z \in \mathbb{C}^k$ (implying that $d_0 = d(0)$). Consequently, the construction of $b \otimes \bar{c}$ implies that

$$r(b \otimes \bar{c})(\text{vec}(\text{vec}(x_1, x_2), \text{vec}(y_1, y_2))) = r(b)(\text{vec}(x_1, y_1))r(c)(\text{vec}(x_2, y_2))$$
$$+ s(b)(\text{vec}(x_1, y_1))s(c)(\text{vec}(x_2, y_2))$$

and

$$s(b \otimes \bar{c})(\text{vec}(\text{vec}(x_1, x_2), \text{vec}(y_1, y_2))) = s(b)(\text{vec}(x_1, y_1))r(c)(\text{vec}(x_2, y_2))$$
$$- r(b)(\text{vec}(x_1, y_1))s(c)(\text{vec}(x_2, y_2))$$

for all $x_1, x_2, y_1, y_2 \in \mathbb{R}^k$. In other words:

$$r(b \otimes \bar{c}) = (r(b) \otimes r(c)) \circ G + (s(b) \otimes s(c)) \circ G \tag{7.1.2}$$

on $\mathbb{R}^{4k}$, and

$$s(b \otimes \bar{c}) = (s(b) \otimes r(c)) \circ G - (r(b) \otimes s(c)) \circ G, \tag{7.1.3}$$

7.1 Multivariate Complex CCP Functions and Their Relation to the Real Case

on $\mathbb{R}^{4k}$, where again $G = G^\top = G^{-1} \in O(4n)$ is the matrix, introduced in (1.2.6). Finally, d is odd (respectively, even) if and only if both, $r(d)$ and $s(d) = r(-i\,d)$ are odd (respectively, even).

Lemma 7.1 *Let* $k \in \mathbb{N}$, $\rho \in [-1, 1]$, $\alpha \in \mathbb{C}$, $\theta \in \mathbb{R}$, $c, d : \mathbb{C}^k \longrightarrow \mathbb{C}$ *and* $\mathrm{vec}(\mathbf{X}, \mathbf{Y}) \sim N_{4k}(0, \Sigma_{4k}(\rho))$. *Then*

(i) $r(d_\alpha) = r(d) \circ R_2(\mathrm{sign}(\alpha)I_k)$ *and* $r(d_\theta) = r(d) \circ \mathrm{sign}(\theta)I_{2k}$.
(ii) *If* $\alpha \neq 0$, *then* $r(d) \in L^2(\mathbb{R}^{2k}, \gamma_{2k})$ *if and only if* $r(d_\alpha) \in L^2(\mathbb{R}^{2k}, \gamma_{2k})$. *In this case, the norms coincide:* $\|r(d)\|_{\gamma_{2k}} = \|r(d_\alpha)\|_{\gamma_{2k}}$.
(iii) *If* $\alpha \neq 0$, $h_{r(d),r(d)}(\rho) = \mathbb{E}[r(d)(\mathbf{X})r(d)(\mathbf{Y})] = \mathbb{E}[r(d_\alpha)(\mathbf{X})r(d_\alpha)(\mathbf{Y})] = h_{r(d_\alpha),r(d_\alpha)}(\rho)$.
(iv) $h_{r(c_r),r(d)}(|r|) = h_{r(c),r(d)}(r)$ *for all* $r \in [-1, 1]$.

Proof

(i) Put $\kappa := \mathrm{sign}(\alpha)$. Fix $x \in \mathbb{R}^{2k}$. Since $J_2^{-1} \circ R_2(\kappa\,I_k) = (\kappa\,I_k) \circ J_2^{-1} = \kappa\,J_2^{-1}$, it follows that $d_\alpha\left(\frac{1}{\sqrt{2}} J_2^{-1}(x)\right) = d\left(\frac{1}{\sqrt{2}} J_2^{-1}(R_2(\kappa\,I_k)x)\right)$, whence $r(d_\alpha) = r(d) \circ R_2(\kappa\,I_k)$.

(ii) Let $\alpha \neq 0$. Since $\det(R_2(\kappa\,I_k)) = 1$ (due to (2.1.6)) and $\|R_2(\kappa\,I_k)a\|^2 = \langle a, R_2(\overline{\kappa}\,I_k)R_2(\kappa\,I_k)a\rangle_{\mathbb{R}_2^{2k}} = \|a\|^2$ for all $a \in \mathbb{R}^{2k}$ (due to (1.2.11)), an application of the change-of-variables formula implies that $r(d_\alpha) \in L^2(\mathbb{R}^{2k}, \gamma_{2k})$ if and only if $r(d) \in L^2(\mathbb{R}^{2k}, \gamma_{2k})$ and $\|r(d_\alpha)\|_2 = \|r(d)\|_2$.

(iii) Let $\alpha \neq 0$. Put $\kappa := \mathrm{sign}(\alpha)$. Then $A_\kappa := R_2(\kappa\,I_k) \neq 0$. Put $\mathbf{X}_\kappa := A_\kappa \mathbf{X}$ and $\mathbf{Y}_\kappa := A_\kappa \mathbf{Y}$. Since

$$\begin{pmatrix} A_\kappa & 0 \\ 0 & A_\kappa \end{pmatrix} \begin{pmatrix} I_{2k} & \rho I_{2k} \\ \rho I_{2k} & I_{2k} \end{pmatrix} \begin{pmatrix} A_{\overline{\kappa}} & 0 \\ 0 & A_{\overline{\kappa}} \end{pmatrix} = \begin{pmatrix} I_{2k} & \rho I_{2k} \\ \rho I_{2k} & I_{2k} \end{pmatrix} = \Sigma_{4k}(\rho),$$

it follows that

$$\mathrm{vec}(\mathbf{X}_\kappa, \mathbf{Y}_\kappa) = \begin{pmatrix} A_\kappa & 0 \\ 0 & A_\kappa \end{pmatrix} \mathrm{vec}(\mathbf{X}, \mathbf{Y}) \sim N_{4k}(0, \Sigma_{4k}(\rho)).$$

Therefore, $\mathrm{vec}(\mathbf{X}, \mathbf{Y}) \stackrel{d}{=} \mathrm{vec}(\mathbf{X}_\kappa, \mathbf{Y}_\kappa)$, whence $\mathbb{E}[r(d_\alpha)(\mathbf{X})r(d_\alpha)(\mathbf{Y})] \stackrel{(i)}{=} \mathbb{E}[r(d)(\mathbf{X}_\kappa)r(d)(\mathbf{Y}_\kappa)] = \mathbb{E}[r(d)(\mathbf{X})r(d)(\mathbf{Y})]$ which proves (iii).

(iv) If $r = 0$, the claim immediately follows from the resulting independence of the standard Gaussian random vectors $\mathbf{X}$ and $\mathbf{Y}$. So, let $r \neq 0$. Let $\mathrm{vec}(\mathbf{X}_1, \mathbf{X}_2) \sim N_{4k}(0, \Sigma_{4k}(|r|))$. Then $\mathrm{vec}(\mathrm{sign}(r)\mathbf{X}_1, \mathbf{X}_2) \sim N_{4k}(0, \Sigma_{4k}(r))$. The second equality of statement (i) therefore implies that

$$h_{r(c_r),r(d)}(|r|) = \mathbb{E}[r(c_r)(\mathbf{X}_1)r(d)(\mathbf{X}_2)]$$
$$= \mathbb{E}[r(c)(\mathrm{sign}(r)\mathbf{X}_1)r(d)(\mathbf{X}_2)] = h_{r(c),r(d)}(r).$$

$\square$

Fix $\text{vec}(\mathbf{Z}, \mathbf{W}) \sim \mathbb{C}N_{2k}(0, \Sigma_{2k}(\zeta))$, where $\zeta \in \overline{\mathbb{D}} \setminus \{0\}$ and $k \in \mathbb{N}$. Let $b, c : \mathbb{C}^k \longrightarrow \mathbb{C}$, such that $b \otimes \overline{c} \in L^1(\mathbb{C}^{2k}, \mathbb{P}_{\text{vec}(\mathbf{Z}, \mathbf{W})})$. We put

$$h_{b,c}^{\mathbb{C}}(\zeta) := h^{\mathbb{C}}(b, c; \zeta) := \mathbb{E}[b(\mathbf{Z})\overline{c(\mathbf{W})}] = \overline{h_{c,b}^{\mathbb{C}}(\overline{\zeta})}.$$

As in the real case (see (6.2.17)), the joint multivariate Gaussian splitting property (3.5.2) of inner products of vectors on the unit sphere and Lemma 2.2 imply the important observation that for any separable $\mathbb{C}$-Hilbert space H, for any $m, n \in \mathbb{N}$, and for any $(u, v) \in S_H^m \times S_H^n$, we have

$$h_{b,c}^{\mathbb{C}}[\Gamma_H(u, v)] = \mathbb{E}\left[\overline{\mathbf{R}_c} \mathbf{S}_b^\top\right] = \mathbb{E}[\Gamma_{\mathbb{C}}(\mathbf{R}_c, \mathbf{S}_b)], \qquad (7.1.4)$$

where the $\mathbb{C}^m$-valued random vector $\mathbf{R}_c$ and the $\mathbb{C}^n$-valued random vector $\mathbf{S}_b$ are defined as $(\mathbf{R}_c)_i := c(\mathbf{Z}_{u_i})$ and $(\mathbf{S}_b)_j := b(\mathbf{Z}_{v_j})$, respectively $((i, j) \in [m] \times [n])$. Fix $f, g \in L^2(\mathbb{R}^k, \gamma_k)$ and put

$$H_{f,g} := h_{f,f} + h_{g,g}.$$

(6.2.9) implies a concrete integral representation of the function $H_{f,g}\big|_{(-1,1)}$, which particularly plays an important role in the complex case (cf. Corollary 7.1):

$$H_{f,g}(\rho) = \frac{1}{(2\pi)^k(1-\rho^2)^{k/2}} \int_{\mathbb{R}^k} \int_{\mathbb{R}^k} \left\langle \begin{pmatrix} f(x) \\ g(x) \end{pmatrix}, \begin{pmatrix} f(y) \\ g(y) \end{pmatrix} \right\rangle_{\mathbb{R}_2^2}$$
$$\exp\left(-\frac{\|x\|^2 + \|y\|^2 - 2\rho\langle x, y\rangle}{2(1-\rho^2)}\right) d^k x \, d^k y \qquad (7.1.5)$$

for all $\rho \in (-1, 1)$. Theorem 6.2 further implies that $H_{f,g}(\rho) = \sum_{\nu=0}^\infty a_\nu \rho^\nu$ for all $\rho \in [-1, 1]$, where $a_\nu := p_\nu(f, f) + p_\nu(g, g) \geq 0$ for all $\nu \in \mathbb{N}_0$. In particular, $H_{f,g}(1) = \|f\|_{\gamma_{2k}}^2 + \|g\|_{\gamma_{2k}}^2$, implying that $H_{f,g}$ is real analytic on $(-1, 1)$, continuous on $[-1, 1]$ and absolutely monotonic on $[0, 1]$. Moreover, $H_{f,g}$ is bounded, and

$$|H_{f,g}(\rho)| \leq \|f\|_{\gamma_{2k}}^2 + \|g\|_{\gamma_{2k}}^2 = H_{f,g}(1) \qquad (7.1.6)$$

for all $\rho \in [-1, 1]$. Hence, if $f \neq 0$ or $g \neq 0$, it follows that

$$H_{f,g} = H_{f,g}(1) \psi_{f,g}, \qquad (7.1.7)$$

where the function $\psi_{f,g} := \frac{H_{f,g}}{H_{f,g}(1)} : [-1, 1] \longrightarrow [-1, 1]$ is CCP (due to Theorem 5.3). In particular, $H_{f,g}\big|_{(-1,1)} \in W_+^\omega((-1, 1))$. Thus, if $b \in L^2(\mathbb{C}^k, \gamma_k^{\mathbb{C}})$,

7.1 Multivariate Complex CCP Functions and Their Relation to the Real Case

then (2.1.11) implies that $r(b) \in L^2(\mathbb{R}^{2k}, \gamma_{2k})$ and $s(b) \in L^2(\mathbb{R}^{2k}, \gamma_{2k})$, and

$$H_{r(b),s(b)}(1) = \|r(b)\|^2_{\gamma_{2k}} + \|s(b)\|^2_{\gamma_{2k}} = \|b\|^2_{\gamma_k^{\mathbb{C}}}. \tag{7.1.8}$$

If—in addition—b is odd and $\|b\|^2_{\gamma_k^{\mathbb{C}}} > 0$, then also $r(b)$ and $s(b)$ are odd functions (such as $H_{r(b),s(b)}$), satisfying $r(b) \neq 0$ or $s(b) \neq 0$. Hence, we may apply Theorem 6.3 to the well-defined odd CCP function $\psi_{r(b),s(b)}$, and it follows that $H_{r(b),s(b)} : [-1, 1] \longrightarrow [-\|b\|^2_{\gamma_k^{\mathbb{C}}}, \|b\|^2_{\gamma_k^{\mathbb{C}}}]$ is a strictly increasing homeomorphism, such as the inverse function $(H_{r(b),s(b)})^{-1} : [-\|b\|^2_{\gamma_k^{\mathbb{C}}}, \|b\|^2_{\gamma_k^{\mathbb{C}}}] \longrightarrow [-1, 1]$ (due to (7.1.8) and (7.1.7)). Similarly, if we—further—assume that $H'_{r(b),s(b)}(0) > 0$, Theorem 6.3 shows that also $(H_{r(b),s(b)})^{-1}$ is real analytic on $(-\|b\|^2_{\gamma_k^{\mathbb{C}}}, \|b\|^2_{\gamma_k^{\mathbb{C}}})$. So, we could apply Lemma 5.3 to the real analytic function $(H_{r(b),s(b)})^{-1}|_{(-\|b\|^2_{\gamma_k^{\mathbb{C}}}, \|b\|^2_{\gamma_k^{\mathbb{C}}})}$ (if the assumptions are given) to check the existence of $(H^{-1}_{r(b),s(b)})_{\text{abs}} : [-\|b\|^2_{\gamma_k^{\mathbb{C}}}, \|b\|^2_{\gamma_k^{\mathbb{C}}}] \longrightarrow \mathbb{R}$ then (a crucial assumption in Theorem 7.5). Equipped with these facts, we arrive at the complex version of Theorem 6.2:

Theorem 7.1 *Let $k \in \mathbb{N}$, $\zeta \in \overline{\mathbb{D}}$, $\mathbf{L} \sim \mathbb{C}N_k(0, I_k)$ and $\text{vec}(\mathbf{Z}, \mathbf{W}) \sim \mathbb{C}N_{2k}(0, \Sigma_{2k}(\zeta))$. Let $b \in L^2(\mathbb{C}^k, \gamma_k^{\mathbb{C}})$ and $c \in L^2(\mathbb{C}^k, \gamma_k^{\mathbb{C}})$. Then $b \otimes \overline{c} \in L^1(\mathbb{C}^k \times \mathbb{C}^k, \mathbb{P}_{\text{vec}(\mathbf{Z}, \mathbf{W})})$. If $\zeta = 0$, then*

$$\overline{h^{\mathbb{C}}_{c,b}(0)} = h^{\mathbb{C}}_{b,c}(0) = \mathbb{E}[b(\mathbf{L})]\,\mathbb{E}[\overline{c}(\mathbf{L})]$$
$$= h_{r(b),r(c)}(0) + h_{s(b),s(c)}(0) + i\,(h_{s(b),r(c)}(0) - h_{r(b),s(c)}(0)). \tag{7.1.9}$$

If $\zeta \neq 0$, then

$$\overline{h^{\mathbb{C}}_{c,b}(\overline{\zeta})} = h^{\mathbb{C}}_{b,c}(\zeta) = h_{r(b_\zeta),r(c)}(|\zeta|) + h_{s(b_\zeta),s(c)}(|\zeta|)$$
$$+ i\,(h_{s(b_\zeta),r(c)}(|\zeta|) - h_{r(b_\zeta),s(c)}(|\zeta|)). \tag{7.1.10}$$

In particular,

$$0 \leq h^{\mathbb{C}}_{b_{\overline{\zeta}},b}(\zeta) = H_{r(b),s(b)}(|\zeta|) \text{ and } h^{\mathbb{C}}_{b,c}(1) = \langle b, c \rangle_{\gamma_k^{\mathbb{C}}}. \tag{7.1.11}$$

$h^{\mathbb{C}}_{b,c} : \overline{\mathbb{D}} \longrightarrow \mathbb{C}$ *is bounded and satisfies*

$$|h^{\mathbb{C}}_{b,c}(\zeta)| \leq \|b\|_{\gamma_k^{\mathbb{C}}} \|c\|_{\gamma_k^{\mathbb{C}}} \text{ for all } \zeta \in \overline{\mathbb{D}}. \tag{7.1.12}$$

Proof Firstly, let $\zeta = 0$, implying that $\Sigma_{2k}(\zeta) = I_{2k}$. Thus, the partitioned random vector parts $\mathbf{Z} \sim \mathbb{C}N_k(0, I_k)$ and $\mathbf{W} \sim \mathbb{C}N_k(0, I_k)$ of the complex Gaussian random vector $\text{vec}(\mathbf{Z}, \mathbf{W}) \sim \mathbb{C}N_{2k}(0, I_{2k})$ are independent (cf. [5, Theorem 2.12]). If $\mathbf{X} \stackrel{d}{=}$

$\sqrt{2}J_2(\mathbf{Z})$ and $\mathbf{Y} \stackrel{d}{=} \sqrt{2}J_2(\mathbf{W})$, then $\mathbf{X} \sim N_{2k}(0, I_{2k})$ and $\mathbf{Y} \sim N_{2k}(0, I_{2k})$. (2.1.9) therefore implies that

$$\begin{aligned}\mathbb{E}[b(\mathbf{Z})\overline{c(\mathbf{W})}] &= \mathbb{E}[b(\mathbf{Z})]\mathbb{E}[\overline{c(\mathbf{W})}] \\ &= \mathbb{E}[r(b)(\mathbf{X})]\,\mathbb{E}[r(c)(\mathbf{Y})] + \mathbb{E}[s(b)(\mathbf{X})]\,\mathbb{E}[s(c)(\mathbf{Y})] \\ &\quad + i(\mathbb{E}[s(b)(\mathbf{X})]\mathbb{E}[r(c)(\mathbf{Y})] - \mathbb{E}[r(b)(\mathbf{X})]\mathbb{E}[s(c)(\mathbf{Y})]) \\ &= h_{r(b),r(c)}(0) + h_{s(b),s(c)}(0) + i(h_{s(b),r(c)}(0) - h_{r(b),s(c)}(0)), \end{aligned}$$
(7.1.13)

where the last equality follows from Theorem 6.2. Next, we consider the remaining case $\zeta \in \overline{\mathbb{D}} \setminus \{0\}$. Put $\sigma := \text{sign}(\overline{\zeta}) = \frac{\overline{\zeta}}{|\zeta|}$ and $\tau := \text{sign}(\zeta) = \frac{1}{\sigma}$. Lemma 7.1 implies that $r(b_\zeta) \in L^2(\mathbb{R}^{2k}, \gamma_{2k})$ and $\|r(b_\zeta)\|_2 = \|r(b)\|_2$. Similarly, it follows that $s(b_\zeta) \in L^2(\mathbb{R}^{2k}, \gamma_{2k})$, and $\|s(b_\zeta)\|_2 = \|s(b)\|_2$. The proof of (7.1.10) is built on Lemmas 2.2 and 7.1. To this end, we have to consider the block matrix

$$G := \begin{pmatrix} I_k & 0 & 0 & 0 \\ 0 & 0 & I_k & 0 \\ 0 & I_k & 0 & 0 \\ 0 & 0 & 0 & I_k \end{pmatrix} = G^\top = G^{-1} \in O(4k)$$

again (cf. (1.2.6)). Lemma 2.2 implies that $\text{vec}(\sigma\,\mathbf{Z}, \mathbf{W}) \sim \mathbb{C}N_{2k}(0, \Sigma_{2k}(|\zeta|))$ and

$$\mathbf{X} := \sqrt{2}\,G\,J_2(\text{vec}(\sigma\,\mathbf{Z}, \mathbf{W})) \stackrel{(1.2.5)}{=} \sqrt{2}\,\text{vec}(J_2(\sigma\,\mathbf{Z}), J_2(\mathbf{W})) \sim N_{4k}(0, \Sigma_{4k}(|\zeta|)).$$
(7.1.14)

Thus, Corollary 2.1, together with the equality $G^2 = I_{4k}$, yields

$$\begin{aligned}\mathbb{E}[b \otimes \overline{c}(\text{vec}(\mathbf{Z}, \mathbf{W}))] &= \mathbb{E}[b_\zeta \otimes \overline{c}(\text{vec}(\sigma\,\mathbf{Z}, \mathbf{W}))] \\ &= \mathbb{E}[r(b_\zeta \otimes \overline{c})(G\,\mathbf{X})] + i\,\mathbb{E}[s(b_\zeta \otimes \overline{c})(G\,\mathbf{X})]. \end{aligned}$$
(7.1.15)

Because of (7.1.2) and (7.1.3), it follows in particular that

$$r(b_\zeta \otimes \overline{c}) \circ G = r(b_\zeta) \otimes r(c) + s(b_\zeta) \otimes s(c)$$

on $\mathbb{R}^{2k} \times \mathbb{R}^{2k} \cong \mathbb{R}^{4k}$, and

$$s(b_\zeta \otimes \overline{c}) \circ G = s(b_\zeta) \otimes r(c) - r(b_\zeta) \otimes s(c),$$

on $\mathbb{R}^{2k} \times \mathbb{R}^{2k} \cong \mathbb{R}^{4k}$. Linearity of the expectation and Theorem 6.2, together with the fact that by construction $\mathbf{X} = \text{vec}(\mathbf{X}_1, \mathbf{X}_2)$, (with $\mathbf{X}_1 := \sqrt{2}\,J_2(\sigma\mathbf{Z})$ and and $\mathbf{X}_2 := \sqrt{2}\,J_2(\mathbf{W})$) now imply (7.1.10). Finally, to achieve the non-trivial

7.1 Multivariate Complex CCP Functions and Their Relation to the Real Case 179

boundedness statement (7.1.12) (which cannot simply be derived by making a standard use of the triangle inequality), we need to implement (a complexification of) the Ornstein-Uhlenbeck semigroup, respectively (6.2.15). Because of (7.1.9), the case $\zeta = 0$ is trivial, though. So, let us fix $0 \neq \zeta \in \overline{\mathbb{D}}$. Let $f \in \{r(b_\zeta), s(b_\zeta)\}$ and $g \in \{r(c), s(c)\}$. Then $h_{f,g}(|\zeta|) \stackrel{(6.2.15)}{=} \langle f, R_\zeta g \rangle_{\gamma_k}$, where $R_\zeta := T_{\vartheta(|\zeta|)}$. Consequently, since $r(R_\zeta g \circ \sqrt{2} J_2) = \mathrm{Re}(R_\zeta g \circ \sqrt{2} J_2) \circ \frac{1}{\sqrt{2}} J_2^{-1} = R_\zeta g = \mathrm{Im}(R_\zeta g \circ \sqrt{2} J_2) \circ \frac{1}{\sqrt{2}} J_2^{-1} = s(R_\zeta g \circ \sqrt{2} J_2)$, (2.1.14) leads to

$$h_{b,c}^{\mathbb{C}}(\zeta) \stackrel{(7.1.10)}{=} h_{r(b_\zeta),r(c)}(|\zeta|) + h_{s(b_\zeta),s(c)}(|\zeta|) + i\,(h_{s(b_\zeta),r(c)}(|\zeta|) - h_{r(b_\zeta),s(c)}(|\zeta|))$$
$$= \langle b_\zeta, \psi(c; \zeta) \rangle_{\gamma_k^{\mathbb{C}}},$$

where $\psi(c; \zeta) := R_\zeta r(c) \circ \sqrt{2} J_2 + i\,(R_\zeta s(c) \circ \sqrt{2} J_2)$. Consequently, we may apply the Cauchy-Schwarz inequality, and Lemma 7.1 implies that

$$|h_{b,c}(\zeta)| \leq \|b_\zeta\|_{\gamma_k^{\mathbb{C}}} \|\psi(c; \zeta)\|_{\gamma_k^{\mathbb{C}}} = \|b\|_{\gamma_k^{\mathbb{C}}} \|\psi(c; \zeta)\|_{\gamma_k^{\mathbb{C}}}.$$

Since $\|T_\zeta\|_{\mathcal{L}(L^2(\gamma_{2k}))} = 1$ (see [20, Theorem 1.4.1]), a two-fold application of (2.1.11) implies that

$$\|\psi(c; \zeta)\|_{\gamma_k^{\mathbb{C}}}^2 = \|R_\zeta r(c)\|_{\gamma_{2k}}^2 + \|R_\zeta s(c)\|_{\gamma_{2k}}^2 \leq \|r(c)\|_{\gamma_{2k}}^2 + \|s(c)\|_{\gamma_{2k}}^2 = \|c\|_{\gamma_k^{\mathbb{C}}}^2.$$

In conclusion, we finally obtain

$$|h_{b,c}(\zeta)| \leq \|b\|_{\gamma_k^{\mathbb{C}}} \|\psi(c; \zeta)\|_{\gamma_k^{\mathbb{C}}} \leq \|b\|_{\gamma_k^{\mathbb{C}}} \|c\|_{\gamma_k^{\mathbb{C}}}.$$

$\square$

By taking into account that $\mathrm{sign}(\zeta) |\zeta|^{2\nu+1} = \zeta \cdot \zeta^\nu \cdot \overline{\zeta}^\nu$ for all $\zeta \in \mathbb{C}$ and $\nu \in \mathbb{N}_0$, Theorem 7.1 also leads to a straightforward generalisation of the Haagerup function (cf. [60, Proof of Theorem 3.1] and Example 7.1). To this end, we introduce a class of complex-valued functions which could be viewed as a transfer of the class of all odd real-valued functions to the complex field and contains the complex signum function $\mathrm{sign} : \mathbb{C} \longrightarrow \mathbb{C}$ as element.

Definition 7.1 Let $\mathbb{F} \in \{\mathbb{R}, \mathbb{C}\}$ and $k \in \mathbb{N}$. A function $b : \mathbb{F}^k \longrightarrow \mathbb{F}$ is circularly symmetric if

$$b(\alpha z) = \alpha b(z) \text{ for all } (\alpha, z) \in S_\mathbb{F} \times \mathbb{F}^k.$$

The set of all circularly symmetric functions is denoted by $CS_k(S_\mathbb{F})$.

Definition 7.1 obviously implies that $CS_k(S_\mathbb{R}) = CS_k(\{-1, 1\})$ coincides with the set of all odd real functions from $\mathbb{R}^k$ to $\mathbb{R}$ and that $CS_k(S_\mathbb{C}) = CS_k(\mathbb{T})$. Moreover,

$(CS_k(S_\mathbb{F}), \circ, \mathrm{id})$ is a monoid (i.e., a semigroup, with unit element), where the binary operation $\circ$ is given by the composition of functions.

Corollary 7.1 *Let $k \in \mathbb{N}$ and $\zeta \in \overline{\mathbb{D}}$. Let $b, c \in L^2(\mathbb{C}^k, \gamma_k^\mathbb{C})$. Suppose that $b \in CS_k(\mathbb{T})$. Then*

(i)
$$h_{b,c}^\mathbb{C}(\zeta) = \mathrm{sign}(\zeta)(h_{r(b),r(c)}(|\zeta|) + h_{s(b),s(c)}(|\zeta|)$$
$$+ i\,(h_{s(b),r(c)}(|\zeta|) - h_{r(b),s(c)}(|\zeta|))). \tag{7.1.16}$$

$$h_{b,b}^\mathbb{C}(\zeta) = \mathrm{sign}(\zeta) H_{r(b),s(b)}(|\zeta|) = \zeta \sum_{\nu=0}^\infty (p_{2\nu+1}(r(b), r(b))$$
$$+ p_{2\nu+1}(s(b), s(b)))\zeta^\nu \overline{\zeta}^\nu. \tag{7.1.17}$$

In particular, we have:

(i-1) $h_{b,b}^\mathbb{C}|_{[-1,1]} = H_{r(b),s(b)} = h_{r(b),r(b)} + h_{s(b),s(b)}$ and $h_{b,b}^\mathbb{C}|_{(-1,1)} \in W_+^\omega((-1,1))$.

(i-2) *If $|\zeta| = 1$, then $h_{b,b}^\mathbb{C}(\zeta) = \zeta \|b\|_{\gamma_k^\mathbb{C}}^2$.*

(i-3) *If $\zeta \in \mathbb{D}$ and $\mathrm{vec}(\mathbf{X}, \mathbf{Y}) \sim N_{2k}(0, \Sigma_{2k}(|\zeta|))$, then*

$$h_{b,b}^\mathbb{C}(\zeta) = \frac{\mathrm{sign}(\zeta)}{(2\pi)^{2k}(1-|\zeta|^2)^k} \int_{\mathbb{R}^{2k}} \int_{\mathbb{R}^{2k}} \langle \binom{r(b)(x)}{s(b)(x)}, \binom{r(b)(y)}{s(b)(y)} \rangle_{\mathbb{R}_2^2}$$
$$\exp\left(-\frac{\|x\|^2 + \|y\|^2 - 2|\zeta|\langle x,y\rangle}{2(1-|\zeta|^2)}\right) d^{2k}x\, d^{2k}y$$
$$= \mathrm{sign}(\zeta)(\mathbb{E}[r(b)(\mathbf{X})r(b)(\mathbf{Y})] + \mathbb{E}[s(b)(\mathbf{X})s(b)(\mathbf{Y})]). \tag{7.1.18}$$

(ii) $h_{b,b}^\mathbb{C}$ *is bounded, and*

$$|h_{b,b}^\mathbb{C}(\zeta)| \le \|b\|_{\gamma_k^\mathbb{C}}^2 \text{ for all } \zeta \in \overline{\mathbb{D}}. \tag{7.1.19}$$

(iii) *If $b \ne 0$, then $\frac{1}{\|b\|_{\gamma_k^\mathbb{C}}^2} H_{r(b),s(b)}$ as well as $\frac{1}{\|b\|_{\gamma_k^\mathbb{C}}^2} h_{b,b}^\mathbb{C}$ are CCP functions. Both, $H_{r(b),s(b)} : [-1,1] \longrightarrow [-\|b\|_{\gamma_k^\mathbb{C}}^2, \|b\|_{\gamma_k^\mathbb{C}}^2]$ and $h_{b,b}^\mathbb{C} : \overline{\mathbb{D}} \longrightarrow \|b\|_{\gamma_k^\mathbb{C}}^2 \overline{\mathbb{D}}$ are circularly symmetric homeomorphisms. $H_{r(b),s(b)}$ is strictly increasing*

7.1 Multivariate Complex CCP Functions and Their Relation to the Real Case

and satisfies $H_{r(b),s(b)}((-1,0)) = (-\|b\|^2_{\gamma^{\mathbb{C}}_k}, 0)$ and $H_{r(b),s(b)}((0,1)) = (0, \|b\|^2_{\gamma^{\mathbb{C}}_k})$. Moreover, $h^{\mathbb{C}}_{b,b}(\mathbb{D}) = \mathbb{D}$ and

$$(h^{\mathbb{C}}_{b,b})^{-1}(w) = \text{sign}(w) H^{-1}_{r(b),s(b)}(|w|) \text{ for all } w \in \|b\|^2_{\gamma^{\mathbb{C}}_k}\overline{\mathbb{D}}. \qquad (7.1.20)$$

(iv) $h'_{r(b),r(b)}(0) > 0$ or $h'_{s(b),s(b)}(0) > 0$ if and only if $H^{-1}_{r(b),s(b)}|_{(-\|b\|^2_{\gamma^{\mathbb{C}}_k}, \|b\|^2_{\gamma^{\mathbb{C}}_k})} = (H_{r(b),s(b)}|_{(-1,1)})^{-1}$ is real analytic on $(-\|b\|^2_{\gamma^{\mathbb{C}}_k}, \|b\|^2_{\gamma^{\mathbb{C}}_k})$.

Proof

(i) Firstly, since $b \in CS_k(\mathbb{T})$, b in particular is an odd function; i.e., $b(-z) = -b(z)$ for all $z \in \mathbb{C}^k$. Thus, also $r(b)$ and $s(b)$ are odd (since $2\,\text{Re}(b) = b + \overline{b}$ and $2i\,\text{Im}(b) = b - \overline{b}$), such as $h_{r(b),r(b)}$, $h_{s(b),s(b)}$ and $H_{r(b),s(b)}$. Fix $\zeta \in \overline{\mathbb{D}}$. Since $b \in CS_k(\mathbb{T})$, it follows that

$$h^{\mathbb{C}}_{b,c}(\zeta) = \text{sign}(\zeta) h^{\mathbb{C}}_{b_{\overline{\zeta}},c}(\zeta)$$

(which also holds for $\zeta = 0$ due to (7.1.13)). (7.1.16) now follows from Lemma 7.1, (2.1.11) and (7.1.10), where the latter is applied to the well-defined function $\overline{\mathbb{D}} \setminus \{0\} \ni z \mapsto h^{\mathbb{C}}_{b_{\overline{\zeta}},c}(z)$. In particular,

$$h^{\mathbb{C}}_{b,b}(\zeta) = \text{sign}(\zeta) H_{r(b),s(b)}(|\zeta|)$$

(due to (7.1.11)). Therefore, Theorem 6.2, (iii), applied to the odd functions $h_{r(b),r(b)}$ and $h_{s(b),s(b)}$ immediately lead to the power series representation (7.1.17) of $h^{\mathbb{C}}_{b,b}$, which holds on $\overline{\mathbb{D}}$. (7.1.18) now follows from (2.2.3) and (7.1.5).

(ii) (7.1.19) is an implication of (7.1.12).

(iii) Assume that $b \neq 0$. Then $H_{r(b),s(b)} = \|b\|^2_{\gamma^{\mathbb{C}}_k} \psi_{r(b),s(b)}$, where $\psi_{r(b),s(b)} := \frac{H_{r(b),s(b)}}{\|b\|^2_{\gamma^{\mathbb{C}}_k}}$ is CCP (due to (7.1.7) and (7.1.8)). (7.1.17), together with Theorem 5.5 (where the latter is applied to the coefficients $a_{kl} := (p_{2\nu+1}(r(b),r(b)) + p_{2\nu+1}(s(b),s(b)))\delta_{k,l+1}$, $k, l \in \mathbb{N}_0$) implies that $\frac{1}{\|b\|^2_{\gamma^{\mathbb{C}}_k}} h^{\mathbb{C}}_{b,b} : \overline{\mathbb{D}} \longrightarrow \overline{\mathbb{D}}$ is a CCP function. (7.1.7) and Theorem 6.3 imply that both, $H_{r(b),s(b)}$ and $h^{\mathbb{C}}_{b,b}$ are homeomorphisms. The representation (7.1.20) of $(h^{\mathbb{C}}_{b,b})^{-1}$ follows at once. (i-2) and Theorem 6.3 obviously conclude the proof of (iii).

(iv) Since $h'_{r(b),r(b)}(0) > 0$ or $h'_{s(b),s(b)}(0) > 0$ if and only if $H'_{r(b),s(b)}(0) > 0$, and since $H_{r(b),s(b)}$ is odd, we may adopt the underlying idea of the proof of Theorem 6.3,(iv), implying statement (iv). □

Remark 7.1 Already the trivial example $b := 1$, $c := 1$ and $\zeta := i$ shows us that the additional assumption $b \in CS_k(\mathbb{T})$ in Corollary 7.1 cannot be dropped.

Recall (2.1.13), including the construction of the function $g^{\mathbb{C}}$ therein. Due to Proposition 6.5 and Corollary 7.1 we obtain

Remark 7.2 Let $k \in \mathbb{N}$ and $f \in L^2(\mathbb{R}^k, \gamma_k)$. Then $r(f^{\mathbb{C}}) = f \otimes 1$ and $s(f^{\mathbb{C}}) = 0$. Moreover,

$$h_{f,f} = h_{f \otimes 1, f \otimes 1} = H_{r(f^{\mathbb{C}}), s(f^{\mathbb{C}})}.$$

If in addition $f^{\mathbb{C}} \in CS_k(\mathbb{T})$, then $h_{f,f} = h_{f^{\mathbb{C}}, f^{\mathbb{C}}}^{\mathbb{C}}\big|_{[-1,1]}$.

At this point, it is very useful to recall Theorem 6.6 and its proof, where we also implemented Abel's theorem on *real* power series. From complex analysis it is well-known that in general Abel's theorem on power series in this form does not hold for $\mathbb{F} = \mathbb{C}$. Moreover, already the structure of the somewhat "simpler" complex-valued odd functions $h_{b,c}$ (cf. (7.1.16) and (7.1.17)) seemingly does not allow a transfer of Theorem 6.6, (ii) and Theorem 6.6, (iii) to the complex case (due to Theorem 5.5). However, since [59, Lemma 1.1.13, statement 2] also holds also for the complex field, the proof of Theorem 6.6 can be easily adapted, as well as the proof of Corollary 6.9, so that at least the following versions of a complex hybrid correlation transform can be stated at once:

Proposition 7.1 *Let* $m, n \in \mathbb{N}$, $0 < r < \infty$ *and*

$$M := \begin{pmatrix} A & S \\ S^* & B \end{pmatrix} \in M_{m+n}(r\overline{\mathbb{D}})^+$$

be positive semidefinite, where any entry of the matrices A, S and B is an element of $r\overline{\mathbb{D}}$. Let $f, g : (-r, r) \longrightarrow \mathbb{R}$ be two functions, such that $(-r, r) \ni x \mapsto f(x) = \sum_{\nu=0}^{\infty} a_\nu x^\nu \in W_+^{\omega}((-r, r))$ and $(-r, r) \ni x \mapsto g(x) = \sum_{\nu=0}^{\infty} b_\nu x^\nu \in W_+^{\omega}((-r, r))$. Let $q : r\mathbb{D} \longrightarrow \mathbb{C}$ be a holomorphic function, such that

$$|c_\nu| \le \sqrt{|a_\nu||b_\nu|} \text{ for all } \nu \in \mathbb{N}_0, \tag{7.1.21}$$

where $c_\nu := \frac{q^{(\nu)}(0)}{\nu!}$. Then $q_r \in W^+(\mathbb{D})$, where $q_r(\zeta) := q(r\zeta)$ for all $\zeta \in \mathbb{D}$. Put $r\overline{\mathbb{D}} \ni z \mapsto \tilde{q}(z) := \sum_{\nu=0}^{\infty} c_\nu z^\nu$, $r\overline{\mathbb{D}} \ni z \mapsto f_{abs}(z) := \sum_{\nu=0}^{\infty} |a_\nu| z^\nu$ and $r\overline{\mathbb{D}} \ni z \mapsto g_{abs}(z) := \sum_{\nu=0}^{\infty} |b_\nu| z^\nu$. Then $\widetilde{q_r}(\frac{z}{r}) = \tilde{q}(z)$ for all $z \in r\overline{\mathbb{D}}$ and $\tilde{q}\big|_{r\mathbb{D}} = q$, where $\widetilde{q_r} \in A(\mathbb{D})$ is the continuous extension of q_r. Moreover, the following properties hold:

(i)

$$\begin{pmatrix} f_{abs}[A] & \tilde{q}[S] \\ \tilde{q}[S]^* & g_{abs}[B] \end{pmatrix} \in M_{m+n}(r\overline{\mathbb{D}})^+$$

is positive semidefinite.

(ii) If $0 < c_* \leq r$ is a root of $f_{abs} - 1$, then

$$\tilde{q}[c_* \, \Gamma] \in \mathcal{Q}_{m,n}(\mathbb{C}) \text{ for all } \Gamma \in \mathcal{Q}_{m,n}(\mathbb{C}).$$

Thus, an application of Lemma 5.3 immediately leads us to

Corollary 7.2 *Let $m, n \in \mathbb{N}$, $A \in \mathbb{M}_{m,n}(\mathbb{C})$ and*

$$\Sigma = \begin{pmatrix} M & S \\ S^\top & N \end{pmatrix} \in C(m+n; \mathbb{C})$$

be an arbitrary complex $(m+n) \times (m+n)$ correlation matrix (with block elements $M \in C(m; \mathbb{C}), N \in C(n; \mathbb{C})$ and $S \in \mathcal{Q}_{m,n}(\mathbb{C})$). Let $r > 0$ and $0 \neq \psi \in W_+^\omega((-r, r))$. Put $r\overline{\mathbb{D}} \ni z \mapsto \tilde{\psi}(z) := \sum_{\nu=0}^\infty a_\nu z^\nu$, where $a_\nu := \frac{\psi^{(\nu)}(0)}{\nu!}$. If $0 < c \leq r$, then

$$\frac{1}{\psi_{abs}(c)} \begin{pmatrix} \psi_{abs}[c\,M] & \tilde{\psi}[c\,S] \\ \tilde{\psi}[c\,S^*] & \psi_{abs}[c\,N] \end{pmatrix} \in C(m+n; \mathbb{C})$$

again is a correlation matrix with complex entries. In particular,

$$\frac{1}{\psi_{abs}(c)} \tilde{\psi}[cS] \in \mathcal{Q}_{m,n}(\mathbb{C}) \text{ for all } S \in \mathcal{Q}_{m,n}(\mathbb{C}). \tag{7.1.22}$$

7.2 On Complex Bivariate Hermite Polynomials

Our next aim is to reveal in detail that it is possible to transfer the content of Theorem 6.8 from the real case to the complex one, while maintaining our constructive proof (including the intended avoidance of the tensor product language). However, we cannot simply copy the proof of Theorem 6.8. Nevertheless, we are going to unfurl that in fact it is possible to transfer (6.4.16) and (6.4.17) from the real case to the complex one. To this end, we are going to work with a particular case of the rich class of complex bivariate Hermite polynomials, first considered by K. Itô while working with complex multiple Wiener integrals (cf. [77]). Similarly to the real case, we need to verify a convenient correlation property of a random version of these polynomials, which to the best of our knowledge have not been published before (see Theorem 7.3 below). A detailed introduction to complex Hermite polynomials (which would exceed the topic of this monograph by far) can be studied in [51, 76]. Firstly, we have to recall the following general construction:

Definition 7.2 (Complex Hermite Polynomial) Let $m, n \in \mathbb{N}_0$ and $z, w \in \mathbb{C}$. The complex Hermite polynomial $H_{m,n} : \mathbb{C}^2 \longrightarrow \mathbb{C}$ is defined as

$$H_{m,n}(z,w) := \frac{1}{\sqrt{m!n!}} \sum_{j=0}^{m \wedge n} (-1)^j j! \binom{m}{j}\binom{n}{j} z^{m-j} w^{n-j}.$$

Within the scope of our work, we need the particular case of Itô's complex Hermite polynomials

$$\mathbb{C} \ni z \mapsto H_{m,n} \circ \kappa(z) := H_{m,n}(z, \bar{z})$$

$$= i^{m+n} \sum_{j=0}^{m}\sum_{k=0}^{n} i^{j+k}(-1)^{j+n} \sqrt{\binom{m}{j}\binom{n}{k}} s(j,k) H_{j+k}(\sqrt{2}\operatorname{Re}(z))$$

$$s(m-j, n-k) H_{m-j+n-k}(\sqrt{2}\operatorname{Im}(z)),$$

where $\kappa(z) := \operatorname{vec}(z, \bar{z})$ and $s(\nu, \mu) := \sqrt{\frac{(\nu+\mu)!}{\nu!\mu!}}$ for all $\nu, \mu \in \mathbb{N}_0$, including the following statements, which we give without proof (cf. [76, 77]).

Theorem 7.2 *Let $m, n \in \mathbb{N}_0$ and $z, w \in \mathbb{C}$. Then*

(i) $\{H_{m,n} \circ \kappa : m, n \in \mathbb{N}_0\}$ *is an orthonormal basis in the complex Hilbert space $L^2(\gamma_1^{\mathbb{C}})$.*

(ii) *The exponential generating function of $\{H_{m,n} \circ \kappa : m, n \in \mathbb{N}_0\}$ is given as*

$$\sum_{m,n=0}^{\infty} H_{m,n}(z, \bar{z}) \frac{u^m}{\sqrt{m!}} \frac{v^n}{\sqrt{n!}} = \exp(uz + v\bar{z} - uv) \text{ for all } u, v \in \mathbb{C}.$$

Lemma 2.1 allows us to transfer (the special case $k = 1$ of) Corollary 6.5 to the complex case. More precisely, we have

Theorem 7.3 *Let $m, n, \nu, \mu \in \mathbb{N}_0$ and $\zeta \in \overline{\mathbb{D}}$. If $\operatorname{vec}(Z, W) \sim \mathbb{C}N_2(0, \Sigma_2(\zeta))$, then*

$$\mathbb{E}\big[H_{m,n}(Z, \bar{Z})\overline{H_{\nu,\mu}(W, \bar{W})}\big] = \delta_{m,\nu}\, \delta_{n,\mu}\, \zeta^m \bar{\zeta}^n = \delta_{(m,n),(\nu,\mu)}\, \zeta^m \bar{\zeta}^n.$$

Proof Let $u, v, a, b \in \mathbb{C}$. Due to Theorem 7.2, a multiplication of the two (random) exponential generating functions leads to

$$\sum_{m,n=0}^{\infty} \sum_{\nu,\mu=0}^{\infty} H_{m,n}(Z, \bar{Z})\overline{H_{\nu,\mu}(W, \bar{W})} \frac{u^m}{\sqrt{m!}}\frac{v^n}{\sqrt{n!}}\frac{\bar{a}^\nu}{\sqrt{\nu!}}\frac{\bar{b}^\mu}{\sqrt{\mu!}}$$

$$= \exp\left(\binom{\bar{u}}{b}^* \binom{Z}{W} + \binom{\bar{v}}{a}^* \overline{\binom{Z}{W}}\right) \exp(-uv - \bar{a}\bar{b}).$$

Thus, if we put $\alpha_{m,n,\nu,\mu}(\zeta) := \mathbb{E}\big[H_{m,n}(Z,\overline{Z})\overline{H_{\nu,\mu}(W,\overline{W})}\big]$, then Lemma 2.1 implies that

$$\sum_{m,n=0}^{\infty}\sum_{\nu,\mu=0}^{\infty} \alpha_{m,n,\nu,\mu}(\zeta) \frac{u^m v^n \overline{a}^\nu \overline{b}^\mu}{\sqrt{m!n!\nu!\mu!}} = \exp\big((u,\overline{b})\Sigma_2(\zeta)\binom{v}{\overline{a}}\big)$$

$$\exp(-uv - \overline{a}\overline{b}) = \exp(u\overline{a}\zeta + \overline{b}v\overline{\zeta}).$$

Since $u, v, a, b \in \mathbb{C}$ were arbitrarily chosen, the multi-index notation shows us, that we actually have proven that

$$\sum_{n \in \mathbb{N}_0^4} \frac{\beta_n(\zeta)}{n!}(z_1,z_2,z_3,z_4)^n = \exp(z_1 z_3\,\zeta + z_2 z_4\,\overline{\zeta})$$

$$= \exp(z_1 z_3\,\zeta)\exp(z_2 z_4\,\overline{\zeta}) \text{ for all } z_1,z_2,z_3,z_4 \in \mathbb{C},$$

where $\beta_n(\zeta) := \sqrt{n!}\,\alpha_n(\zeta)$. Consequently, it follows that

$$\sum_{n \in \mathbb{N}_0^4} \big(\frac{\beta_n(\zeta) - \sqrt{n!}\,\delta_{n_1,n_3}\delta_{n_2,n_4}\,\zeta^{n_1}\overline{\zeta}^{n_2}}{n!}\big) z^n = 0 \text{ for all } z \in \mathbb{C}^4,$$

wherefrom the claim follows (by uniqueness of the multi-dimensional power series expansion for the entire function $f = 0$ around 0 [132, Ch. 1.2.2.]). □

7.3 Upper Bounds of $K_G^{\mathbb{C}}$ and Inversion of Complex CCP Functions

Equipped with the complex Hermite polynomials and Theorem 7.3, it is possible to transfer Theorem 6.4 from the real field $\mathbb{R}$ to the complex field $\mathbb{C}$; at least in the odd case. We "just" have to construct the mappings $\alpha_1^{\psi,c} \in S_{L^2(\gamma_1^{\mathbb{C}})}$ and $\beta_1^{\psi,c} \in S_{L^2(\gamma_1^{\mathbb{C}})}$ properly.

Theorem 7.4 *Let $k \in \mathbb{N}$ and $0 < c \le 1$. Let $0 \ne \psi \in W_+^\omega((-1,1))$ be odd. Then there exist $\alpha \equiv \alpha_{\psi,c}, \beta \equiv \beta_{\psi,c} \in S_{L^2(\gamma_1^{\mathbb{C}})}$, which satisfy the following properties:*

(i) $\mathbb{E}[\alpha(Z)] = \mathbb{E}[\beta(Z)] = 0$ *for all* $Z \sim \mathbb{C}N_1(0,1)$.
(ii) *If* $c\zeta \in \mathbb{D}$, *then*

$$\text{sign}(\zeta)\psi(c|\zeta|) = \psi_{abs}(c)\,h_{\alpha,\beta}^{\mathbb{C}}(\zeta)$$

and
$$sign(\zeta)\psi_{abs}(c|\zeta|) = h_{\alpha,\alpha}^{\mathbb{C}}(\zeta)\psi_{abs}(c) = h_{\beta,\beta}^{\mathbb{C}}(\zeta)\psi_{abs}(c).$$

In particular,
$$\psi(c) = \psi_{abs}(c)\,\langle\alpha,\beta\rangle_{\gamma_1^{\mathbb{C}}}. \tag{7.3.1}$$

(iii) *If $c \neq 1$ and H is an arbitrary $\mathbb{C}$-Hilbert space, then*
$$sign(\langle u, v\rangle_H)\psi(c|\langle u, v\rangle_H|) = \psi_{abs}(c)\,h_{\alpha,\beta}^{\mathbb{C}}(\langle u, v\rangle_H)$$

and
$$sign(\langle u, v\rangle_H)\psi_{abs}(c|\langle u, v\rangle_H|) = \psi_{abs}(c)\,h_{\alpha,\alpha}^{\mathbb{C}}(\langle u, v\rangle_H)$$
$$= \psi_{abs}(c)\,h_{\beta,\beta}^{\mathbb{C}}(\langle u, v\rangle_H)$$

for all $u, v \in S_H$.

Proof Let $0 < c \leq 1$ and $\zeta \in \overline{\mathbb{D}}$. Since $0 \neq \psi \in W_+^\omega((-1, 1))$, it follows that $\psi^{(n_0)}(0) \neq 0$ for some $n_0 \in \mathbb{N}_0$, implying that $\psi_{abs}(x) > 0$ for all $x \in (0, 1]$. Thus, $\psi_{abs}(c) > 0$. Put $\widetilde{b}_n \equiv \widetilde{b}_{n_1,n_2} := \frac{\psi^{(2n_1+1)}(0)}{(2n_1+1)!}\delta_{n_1,n_2}$, where $n = (n_1, n_2) \in \mathbb{N}_0^2$. Because of (3.5.2) and (3.5.3), we may choose complex random variables $Z_\zeta \sim \mathbb{C}N_1(0, 1)$ and $Z_1 \sim \mathbb{C}N_1(0, 1)$, such that $\text{vec}(Z_\zeta, Z_1) \sim \mathbb{C}N_2(0, \Sigma_2(\zeta))$, $\text{vec}(Z_\zeta, Z_\zeta) \sim \mathbb{C}N_2(0, \Sigma_2(1))$ and $\text{vec}(Z_1, Z_1) \sim \mathbb{C}N_2(0, \Sigma_2(\zeta))$. Theorem 7.3 therefore in particular implies that

$$\mathbb{E}\big[H_{k+1,l}(Z_\zeta, \overline{Z_\zeta})\overline{H_{\nu+1,\mu}(Z_1, \overline{Z_1})}\big] = \delta_{(k,l),(\nu,\mu)}\,\zeta^k\overline{\zeta}^l. \tag{7.3.2}$$

and
$$\mathbb{E}\big[H_{k+1,l}(Z_\zeta, \overline{Z_\zeta})\overline{H_{\nu+1,\mu}(Z_\zeta, \overline{Z_\zeta})}\big] = \delta_{(k,l),(\nu,\mu)}$$
$$= \mathbb{E}\big[H_{k+1,l}(Z_1, \overline{Z_1})\overline{H_{\nu+1,\mu}(Z_1, \overline{Z_1})}\big] \tag{7.3.3}$$

for all $k, l, \nu, \mu \in \mathbb{N}_0$. Consequently, since ψ is odd by assumption, a straightforward calculation (including the definition of the real numbers $\widetilde{b}_n$ and (2.1.11) and (7.3.3)) implies that

$$\mathbb{C} \ni z \mapsto \alpha(z) \equiv \alpha_{\psi,c}(z) := \frac{1}{\sqrt{\psi_{abs}(c)}} \sum_{n \in \mathbb{N}_0^2} \text{sign}(\widetilde{b}_n)\sqrt{|\widetilde{b}_n|}\,H_{n_1+1,n_2}(z, \overline{z})(\sqrt{c})^{|n|+1}$$

7.3 Upper Bounds of $K_G^{\mathbb{C}}$ and Inversion of Complex CCP Functions

and

$$\mathbb{C} \ni z \mapsto \beta(z) \equiv \beta_{\psi,c}(z) := \frac{1}{\sqrt{\psi_{\text{abs}}(c)}} \sum_{n \in \mathbb{N}_0^2} \sqrt{|\tilde{b}_n|} H_{m_1+1,m_2}(z,\overline{z})(\sqrt{c})^{|m|+1}$$

both lead to well-defined elements $\alpha \in L^2(\gamma_1^{\mathbb{C}})$ and $\beta \in L^2(\gamma_1^{\mathbb{C}})$.

(i) Let $Z \sim \mathbb{C}N_1(0,1)$. Then $(Z,Z)^\top \sim \mathbb{C}N_2(0, \Sigma_2(1))$ (due to (2.1.7)). Since $H_{0,0} = 1$, Theorem 7.3 therefore implies that

$$\mathbb{E}[H_{l_1+1,l_2}(Z,\overline{Z})] = \mathbb{E}[H_{l_1+1,l_2}(Z,\overline{Z})\overline{H_{0,0}(Z,\overline{Z})}] = 0 \text{ for all } (l_1,l_2) \in \mathbb{N}_0^2,$$

and (i) follows.

(ii) Similarly, as explained above, (7.3.2), together with the construction of the real numbers $\tilde{b}_n$ and the fact that $\zeta^{l+1}\overline{\zeta}^l = |\zeta|^{2l}\zeta = \text{sign}(\zeta)|\zeta|^{2l+1}$ for all $l \in \mathbb{N}_0$ implies that

$$\psi_{\text{abs}}(c)\mathbb{E}[\alpha(Z_\zeta)\overline{\alpha(Z_1)}] = \text{sign}(\zeta)\psi_{\text{abs}}(c|\zeta|) = \psi_{\text{abs}}(c)\mathbb{E}[\beta(Z_\zeta)\overline{\beta(Z_1)}].$$

and

$$\psi_{\text{abs}}(c)\mathbb{E}[\alpha(Z_\zeta)\overline{\beta(Z_1)}] = \text{sign}(\zeta)\psi(|c\,\zeta|).$$

However, $\mathbb{E}[\alpha(Z_\zeta)\overline{\beta(Z_1)}] = h_{\alpha,\beta}^{\mathbb{C}}(\zeta)$, $\mathbb{E}[\alpha(Z_\zeta)\overline{\alpha(Z_1)}] = h_{\alpha,\alpha}^{\mathbb{C}}(\zeta)$ and $\mathbb{E}[\beta(Z_\zeta)\overline{\beta(Z_1)}] = h_{\beta,\beta}^{\mathbb{C}}(\zeta)$, which proves (ii).

(iii) We just have to apply (ii) to $\overline{\mathbb{D}} \ni \zeta := \langle u,v \rangle_H$. □

Recall Corollary 7.1 including the structure of $h_{b,b}^{\mathbb{C}}$, strongly built on the odd homeomorphic real CCP function $\frac{1}{\|b\|_{\gamma_k^{\mathbb{C}}}^2} H_{r(b),s(b)} : [-1,1] \longrightarrow [-1,1]$. Thus, if we link (7.1.17), (7.1.20) and Theorem 7.4 (where the latter is applied to the function $\psi := \left(\frac{1}{\|b\|_{\gamma_k^{\mathbb{C}}}^2} H_{r(b),s(b)}\right)^{-1} = H_{r(b),s(b)}^{-1}(\|b\|_{\gamma_k^{\mathbb{C}}}^2 \cdot)$), we are able to prove

Theorem 7.5 (Complex Inner Product Rounding) *Let $k \in \mathbb{N}$ and $b \in S_{L^2(\gamma_k^{\mathbb{C}})}$. If $b \in CS_k(\mathbb{T})$ and $(H_{r(b),s(b)})^{-1}|_{(-1,1)} \in W_+^\omega((-1,1))$, then $\left(H_{r(b),s(b)}^{-1}\right)_{\text{abs}}(1) > 1$, and there exist $\alpha_b \in L^2(\gamma_1^{\mathbb{C}})$ and $\beta_b \in L^2(\gamma_1^{\mathbb{C}})$, satisfying $0 < \langle \alpha_b, \beta_b \rangle_{\gamma_1^{\mathbb{C}}} < 1$, such that for all $\mathbb{C}$-Hilbert spaces H and $u,v \in S_H$ the following properties are satisfied:*

(i)

$$\langle u,v \rangle_H = \frac{1}{c^*} h_{b,b}^{\mathbb{C}}(\zeta_{u,v}(b)), \tag{7.3.4}$$

where $0 < c^* \equiv c^*(b) := H_{r(b),s(b)}^{hyp}(1) < 1$ and $\zeta_{u,v}(b) := h_{\alpha_b,\beta_b}^{\mathbb{C}}(\langle u,v \rangle_H) \in \mathbb{D}$.

(ii)
$$c^* = H_{r(b),s(b)}(\langle \alpha_b, \beta_b \rangle_{\gamma_1^{\mathbb{C}}}) = h_{b,b}^{\mathbb{C}}(\langle \alpha_b, \beta_b \rangle_{\gamma_1^{\mathbb{C}}}).$$

(iii) *If* $vec(\mathbf{X}, \mathbf{Y}) \sim N_{2k}(0, \Sigma_{2k}(|\zeta_{u,v}(b)|))$, *then*

$$c^* \langle u,v \rangle_H = \frac{sign(\zeta_{u,v})}{(2\pi)^{2k}(1-|\zeta_{u,v}|^2)^k} \int_{\mathbb{R}^{2k}} \int_{\mathbb{R}^{2k}} \langle \binom{r(b)(x)}{s(b)(x)}, \binom{r(b)(y)}{s(b)(y)} \rangle_{\mathbb{R}_2^2}$$
$$\exp\left(-\frac{\|x\|^2 + \|y\|^2 - 2|\zeta_{u,v}(b)| \langle x,y \rangle}{2(1-|\zeta_{u,v}(b)|^2)}\right) d^{2k}x \, d^{2k}y$$
$$= sign(\zeta_{u,v}) \big(\mathbb{E}[r(b)(\mathbf{X})r(b)(\mathbf{Y})] + \mathbb{E}[s(b)(\mathbf{X})s(b)(\mathbf{Y})]\big),$$

(iv) *If* $m, n \in \mathbb{N}$ *and* $(z, w) \in S_H^m \times S_H^n$, *then there exist* $m + n$ $\mathbb{C}^k$-*valued random vectors* $\mathbf{Z}_1, \ldots, \mathbf{Z}_m, \mathbf{W}_1, \ldots, \mathbf{W}_n$, *such that* $vec(\mathbf{Z}_i, \mathbf{W}_j) \sim \mathbb{C}N_{2k}(0, \Sigma_{2k}(\zeta_{ij}(b)))$ *for all* $(i, j) \in [m] \times [n]$, *and*

$$\Gamma_H(z, w) = \frac{1}{c^*} \mathbb{E}[\overline{\mathbf{P}_b} \mathbf{Q}_b^\top], \qquad (7.3.5)$$

where $(\mathbf{P}_b)_i := b(\mathbf{Z}_i)$, $(\mathbf{Q}_b)_j := b(\mathbf{W}_j)$ *and* $\zeta_{ij}(b) := h_{\alpha_b,\beta_b}^{\mathbb{C}}(\langle z_i, w_j \rangle_H) \in \mathbb{D}$, $(i, j) \in [m] \times [n]$.

Proof

(i) Fix $u, v \in S_H$. Put $\psi := H_{r(b),s(b)}^{-1}$ and $\zeta := \langle u, v \rangle_H$. Since $b \in CS_k(\mathbb{T})$, it follows that both, $r(b)$ and $s(b)$ are odd. Hence, the CCP function $H_{r(b),s(b)} = h_{r(b),r(b)} + h_{s(b),s(b)}$ is odd as well (Corollary 7.1), implying that also its inverse ψ is an odd function. So, we may apply Proposition 6.4, together with Theorem 6.5, and it follows that $0 < c^* = H_{r(b),s(b)}^{hyp}(1) < 1$ is well-defined and satisfies $\psi_{abs}(c^*) = 1$. (7.1.20), together with Theorem 7.4 therefore implies that

$$(h_{b,b}^{\mathbb{C}})^{-1}(c^* \zeta) = sign(c^* \zeta) \psi(|c^* \zeta|) = sign(\zeta) \psi(c^* |\zeta|)$$
$$= \psi_{abs}(c^*) \zeta_{u,v}(b) = \zeta_{u,v}(b),$$

where $\alpha_b := \alpha_{\psi,c^*}$, $\beta_b := \beta_{\psi,c^*}$ and $\zeta_{u,v}(b) := h_{\alpha_b,\beta_b}^{\mathbb{C}}(\zeta) \in \overline{\mathbb{D}}$ (due to (7.1.12)). Hence, $c^* \zeta = h_{b,b}^{\mathbb{C}}(\zeta_{u,v}(b))$. However, since $c^* \zeta \in \mathbb{D}$, it follows that $\zeta_{u,v}(b) \in \mathbb{D}$ (due to Corollary 7.1, (i)).

7.3 Upper Bounds of $K_G^{\mathbb{C}}$ and Inversion of Complex CCP Functions

(ii) Since $H_{r(b),s(b)}(0) = 0 < c^* < 1 = \|b\|_{\gamma_k^{\mathbb{C}}}^2 = H_{r(b),s(b)}(1)$ and $\psi_{\mathrm{abs}}(c^*) = 1$, the strict monotonicity of the odd function $\psi = H_{r(b),s(b)}^{-1}$ implies that $0 = \psi(0) < \psi(c^*) \stackrel{(7.3.1)}{=} \langle \alpha_b, \beta_b \rangle_{\gamma_1^{\mathbb{C}}} < \psi(1) = 1$. Thus, $\mathrm{sign}(\langle \alpha_b, \beta_b \rangle_{\gamma_1^{\mathbb{C}}}) = 1$, and we obtain

$$c^* = H_{r(b),s(b)}(\langle \alpha_b, \beta_b \rangle_{\gamma_1^{\mathbb{C}}}) \stackrel{(7.1.17)}{=} h_{b,b}^{\mathbb{C}}(\langle \alpha_b, \beta_b \rangle_{\gamma_1^{\mathbb{C}}}).$$

(iii) This is an immediate application of (i) and Corollary 7.1 (since $\zeta_{u,v}(b) \in \mathbb{D}$).

(iv) Let $(z,w) \in S_H^m \times S_H^n$ and $(i,j) \in [m] \times [n]$. Because of (i), there exists $\mathrm{vec}(\mathbf{Z}_i, \mathbf{W}_j) \sim \mathbb{C}N_{2k}(0, \Sigma_{2k}(\zeta_{ij}(b)))$ (where $\zeta_{ij}(b) \equiv \zeta_{z_i,w_j}(b)$), such that

$$\Gamma_H(z,w)_{ij} = \langle w_j, z_i \rangle_H = \overline{\langle z_i, w_j \rangle_H} \stackrel{(i)}{=} \frac{1}{c^*} \overline{h_{b,b}(\zeta_{ij}(b))}$$

$$= \frac{1}{c^*} \mathbb{E}[\overline{b(\mathbf{Z}_i)} b(\mathbf{W}_j)] = \frac{1}{c^*} \mathbb{E}[\overline{(\mathbf{P}_b)_i} (\mathbf{Q}_b)_j] = \frac{1}{c^*} \mathbb{E}[\overline{\mathbf{P}_b} \mathbf{Q}_b^\top]_{ij}.$$

$\square$

Our next result shows that in fact also Theorem 6.9, respectively Corollary 6.12 can be transferred to the complex case:

Theorem 7.6 *Let $k,m,n \in \mathbb{N}$ and $b,c \in L^\infty(\mathbb{C}^k)$. Then $0 \leq r \equiv r_k(b) := \|b\|_{\gamma_k^{\mathbb{C}}}^2 < \infty$ and $\|b\|_\infty \geq \sqrt{r}$. Moreover, the following statements hold:*

(i)

$$|\mathrm{tr}(A^* h_{b,c}^{\mathbb{C}}[S])| \leq \|b\|_\infty \|c\|_\infty \|A\|_{\infty,1}^{\mathbb{C}} \text{ for all } A \in \mathbb{M}_{m,n}(\mathbb{C}) \text{ and } S \in \mathcal{Q}_{m,n}(\mathbb{C}).$$

(ii) *Assume that $b \in CS_k(\mathbb{T}) \setminus \{0\}$ and $H_{r(b),s(b)}^{-1}|_{(-r,r)} \in W_+^\omega((-r,r))$. Put $c^* \equiv c^*(b) := H_{\frac{r(b)}{\sqrt{r}}, \frac{s(b)}{\sqrt{r}}}^{hyp}(1)$. Then $c^* \in (0,1)$, and the following properties are satisfied:*

(ii-1)

$$r K_G^{\mathbb{C}} \leq \frac{\|b\|_\infty^2}{c^*}. \qquad (7.3.6)$$

In particular, if $\mathrm{sign}\big((H_{\frac{r(b)}{\sqrt{r}}, \frac{s(b)}{\sqrt{r}}}^{-1}|_{(-1,1)})^{(2n+1)}(0)\big) = (-1)^n$ for all $n \in \mathbb{N}_0$, then

$$r K_G^{\mathbb{C}} \leq i \frac{\|b\|_\infty^2}{F_b(i)},$$

where $F_b := \widetilde{H_{\frac{r(b)}{\sqrt{r}}, \frac{s(b)}{\sqrt{r}}}}|_{(-1,1)} : \overline{\mathbb{D}} \longrightarrow \overline{\mathbb{D}}$ is defined as in Lemma 5.3.

(ii-2) Let $1 \leq \kappa_* < K_G^{\mathbb{C}}$. If $\|b\|_\infty = \sqrt{r}$, then $0 < H_{\frac{r(b)}{\sqrt{r}}, \frac{s(b)}{\sqrt{r}}}^{hyp}(\kappa_*) < 1$ and there is exactly one number $\gamma^* \equiv \gamma^*(b) \in (H_{\frac{r(b)}{\sqrt{r}}, \frac{s(b)}{\sqrt{r}}}^{hyp}(\kappa_*), 1]$, such that

$$K_G^{\mathbb{C}} = \left(H_{r(b),s(b)}^{-1}\big|_{(-r,r)}\right)_{abs}(r\gamma^*) \leq \min\left\{\frac{1}{c^*}, \left(H_{r(b),s(b)}^{-1}\big|_{(-r,r)}\right)_{abs}(r)\right\}. \tag{7.3.7}$$

Proof Without loss of generality, as in the proof of the real case (Theorem 6.8), we may assume throughout the proof that $\|b\|_\infty = 1$ and $\|c\|_\infty = 1$ if $\|b\|_\infty > 0$ and $\|c\|_\infty > 0$ (else, we just have to rescale the pair (b, c) to the pair $\left(\frac{1}{\|b\|_\infty}b, \frac{1}{\|c\|_\infty}c\right)$). In particular, $\|b\|_\infty^2 = 1$.

(i) Because of (7.1.4), we may fully adopt the proof of Theorem 6.8.
(ii) Recall from Corollary 7.1 that both, $H_{r(b),s(b)} : [-1,1] \longrightarrow [-r,r]$ and $h_{b,b}^{\mathbb{C}} : \overline{\mathbb{D}} \longrightarrow r\overline{\mathbb{D}}$ are circularly symmetric homeomorphisms. Put $\psi_b := H_{r(b),s(b)}^{-1} : [-r, r] \longrightarrow [-1, 1]$ and $a_m := \frac{|\psi_b^{(m)}(0)|}{m!}$, $m \in \mathbb{N}$.

(ii-1) Given our assumptions, we may apply Theorem 7.5 to the function $H_{r(b^\circ),s(b^\circ)}^{-1} = \left(\frac{1}{r}H_{r(b),s(b)}\right)^{-1} = \psi_b(r\,\cdot)$, where $b^\circ := \frac{b}{\sqrt{r}} \in S_{L^2(\gamma_k^{\mathbb{C}})}$ is an element of the unit sphere of $L^2(\gamma_k^{\mathbb{C}})$. Put $\alpha^* := rc^*$. Thus, if $S \in \mathcal{Q}_{m,n}(\mathbb{C})$ is arbitrarily given, then

$$S \stackrel{(7.3.5)}{=} \frac{1}{c^*}\mathbb{E}[\overline{\mathbf{P}_{b^\circ}}\mathbf{Q}_{b^\circ}^\top] = \frac{1}{\alpha^*}\mathbb{E}[\overline{\mathbf{P}_b}\mathbf{Q}_b^\top],$$

where the m-dimensional complex random vector $\mathbf{P}_b$ maps into $\overline{\mathbb{D}}^m$ and the n-dimensional complex random vector $\mathbf{Q}_b$ maps into $\overline{\mathbb{D}}^n$ (since $\|b\|_\infty = 1$). Hence,

$$|\mathrm{tr}(A^*S)| = \frac{1}{\alpha^*}|\mathrm{tr}(A^*\mathbb{E}[\overline{\mathbf{P}_b}\mathbf{Q}_b^\top])| \leq \frac{1}{\alpha^*}\mathbb{E}[|\mathrm{tr}(A^*\overline{\mathbf{P}_b}\mathbf{Q}_b^\top)|] \leq \frac{1}{\alpha^*}\|A\|_{\infty,1},$$

and (7.3.6) follows. (6.2.26) concludes the proof of (ii-1).
(ii-2) Since $r = \|b\|_\infty^2 = 1$ (by assumption), it follows that $\sum_{\nu=0}^\infty a_{2\nu+1} = (\psi_b|_{(-1,1)})_{abs}(1) < \infty$. So, we may apply Lemma 5.3 to the odd real analytic function $\psi_b|_{(-1,1)}$, whence

$$\psi_b(|w|) = \sum_{\nu=0}^\infty a_{2\nu+1}|w|^{2\nu+1}$$

$$= |w|\sum_{\nu=0}^\infty a_{2\nu+1}w^\nu \overline{w}^\nu \text{ for all } w \in \overline{\mathbb{D}} = h_{b,b}^{\mathbb{C}}(\mathbb{D}).$$

7.3 Upper Bounds of $K_G^{\mathbb{C}}$ and Inversion of Complex CCP Functions

Thus, (7.1.20) implies that

$$\overline{\mathbb{D}} \ni (h_{b,b}^{\mathbb{C}})^{-1}(w) = \mathrm{sign}(w)\psi_b(|w|) = w \sum_{\nu=0}^{\infty} a_{2\nu+1} w^{\nu} \overline{w}^{\nu} \text{ for all } w \in \overline{\mathbb{D}}.$$

Consequently, the matrix equality (7.1.4) leads to

$$S = (h_{b,b}^{\mathbb{C}})^{-1}[h_{b,b}^{\mathbb{C}}[S]] = (h_{b,b}^{\mathbb{C}})^{-1}[\mathbb{E}[\mathbf{\Lambda}_b]] = \sum_{\nu=0}^{\infty} a_{2\nu+1} \mathbb{E}[\mathbf{\Lambda}_b]^{*\nu} * \mathbb{E}[\overline{\mathbf{\Lambda}_b}]^{*\nu} * \mathbb{E}[\mathbf{\Lambda}_b],$$

where $*$ denotes entrywise multiplication on $M_n(\mathbb{F})$ (i.e., the Hadamard product of matrices) and $\mathbf{\Lambda}_b := \overline{\mathbf{R}_b}\,\mathbf{S}_b^{\top}$. Put $M_{\nu} := \mathbb{E}[\mathbf{\Lambda}_b]^{*\nu} * \mathbb{E}[\overline{\mathbf{\Lambda}_b}]^{*\nu} = \overline{M_{\nu}}$. Since $\|b\|_{\infty} = 1$, we may apply Lemma 6.2 to the random matrix $\mathbf{\Lambda}_b$, and it follows that

$$|\mathrm{tr}(A^*S)| \leq \sum_{\nu=0}^{\infty} a_{2\nu+1} |\langle M_{\nu} * \mathbb{E}[\mathbf{\Lambda}_b], A\rangle_2| = \sum_{\nu=0}^{\infty} a_{2\nu+1} |\langle \mathbb{E}[\mathbf{\Lambda}_b], A * M_{\nu}\rangle_2|$$

$$\leq \sum_{\nu=1}^{\infty} a_{2\nu+1} \|A * M_{\nu}\|_{\infty,1},$$

whereby the last inequality follows from the construction of the random Gram matrix $\mathbf{\Lambda}_b$. Next, we may apply Lemma 6.2 twice, so that

$$\|A * M_{\nu}\|_{\infty,1} \leq \|A * \mathbb{E}[\mathbf{\Lambda}_b]^{*\nu}\|_{\infty,1} \leq \|A\|_{\infty,1}.$$

Altogether, it follows that

$$|\mathrm{tr}(A^*S)| \leq \|A\|_{\infty,1} \sum_{\nu=0}^{\infty} a_{2\nu+1} = (\psi_b|_{(-1,1)})_{\mathrm{abs}}(1) \|A\|_{\infty,1}.$$

Consequently, $K_G^{\mathbb{C}} \leq (\psi_b|_{(-1,1)})^{\mathrm{abs}}(1) = \left(H_{r(b),s(b)}^{-1}\big|_{(-r,r)}\right)_{\mathrm{abs}}(r)$. Hence, if we put $\gamma^* := H_{\frac{r(b)}{\sqrt{r}},\frac{s(b)}{\sqrt{r}}}^{\mathrm{hyp}}(K_G^{\mathbb{C}})$, then $0 < H_{\frac{r(b)}{\sqrt{r}},\frac{s(b)}{\sqrt{r}}}^{\mathrm{hyp}}(1) \leq H_{\frac{r(b)}{\sqrt{r}},\frac{s(b)}{\sqrt{r}}}^{\mathrm{hyp}}(\kappa_*) < \gamma^* \leq 1$, and (ii-1) concludes the proof of (7.3.7). □

Recall again from Theorem 5.5 that in general complex CCP functions need not be analytic. Due to this fact and the structure of the complex-valued (circularly symmetric) functions $h_{b,b}$, which is built on absolute values and signs of complex numbers (cf. Corollary 7.1, including (7.1.17)), it seems that Proposition 6.12 cannot be easily transferred to the complex case (if at all).

Example 7.1 (Haagerup Function) Both, Corollary 7.1 and Proposition 6.7 enable us to recover quickly a direct power series representation of Haagerup's

CCP function $h_{b,b} : \mathbb{D} \longrightarrow \mathbb{D}$, where $b :=$ sign. (Retranslated into the terminology of Haagerup, $h_{b,b} = \Phi$ and $H_{r(b),s(b)} = \varphi$ (cf. [60, Lemma 3.5 and Theorem 3.1])). To this end, let $0 \neq \zeta \in \overline{\mathbb{D}}$ and $\rho := |\zeta|$. As shown on [60, p. 200], the function $h_{b,b}$ can then be written in terms of $\mathrm{sign}(\zeta)$ and the two complete elliptic integrals $E(\rho)$ and $K(\rho)$. However, in our approach (by which Haagerup's CCP function is obtained as a special case), we don't have to work with elliptic integration. Firstly note that

$$r(b)(x_1, x_2) = r(b)(x) = \frac{x_1}{\|x\|_2} \text{ and } s(b)(x) = r(b)(x_2, x_1) = \frac{x_2}{\|x\|_2}$$

for all $x = \mathrm{vec}(x_1, x_2) \in \mathbb{R}^2 \setminus \{0\}$. Thus, if $\tau \in [-1, 1]$ and $\mathrm{vec}(\mathbf{X}, \mathbf{Y}) \sim N_{2k}(0, \Sigma_{2k}(\tau))$, $k := 2$ are given, then

$$\begin{aligned} H_{r(b),s(b)}(\tau) &= h_{r(b),r(b)}(\tau) + h_{s(b),s(b)}(\tau) \\ &= \mathbb{E}[r(b)(\mathbf{X})r(b)(\mathbf{Y})] + \mathbb{E}[s(b)(\mathbf{X})s(b)(\mathbf{Y})] \\ &= \mathbb{E}[r(b)(X_1, X_2)r(b)(Y_1, Y_2)] + \mathbb{E}[r(b)(X_2, X_1)r(b)(Y_2, Y_1)] \\ &= \mathbb{E}\left[\frac{X_1 Y_1}{\|\mathbf{X}\|_2 \|\mathbf{Y}\|_2}\right] + \mathbb{E}\left[\frac{X_2 Y_2}{\|\mathbf{X}\|_2 \|\mathbf{Y}\|_2}\right] \\ &= \mathbb{E}\left[\langle \frac{\mathbf{X}}{\|\mathbf{Y}\|_2}, \frac{\mathbf{Y}}{\|\mathbf{X}\|_2} \rangle\right] \\ &= \frac{\pi}{4} \tau \, {}_2F_1\left(\frac{1}{2}, \frac{1}{2}, 2; \tau^2\right) = h_{f_2, f_2}(\tau), \end{aligned}$$

whereby the equalities in the last line follow from Proposition 6.7 (applied to $k = 2$). An application of Corollary 7.1 (or the original proof of [60, Theorem 3.1]) therefore implies

$$\mathrm{sign}(\overline{\zeta}) h_{b,b}(\zeta) = H_{r(b),s(b)}(\rho) = \frac{\pi}{4} \rho \, {}_2F_1\left(\frac{1}{2}, \frac{1}{2}, 2; \rho^2\right) = h_{f_2, f_2}(\rho).$$

Hence,

$$\begin{aligned} h_{b,b}(\zeta) &= \mathrm{sign}(\zeta) \, \mathbb{E}\left[\langle \frac{\mathbf{X}}{\|\mathbf{X}\|_2}, \frac{\mathbf{Y}}{\|\mathbf{X}\|_2} \rangle\right] = \frac{\pi}{4} \zeta \, {}_2F_1\left(\frac{1}{2}, \frac{1}{2}, 2; |\zeta|^2\right) \\ &= \mathrm{sign}(\zeta) \, h_{f_2, f_2}(|\zeta|). \end{aligned}$$

A highly non-trivial part in [60] consists of a multi-page proof of the fact that $h_{f_2, f_2}^{-1} = H_{r(b),s(b)}^{-1} \in W_+^\omega((-1, 1))$, implying that $\left(h_{f_2, f_2}^{-1}\right)_{\mathrm{abs}}$ is well-defined

7.3 Upper Bounds of $K_G^{\mathbb{C}}$ and Inversion of Complex CCP Functions

(cf. [60, Lemma 2.6]). The Maclaurin series of $\left(h_{f_2,f_2}^{-1}\right)_{\text{abs}}$ can e.g. be approximated by the Taylor polynomial of degree 7 as:

$$\left(h_{f_2,f_2}^{-1}\right)_{\text{abs}}(s) = \frac{4}{\pi}s + \frac{8}{\pi^3}s^3 + 0 \cdot s^5 + \frac{16}{\pi^7}s^7 + o(|s|^7). \tag{7.3.8}$$

Hence (cf. Proposition 6.7 and (7.3.7)),

$$\frac{K_G^{\mathbb{R}}}{\sqrt{2}} \overset{(1.1.1)}{\leq} K_G^{\mathbb{C}} \overset{(7.3.7)}{\leq} \min\left\{\frac{1}{h_{f_2,f_2}^{\text{hyp}}(1)}, \left(h_{f_2,f_2}^{-1}\right)_{\text{abs}}(1)\right\} \tag{7.3.9}$$

$$\leq \left(h_{f_2,f_2}^{-1}\right)_{\text{abs}}(1) = \frac{4}{\pi} + \frac{8}{\pi^3} + \frac{16}{\pi^7} + o(1) \approx 1,53655 + o(1). \tag{7.3.10}$$

These facts follow from [60], respectively (6.3.1) and (9.1.8), where the latter is applied to

$$\alpha_\nu := \begin{cases} 0 & \text{if } \nu \text{ is even} \\ \frac{\pi}{2}\frac{((\nu-2)!!)^2}{((\nu-1)!!)^2 (\nu+1)} & \text{if } \nu \text{ is odd} \end{cases} \quad \text{and} \quad \alpha_\nu^\times := \frac{\alpha_\nu}{\alpha_1} = \frac{\alpha_\nu}{c_2^2} = \frac{4}{\pi}\alpha_\nu.$$

Already a numerical calculation of the single root $0 < c^* < \frac{\pi}{4}$ of the polynomial $s \mapsto \frac{4}{\pi}s + \frac{8}{\pi^3}s^3 + \frac{16}{\pi^7}s^7 - 1$ leads to the number $\frac{1}{c^*} \approx 1.40449$. The latter outcome should now be compared with the result of Haagerup in [60].

Chapter 8
A Summary Scheme of the Main Result

To highlight and summarise our approach, it completely suffices to list in detail the single steps and assumptions in the form of a "flowchart", possibly leading to a computer-aided approach regarding the implementation of an approximation to the lowest upper bound of the Grothendieck constant $K_G^\mathbb{F}$ as a next step. *Very likely, supercomputers are required to perform these approximations. That (technical) implementation would go however far beyond the scope of our groundwork; especially since we have no access to equipment of this type.*

Fix $\mathbb{F} \in \{\mathbb{R}, \mathbb{C}\}$ and $k \in \mathbb{N}$.

(SIGN) Choose a function $0 \neq b : \mathbb{F}^k \longrightarrow \mathbb{F}$, which satisfies the following conditions:
 (a) b is circularly symmetric (*cf. Definition 7.1*);
 (b) $b \in L^\infty(\mathbb{F}^k)$.

(CCP) Consider the function $b_\mathbb{F}^\circ : \mathbb{C}^k \longrightarrow \mathbb{C}$, defined as

$$b^\circ \equiv b_\mathbb{F}^\circ := \begin{cases} \left(\frac{b}{\|b\|_{\gamma_k}}\right)^\mathbb{C} & \text{if } \mathbb{F} = \mathbb{R} \\ \frac{b}{\|b\|_{\gamma_k^\mathbb{C}}} & \text{if } \mathbb{F} = \mathbb{C}. \end{cases}$$

Then $\|b^\circ\|_{\gamma_k^\mathbb{C}} = 1$ and $1 \leq \|b^\circ\|_\infty = \frac{\|b\|_\infty^\mathbb{F}}{\|b\|_{\gamma_k^\mathbb{F}}}$ (*cf.* (2.1.13)). Construct its allocated homeomorphic real CCP function $H_{r(b^\circ), s(b^\circ)} = h_{r(b^\circ), r(b^\circ)} + h_{s(b^\circ), s(b^\circ)}$ (*cf. Corollary 7.1 and Remark 7.2*).

(CRA) Assume that $H_{r(b^\circ), s(b^\circ)}^{-1}\big|_{(-1,1)} \in W_+^\omega((-1, 1))$ (*cf. Definition 5.3*).

An application of Lemma 6.1, Corollary 6.12 and Theorem 7.6 consequently leads to the following result which holds for both, $\mathbb{R}$ and $\mathbb{C}$ simultaneously:

Assume that (SIGN), (CCP) and (CRA) are satisfied. Then $\left(H^{-1}_{r(b^\circ),s(b^\circ)}\big|_{(-1,1)}\right)_{\text{abs}}(1) > 1$. Put $c^*_\mathbb{F} := H^{\text{hyp}}_{r(b^\circ),s(b^\circ)}(1)$ (cf. Lemma 6.1). Then $0 < c^*_\mathbb{F} < 1$, and the following statements hold:

(i)
$$K^\mathbb{F}_G \leq \frac{1}{c^*_\mathbb{F}} \|b^\circ\|^2_\infty.$$

In particular, if $\text{sign}\left((H^{-1}_{r(b^\circ),s(b^\circ)}\big|_{(-1,1)})^{(2n+1)}(0)\right) = (-1)^n$ for all $n \in \mathbb{N}_0$, then

$$K^\mathbb{F}_G \leq i \, \frac{\|b^\circ\|^2_\infty}{\widetilde{\psi}_b(i)},$$

where $\psi_b := H_{r(b^\circ),s(b^\circ)}\big|_{(-1,1)}$ and $\widetilde{\psi}_b : \overline{\mathbb{D}} \longrightarrow \overline{\mathbb{D}}$ is defined as in Lemma 5.3.

(ii) Let $1 \leq \kappa_* < K^\mathbb{F}_G$. If $\|b\|^\mathbb{F}_\infty = \|b\|_{\gamma^\mathbb{F}_k}$, then $0 < H^{\text{hyp}}_{r(b^\circ),s(b^\circ)}(\kappa_*) < 1$ and there is exactly one number $\gamma^*_\mathbb{F} \equiv \gamma^*_\mathbb{F}(b) \in (H^{\text{hyp}}_{r(b^\circ),s(b^\circ)}(\kappa_*), 1]$, such that

$$K^\mathbb{F}_G = \left(H^{-1}_{r(b^\circ),s(b^\circ)}\big|_{(-1,1)}\right)_{\text{abs}}(\gamma^*_\mathbb{F}) \leq \min\left\{\frac{1}{c^*_\mathbb{F}}, \left(H^{-1}_{r(b^\circ),s(b^\circ)}\big|_{(-1,1)}\right)_{\text{abs}}(1)\right\}.$$

Again, we recognise that Maclaurin series representation (or at least its approximation by the Taylor polynomial of a given degree) and Maclaurin series inversion of CCP functions play the key role regarding the search for the lowest upper bound of the Grothendieck constant $K^\mathbb{F}_G$. Unfortunately, a closed form representation of the coefficients of the inverse of a Taylor series runs against a well-known combinatorial complexity issue (due to the presence of ordinary partial Bell polynomials as building blocks of these coefficients—cf. Sect. 9.1 below for details), which in general does not allow a closed form representation of these coefficients. The inverse of the real function factor of the Haagerup function is one such example. It is given by

$$[-1,1] \ni \tau \mapsto H_{r(\text{sign}),s(\text{sign})}(\tau) = h_{f_2,f_2}(\tau) = \frac{\pi}{4} \tau \, _2F_1\left(\frac{1}{2}, \frac{1}{2}, 2; \tau^2\right)$$

8 A Summary Scheme of the Main Result

in the complex case (cf. [60], Remark on page 216 and Example 7.1), as opposed to the Grothendieck function

$$[-1, 1] \ni \rho \mapsto H_{r(\text{sign}^{\mathbb{C}}), s(\text{sign}^{\mathbb{C}})}(\rho) = h_{\text{sign}, \text{sign}}(\rho) = \frac{2}{\pi} \arcsin(\rho)$$

in the real case.

Chapter 9
Concluding Remarks and Open Problems

Not very surprisingly, the long-standing, intensive and technically quite demanding attempts to compute the—still not available—value of the real and complex Grothendieck constants (an open problem since 1953) leads to further projects and open problems, such as the following ones; addressed in particular to researchers who also wish to get a better understanding of the reasons underlying these topics.

9.1 Open Problem 1: Grothendieck Constant Versus Taylor Series Inversion

Only between 2011 and 2013 it was shown that $K_G^{\mathbb{R}}$ is strictly smaller than Krivine's upper bound, stating that $K_G^{\mathbb{R}} < \frac{\pi}{2\ln(1+\sqrt{2})}$ (cf. [22] and Example 6.10). Consequently, in the real case sign is not the "optimal" function to choose (answering a question of H. König to the negative—cf. [87]). So, if we wish to reduce the value of the upper bound of the real Grothendieck constant we have to look for functions $b : \mathbb{R}^k \longrightarrow \mathbb{R}$ which are different from sign $: \mathbb{R} \longrightarrow \{-1, 1\}$. However, these functions are required to satisfy any of the conditions in the workflow, listed in Chap. 8. In particular, we have to look for both, the Fourier-Hermite coefficients of the Taylor series (respectively the approximating Taylor polynomial) of $H_{r(b),s(b)} = h_{r(b),r(b)} + h_{s(b),s(b)}$ and the coefficients of the Taylor series of both, the *inverse* function $H_{r(b),s(b)}^{-1}$ and $\left(H_{r(b),s(b)}^{-1}\right)_{\text{abs}}$. It is well-known that the latter task increases rapidly in computational complexity if we want to calculate such Taylor coefficients of a higher degree, leading to the involvement of highly non-trivial combinatorial aspects, concretised by the use of partitions of positive integers and partial exponential Bell polynomials as part of the Taylor coefficients of the inverse Taylor series (a thorough introduction to this framework including the related Lagrange-Bürmann inversion formula is given in [29, 85]).

To reveal the origin of these difficulties let us focus on the one-dimensional real case. Let $b \in L^2(\mathbb{R}, \gamma_1)$ be given. Assume that $\alpha_0 := h_{b,b}(0) = 0$ (which is the case if b were odd). First recall from (6.2.3) that

$$h_{b,b}(\rho) = \sum_{n=1}^{\infty} \langle b, H_n \rangle_{\gamma_1}^2 \rho^n$$

for all $\rho \in [-1, 1]$, where for $n \in \mathbb{N}$ and $x \in \mathbb{R}$

$$H_n(x) := \frac{1}{\sqrt{n!}} (-1)^n \exp\left(\frac{x^2}{2}\right) \frac{d^n}{dx^n} \exp\left(-\frac{x^2}{2}\right)$$

denotes the (probabilistic version of the) n-th Hermite polynomial. Put $\alpha_n := \langle b, H_n \rangle_{\gamma_1}^2$. If $h'_{b,b}(0) = \alpha_1 > 0$, we know that the real analytic function $h_{b,b}|_{(-1,1)}$ is invertible around $0 = h_{b,b}(0)$. Its inverse is also expressible as a power series there; i.e., around 0, $\left(h_{b,b}|_{(-1,1)}\right)^{-1}$ is real analytic, too. Hence, given the assumption (CRA), listed in Chap. 8, it follows that $g_b(y) := h_{b,b}^{-1}(y) = \sum_{n=1}^{\infty} \beta_n y^n$ for all $y \in [-1, 1]$, where $\beta_1 = \frac{1}{\alpha_1}$ and

$$\beta_n = \frac{1}{n} \sum_{k=1}^{n-1} \frac{1}{\alpha_1^{n+k}} (-1)^k \binom{n-1+k}{k} B_{n-1,k}^{\circ}(\alpha_2, \alpha_3, \ldots, \alpha_{n-k+1})$$

$$= \frac{1}{n \alpha_1^n} \sum_{k=1}^{n-1} (-1)^k \binom{n-1+k}{k} B_{n-1,k}^{\circ}\left(\frac{\alpha_2}{\alpha_1}, \frac{\alpha_3}{\alpha_1}, \ldots, \frac{\alpha_{n-k+1}}{\alpha_1}\right) \quad (9.1.1)$$

for all $n \in \mathbb{N}_2$. In this context,

$$B_{n,k}^{\circ}(x_1, x_2, \ldots, x_{n-k+1}) := \sum_{v \in P(n,k)} \binom{k}{v_1, v_2, \ldots, v_{n-k+1}} \prod_{i=1}^{n-k+1} x_i^{v_i}$$

$$= \sum_{v \in P(n,k)} k! \frac{x^v}{v!}$$

denotes the ordinary partial Bell polynomial, where the multinomial coefficient $\binom{\sum_{i=1}^{n-k+1} v_i}{v_1, v_2, \ldots, v_{n-k+1}} := \frac{\left(\sum_{i=1}^{n-k+1} v_i\right)!}{\prod_{i=1}^{n+1-k} v_i!}$ represents the number of ways of allocating $\sum_{i=1}^{n-k+1} v_i$ distinct objects into $n - k + 1$ distinct bins, with v_i objects in the i'th

9.1 Open Problem 1: Grothendieck Constant Versus Taylor Series Inversion

bin and $P(n, k)$ indicates the set of all multi-indices $v \equiv (v_1, v_2, \ldots, v_{n+1-k}) \in \mathbb{N}_0^{n-k+1}$ ($k \leq n$) which satisfy the Diophantine equations

$$\sum_{i=1}^{n+1-k} v_i = k \text{ and } \sum_{i=1}^{n+1-k} i\, v_i = n;$$

i.e., summation is extended over all partitions of the number n into k positive (non-zero) integers (cf. e.g. [29, 31, 101, 146]). Observe that (dependent on the choice on n and k, of course) these Diophantine equations may have an extremely large, if not even an unmanageable set of solutions! Already that definition implies the well-known and important fact that

$$B_{n,k}^{\circ}(abx_1, ab^2 x_2, \ldots, ab^{n-k+1} x_{n-k+1}) = a^k\, b^n\, B_{n,k}^{\circ}(x_1, x_2, \ldots, x_{n-k+1}), \tag{9.1.2}$$

for all $a, b, x_1, \ldots, x_{n-k+1} \in \mathbb{C}$. If the element $(x_1, \ldots, x_{n-k+1}) \in \mathbb{N}_0^{n-k+1}$ consists of at most 2 non-zero elements, x_{i_1} and x_{i_2}, say ($i_1 < i_2$), we only have to sum over the set of all $v \in P(n, k)$, such that $v_i = 0$ for all $i \notin \{i_1, i_2\}$ (since $0^0 = 1$). In this case, the Diophantine equations reduce to a simple two-dimensional linear equation system. The latter has a solution $(v_{i_1}, v_{i_2}) \in \mathbb{N}_0^2$ if and only if

$$i_1 \leq \frac{n}{k} \leq i_2 \text{ and } i_2 - i_1 \text{ divides both, } k i_2 - n \in \mathbb{N}_0 \text{ and } n - k i_1 \in \mathbb{N}_0. \tag{9.1.3}$$

The solution $(v_{i_1}, v_{i_2}) \in \mathbb{N}_0^2$ is then unique and given as

$$v_{i_1} = \frac{k i_2 - n}{i_2 - i_1} \text{ and } v_{i_2} = \frac{n - k i_1}{i_2 - i_1}.$$

Hence, if (9.1.3) is satisfied, we immediately recognise that

$$B_{n,k}^{\circ}(0, 0, \ldots, 0, x_{i_1}, 0, 0, \ldots, 0, x_{i_2}, 0, 0, \ldots, 0) = \binom{k}{v_{i_1}} x_{i_1}^{v_{i_1}} x_{i_2}^{v_{i_2}}.$$

In particular, if $k \leq n \leq 2k$, we reobtain the well-known special case

$$B_{n,k}^{\circ}(x, y, 0, \ldots 0) = \binom{k}{2k - n} x^{2k-n} y^{n-k} \text{ for all } x, y \in \mathbb{C}.$$

In the literature, one frequently finds the so-called *exponential partial Bell polynomials* $B_{n,k}$, characterised as (cf. e.g. [29, Remark on page 136])

$$B_{n,k}(x_1, \ldots, x_{n-k+1}) := \frac{n!}{k!} B_{n,k}^{\circ}\left(\frac{x_1}{1!}, \frac{x_2}{2!}, \ldots, \frac{x_{n-k+1}}{(n-k+1)!}\right), \tag{9.1.4}$$

implying that any result about $B_{n,k}$ can be directly converted into a result about $B_{n,k}^\circ$ and conversely. For example, if $n > k$, then [29, formula (31)] transforms very pleasantly and reiteratively into

$$B_{n,k}^\circ(x_1, x_2, \ldots x_{n-k+1}) = \sum_{l=k-\alpha(n,k)}^{k-1} \binom{k}{l} x_1^l B_{n-k,k-l}^\circ(x_2, \ldots, x_{n-2k+l+1})$$

$$= x_1^{k-\alpha(n,k)} \sum_{l=0}^{\alpha(n,k)-1} \binom{k}{\alpha(n,k)-l}$$

$$x_1^l B_{n-k,\alpha(n,k)-l}^\circ(x_2, \ldots, x_{(n-2k)^+ + l + 2})$$

for all $x_1, x_2, \ldots, x_{n-k+1} \in \mathbb{C}$, where $\alpha(n, k) := \min\{n-k, k\}$. Another application of (9.1.4) implies the well-known fact that (9.1.1) is equivalent to

$$\left(h_{b,b}^{-1}\right)^{(n)}(0) = \sum_{k=1}^{n-1} \frac{1}{\delta_1^{n+k}} (-1)^k \frac{(n-1+k)!}{(n-1)!} B_{n-1,k}(\delta_2, \delta_3, \ldots, \delta_{n-k+1}),$$

(9.1.5)

where $\delta_l := \frac{h_{b,b}^{(l)}(0)}{l}, l \in [n-k+1]$ (cf. [29, Theorem E on p. 150]).

Regarding an explicit recursive construction of these polynomials in full generality, yet without having to know the sets $P(n, k)$ beforehand, we recall the important fact that any ordinary partial Bell polynomial $B_{n,k}^\circ$ actually arises as a (kind of) discrete convolution of two ordinary partial Bell polynomial series. More precisely, we have:

Lemma 9.1 *Let $m \in \mathbb{N}_0$, $k \in \mathbb{N}$, $n \in \mathbb{N}_k$ and $x_1, \ldots, x_{n-k+1} \in \mathbb{C}$. Then the following equalities are satisfied:*

(i) $B_{m,0}^\circ(x_1, \ldots, x_{m+1}) = \delta_{m0}$ and

$$B_{n,k}^\circ(x_1, \ldots, x_{n-k+1}) = \sum_{i=k-1}^{n-1} x_{n-i} B_{i,k-1}^\circ(x_1, \ldots, x_{i-k+2})$$

$$= \sum_{i=1}^{n-k+1} x_i B_{n-i,k-1}^\circ(x_1, \ldots, x_{n-k+2-i}).$$

9.1 Open Problem 1: Grothendieck Constant Versus Taylor Series Inversion

(ii)

$$n B_{n,k}^{\circ}(x_1, \ldots, x_{n-k+1}) = k \sum_{i=k-1}^{n-1} (n-i) x_{n-i}\, B_{i,k-1}^{\circ}(x_1, x_2, \ldots, x_{i-k+2})$$

$$= k \sum_{i=1}^{n-k+1} i x_i\, B_{n-i,k-1}^{\circ}(x_1, x_2, \ldots, x_{n-k+2-i}).$$

Proof

(i) Follows from [29, page 136, formula (3k)], respectively [117, page 366, formula (13)].
(ii) See [31, formula (2.3) and its equivalent (unnumbered) representation on page 1546 (line 3)].

□

Lemma 9.1 obviously implies the following multiple-sum representation of the ordinary partial Bell polynomials:

$$B_{n,k+1}^{\circ}(x_1, x_2, \ldots, x_{n-k}) = \sum_{i_1=k}^{n-1} \sum_{i_2=k-1}^{i_1-1} \cdots \sum_{i_k=1}^{i_{k-1}-1} x_{n-i_1} \prod_{\nu=2}^{k} x_{i_{\nu-1}-i_\nu}\, x_{i_k} \quad (9.1.6)$$

for all $k \in \mathbb{N}$, $n \in \mathbb{N}_{k+1}$ and $x_1, x_2, \ldots, x_{n-k} \in \mathbb{C}$. For the convenience of the readers, we list a few examples of ordinary partial Bell polynomials that can be displayed in closed form. For review, we refer to the widely comprehensive table of these polynomials on page 309 of [29] (enumerating all polynomials $B_{n,m}^{\circ}$ for which $10 \geq n \geq m \geq 1$). To this end, fix $k \in \mathbb{N}$ and $x_1, \ldots, x_{k+1} \in \mathbb{C}$. Then

(i) $B_{0,0}^{\circ}(x_1) = 1$ and $B_{k,0}^{\circ}(x_1, \ldots, x_{k+1}) = 0$.
(ii) $B_{k,1}^{\circ}(x_1, \ldots, x_k) = x_k$ and $B_{k,k}^{\circ}(x_1) = x_1^k$.
(iii) $B_{k+1,k}^{\circ}(x_1, x_2) = k\, x_1^{k-1}\, x_2$.
(iv) $B_{k+2,k}^{\circ}(x_1, x_2, x_3) = \binom{k}{2} x_1^{k-2} x_2^2 + k\, x_1^{k-1}\, x_3$ if $k \geq 2$.
(v) $B_{k+3,k}^{\circ}(x_1, x_2, x_3, x_4) = \binom{k}{3} x_1^{k-3} x_2^3 + k(k-1) x_1^{k-2} x_2 x_3 + k\, x_1^{k-1}\, x_4$ if $k \geq 3$.
(vi) $B_{k+4,k}^{\circ}(x_1, \ldots, x_5) = \binom{k}{4} x_1^{k-4} x_2^4 + \binom{k}{3} x_1^{k-3} (3\, x_2^2\, x_3) + \binom{k}{2} x_1^{k-2} (x_3^2 + 2 x_2 x_4) + k\, x_1^{k-1}\, x_5$ if $k \geq 4$.

Moreover, we have

$$B_{k,2}^\circ(x_1,\ldots,x_{k-1}) = \sum_{i=1}^{k-1} x_i\, x_{k-i} \text{ if } k \geq 2.$$

Since the Taylor series of the inverse of the "standardised" Taylor series $\sum_{n=1}^{\infty} \alpha_n^\times \rho^n = \frac{1}{\alpha_1} h_{b,b}(\rho)$ of the function $h_{b,b}$ obviously is given by $\left(\frac{1}{\alpha_1} h_{b,b}\right)^{-1}(y) = h_{b,b}^{-1}(\alpha_1 y) = \sum_{n=1}^{\infty} (\beta_n \alpha_1^n) y^n$ for all $y \in [-1,1]$, where $\alpha_n^\times := \frac{\alpha_n}{\alpha_1}$ ($n \in \mathbb{N}$), it follows that the n-th Taylor series coefficient $\beta_n^\times$ of the Taylor series of $\left(\frac{1}{\alpha_1} h_{b,b}\right)^{-1}$ is given by $\beta_n^\times = \beta_n \alpha_1^n$. Consequently,

$$\beta_n \alpha_1^n = \beta_n^\times = \frac{1}{n} \sum_{k=1}^{n-1} (-1)^k \binom{n-1+k}{k} B_{n-1,k}^\circ(\alpha_2^\times, \alpha_3^\times, \ldots, \alpha_{n-k+1}^\times) \tag{9.1.7}$$

(due to (9.1.1)). In the odd case, i.e., if in addition $\alpha_{2n} = 0$ for all $n \in \mathbb{N}$, the intrinsic combinatorical complexity of (9.1.1) can be even further reduced, possibly allowing a non-negligible saving of computing time (see Corollary 9.1).

We explicitly list $\beta_1, \beta_2, \beta_3, \beta_4, \beta_5, \beta_6$ and β_7 in full generality. To this end, as discussed above, if $\alpha_1 \neq 0$, we have to consider the Taylor coefficients $\alpha_n^\times := \frac{\alpha_n}{\alpha_1}, n \in \mathbb{N}$ of the "standardised" Taylor series $\frac{1}{\alpha_1} h_{b,b}$. Note again that $\alpha_1^\times = 1$. (9.1.1) therefore implies that

$$\beta_1 \alpha_1 = 1 = \alpha_1^\times$$

$$\beta_2 \alpha_1^2 = -\alpha_2^\times$$

$$\beta_3 \alpha_1^3 = -\alpha_3^\times + 2(\alpha_2^\times)^2$$

$$\beta_4 \alpha_1^4 = -\alpha_4^\times + 5\alpha_2^\times \alpha_3^\times - 5(\alpha_2^\times)^3$$

$$\beta_5 \alpha_1^5 = -\alpha_5^\times + 6\alpha_2^\times \alpha_4^\times + 3(\alpha_3^\times)^2 - 21(\alpha_2^\times)^2 \alpha_3^\times + 14(\alpha_2^\times)^4$$

$$\beta_6 \alpha_1^6 = -\alpha_6^\times + 7\alpha_2^\times \alpha_5^\times + 7\alpha_3^\times \alpha_4^\times - 28\alpha_2^\times (\alpha_3^\times)^2$$
$$\qquad - 28(\alpha_2^\times)^2 \alpha_4^\times + 84(\alpha_2^\times)^3 \alpha_3^\times - 42(\alpha_2^\times)^5$$

$$\beta_7 \alpha_1^7 = -\alpha_7^\times + 8\alpha_2^\times \alpha_6^\times + 8\alpha_3^\times \alpha_5^\times + 4(\alpha_4^\times)^2 - 36(\alpha_2^\times)^2 \alpha_5^\times$$
$$\qquad - 72\alpha_2^\times \alpha_3^\times \alpha_4^\times - 12(\alpha_3^\times)^3 + 120(\alpha_2^\times)^3 \alpha_4^\times$$
$$\qquad + 180(\alpha_2^\times)^2 (\alpha_3^\times)^2 - 330(\alpha_2^\times)^4 \alpha_3^\times + 132(\alpha_2^\times)^6. \tag{9.1.8}$$

In fact, if we make use of the key result, listed in [155], paired with the general Theorem [26, p. 222], we are able to present a (purely linear algebraic and algorith-

9.1 Open Problem 1: Grothendieck Constant Versus Taylor Series Inversion

mic) representation of each coefficient β_n, which avoids an explicit use of ordinary partial Bell polynomials (where no closed form seems to be available). To the best of our knowledge, in this context, that representation had not been published before. Instead of ordinary partial Bell polynomials, we have to calculate determinants of leading principal submatrices. Of course, the computational complexity induced by the increasing size of partial Bell polynomials transforms into the rapidly increasing computing time, induced by the increasing size of the determinants including the need to sum proper parts of determinants of different size. However, that summation is a recurrence relation (see (9.1.11) and the instructive Example 9.1).

Firstly, if $\alpha_1 \neq 0$, an enhancement of [155] reveals the following explicit representation of each coefficient β_n $(n \in \mathbb{N}_2)$:

$$\beta_n = \frac{(-1)^{n-1}}{n!\,\alpha_1^n} \det(A_n * T_n(1, \alpha_2^\times, \alpha_3^\times, \ldots, \alpha_n^\times))$$

$$= \frac{(-1)^{n-1}}{n!\,\alpha_1^{2n-1}} \det(A_n * T_n(\alpha_1, \alpha_2, \ldots, \alpha_n))$$

$$= \frac{1}{n!\,\alpha_1^{2n-1}} \det(-(A_n * T_n(\alpha_1, \alpha_2, \ldots, \alpha_n))), \qquad (9.1.9)$$

where $*$ again denotes the Hadamard product and the matrices $A_n \in \mathbb{M}_{n-1}(\mathbb{R})$ and $T_n \equiv T_n(x_1, x_2, \ldots, x_n) \in \mathbb{M}_{n-1}(\mathbb{R})$ $(x_1, \ldots, x_n \in \mathbb{C})$ are respectively defined as

$$A_n := \begin{pmatrix} n & 1 & 0 & 0 & 0 & \cdots & 0 & 0 \\ 2n & n+1 & 2 & 0 & 0 & \cdots & 0 & 0 \\ 3n & 2n+1 & n+2 & 3 & 0 & \cdots & 0 & 0 \\ 4n & 3n+1 & 2n+2 & n+3 & 4 & \cdots & 0 & 0 \\ \vdots & \vdots & \vdots & \vdots & \vdots & & \vdots & \vdots \\ (n-3)n & (n-4)n+1 & (n-5)n+2 & \cdots & \cdots & & n-3 & 0 \\ (n-2)n & (n-3)n+1 & (n-4)n+2 & \cdots & \cdots & & n+(n-3) & n-2 \\ (n-1)n & (n-2)n+1 & (n-3)n+2 & \cdots & \cdots & & 2n+(n-3) & n+(n-2) \end{pmatrix}$$

and

$$T_n \equiv T_n(x_1, x_2, \ldots, x_n) := \begin{pmatrix} x_2 & x_1 & 0 & 0 & 0 & \cdots & 0 & 0 \\ x_3 & x_2 & x_1 & 0 & 0 & \cdots & 0 & 0 \\ x_4 & x_3 & x_2 & x_1 & 0 & \cdots & 0 & 0 \\ x_5 & x_4 & x_3 & x_2 & x_1 & \ddots & 0 & 0 \\ \vdots & \vdots & \vdots & \vdots & \vdots & \ddots & \ddots & \vdots \\ x_{n-2} & x_{n-3} & \cdots & \cdots & x_3 & x_2 & x_1 & 0 \\ x_{n-1} & x_{n-2} & \cdots & \cdots & \cdots & x_3 & x_2 & x_1 \\ x_n & x_{n-1} & x_{n-2} & \cdots & \cdots & x_4 & x_3 & x_2 \end{pmatrix}.$$

Obviously, the Toeplitz matrix $T_n(x_1, x_2, \ldots, x_n)$ is well-defined for any $x_1, \ldots, x_n \in \mathbb{C}$. Observe that the appearance of the rather uncommon factor $\frac{1}{\alpha_1^{2n-1}} = \frac{1}{\alpha_1^n} \cdot \frac{1}{\alpha_1^{n-1}}$ in (9.1.9) actually originates from the simple, yet important, transformation

$$T_n \equiv T_n(\alpha_1, \alpha_2, \ldots, \alpha_n) = \alpha_1 T_n(1, \alpha_2^\times, \alpha_3^\times, \ldots, \alpha_n^\times). \tag{9.1.10}$$

More precisely, if $n \in \mathbb{N}$ and $(i, j) \in [n-1] \times [n-1]$, we have:

$$(A_n)_{ij} := \begin{cases} 0 & \text{if } j \geq i+2 \\ i & \text{if } j = i+1 \\ (i-j+1)n + j - 1 & \text{if } j \leq i \end{cases}$$

and $(T_n)_{ij} := \begin{cases} 0 & \text{if } j \geq i+2 \\ \alpha_1 & \text{if } j = i+1 \\ x_{i-j+2} & \text{if } j \leq i \end{cases}$.

Comparing (9.1.1) and (9.1.9), it follows that for all $\alpha_1 \in \mathbb{C}^*$ and $\alpha_2, \ldots, \alpha_n \in \mathbb{C}$

$$\det((A_n * T_n(\alpha_1, \ldots, \alpha_n))) = \alpha_1^{n-1} \sum_{k=1}^{n-1} (-1)^{n-1+k} \frac{(n-1+k)!}{k!}$$
$$B_{n-1,k}^\circ(\alpha_2^\times, \alpha_3^\times, \ldots, \alpha_{n-k+1}^\times).$$

Hence,

$$\det((A_n * T_n(1, \alpha_2^\times, \alpha_3^\times, \ldots, \alpha_n^\times))) = \sum_{k=1}^{n-1} (-1)^{n-1+k} \frac{(n-1+k)!}{k!}$$
$$B_{n-1,k}^\circ(\alpha_2^\times, \alpha_3^\times, \ldots, \alpha_{n-k+1}^\times).$$

Remark 9.1 (Connection to Apostol's Approach in [7]) In fact, it can be shown that

$$\det\left(-\left(A_n * T_n\left(x_1, \frac{x_2}{2!}, \frac{x_3}{3!}, \ldots, \frac{x_n}{n!}\right)\right)\right) = P_n(x_1, \ldots, x_n) \text{ for all } x_1, \ldots, x_n \in \mathbb{C}$$

precisely coincides with the function P_n, introduced in [7] (due to the convolution representation and the partial derivative structure of ordinary partial Bell polynomials)! In particular,

$$P_n(1, x_2, \ldots, x_n) = \sum_{k=1}^{n-1} (-1)^k \frac{(n-1+k)!}{k!} B_{n-1,k}^\circ\left(\frac{x_2}{2!}, \frac{x_3}{3!}, \ldots, \frac{x_{n-k+1}}{(n-k+1)!}\right)$$

$$= \det\left(-\left(A_n * T_n\left(1, \frac{x_2}{2!}, \frac{x_3}{3!}, \ldots, \frac{x_n}{n!}\right)\right)\right).$$

9.1 Open Problem 1: Grothendieck Constant Versus Taylor Series Inversion

Due to (9.1.10), it follows that

$$A_n * T_n(\alpha_1, \ldots, \alpha_n) = \alpha_1(A_n * T_n(1, \alpha_2^\times, \ldots, \alpha_n^\times)) = \alpha_1 B_n[n-1](\alpha_2^\times, \ldots, \alpha_n^\times),$$

where for any $p \in [n-1]$ and $x_1, \ldots, x_p \in \mathbb{C}$, the matrix $B_n[p] \equiv B_n[p](x_1, \ldots, x_p) \in \mathbb{M}_p(\mathbb{C})$ is defined as

$$B_n[p] := \begin{pmatrix} nx_1 & 1 & 0 & 0 & 0 & \cdots & 0 & 0 \\ 2nx_2 & (n+1)x_1 & 2 & 0 & 0 & \cdots & 0 & 0 \\ 3nx_3 & (2n+1)x_2 & (n+2)x_1 & 3 & 0 & \cdots & 0 & 0 \\ 4nx_4 & (3n+1)x_3 & (2n+2)x_2 & (n+3)x_1 & 4 & \cdots & 0 & 0 \\ \vdots & \vdots & \vdots & \vdots & \vdots & & \vdots & \vdots \\ (p-2)nx_{p-2} & ((p-3)n+1)x_{p-3} & ((p-4)n+2)x_{p-4} & \cdots & \cdots & \cdots & p-2 & 0 \\ (p-1)nx_{p-1} & ((p-2)n+1)x_{p-2} & ((p-3)n+2)x_{p-3} & \cdots & \cdots & \cdots & (n+(p-2))x_1 & p-1 \\ pnx_p & ((p-1)n+1)x_{p-1} & ((p-2)n+2)x_{p-2} & \cdots & \cdots & \cdots & (2n+(p-2))x_2 & (n+(p-1)x_1) \end{pmatrix}$$

$B_n[p]$ therefore denotes the p-th leading principal submatrix of the matrix $B_n[n-1]$ if $p \in [n-1]$ (cf. e.g. [70, 0.7.1]). More precisely, if $p \in [n-1]$ and $i, j \in [p]$, then:

$$B_n[p](x_1, x_2, \ldots, x_p)_{ij} := \begin{cases} 0 & \text{if } j \geq i+2 \\ i & \text{if } j = i+1 \\ ((i-j+1)n + j - 1)x_{i-j+1} & \text{if } j \leq i. \end{cases}$$

Equipped with all p leading principal submatrices $B_n[1], B_n[2], \ldots, B_n[p-1]$ of $B_n[p]$ and the "incipient matrix" $B_n[0] := (1)$, we may apply the main (unnumbered) theorem on page 222 of [26] to the matrix $B_n[p](x_1, \ldots, x_p)$, and it follows that

$$\det(B_n[p](x_1, \ldots, x_p)) = (p-1)! \sum_{k=1}^{p} (-1)^{p-k} \frac{pn - (k-1)(n-1)}{(k-1)!}$$

$$x_{p-k+1} \det(B_n[k-1]). \tag{9.1.11}$$

Consequently, if $p = n - 1$, it follows that

$$\det(A_n * T_n(\alpha_1, \ldots, \alpha_n)) = \alpha_1^{n-1} \det(B_n[n-1](\alpha_2^\times, \ldots, \alpha_n^\times))$$

$$= \alpha_1^{n-1} (n-2)! \sum_{k=1}^{n-1} (-1)^{n-1-k} \frac{(n-1)(n-k+1)}{(k-1)!}$$

$$\alpha_{n-k+1}^\times \det(B_n[k-1](\alpha_2^\times, \ldots, \alpha_k^\times))$$

$$= \alpha_1^{n-1}(n-1)! \sum_{k=1}^{n-1}(-1)^{n-k-1} \frac{n-k+1}{(k-1)!}$$

$$\alpha_{n-k+1}^\times \det(B_n[k-1](\alpha_2^\times, \ldots, \alpha_k^\times)). \qquad (9.1.12)$$

Thus,

$$\beta_n \alpha_1^n \stackrel{(9.1.9)}{=} \frac{1}{n} \sum_{k=1}^{n-1}(-1)^k \frac{n-k+1}{(k-1)!} \alpha_{n-k+1}^\times \det(B_n[k-1](\alpha_2^\times, \ldots, \alpha_k^\times)).$$

In particular, if $n = 2m+1 \in \mathbb{N}_3$ is odd ($m \in \mathbb{N}$) and $\alpha_{2l} := 0$ for all $l \in \mathbb{N}$, then

$$\beta_n \alpha_1^n = \frac{1}{n} \sum_{\substack{k=1 \\ k \text{ odd}}}^{n-1}(-1)^k \frac{n-k+1}{(k-1)!} \alpha_{n-k+1}^\times \det(B_n[k-1](0, \alpha_3^\times, 0 \ldots, 0, \alpha_k^\times))$$

$$= -\alpha_{2m+1}^\times - \frac{1}{2m+1} \sum_{r=1}^{m-1} \frac{2(m-r)+1}{(2r)!}$$

$$\alpha_{2(m-r)+1}^\times \det(B_{2m+1}[2r](0, \alpha_3^\times, 0 \ldots, 0, \alpha_{2r+1}^\times)).$$

Note also that (9.1.11) implies that

$$\det(B_{2m+1}[2r](0, x_1, 0, x_2, 0, \ldots, 0, x_r))$$

$$= -2(2r-1)! \left((2m+1)r\, x_r + \sum_{k=1}^{r-1} p_k(r,m) x_{r-k}\right) \qquad (9.1.13)$$

for all $m \in \mathbb{N}$, $r \in [m]$ and $x_1, \ldots, x_r \in \mathbb{C}$, where

$$p_k(r,m) := \frac{2m(r-k)+r}{(2k)!} \det(B_{2m+1}[2k](0, x_1, 0 \ldots, 0, x_k)) \qquad (k \in [r-1]).$$

Example 9.1 Fix $m \in \mathbb{N}$. Assume for simplification that $\alpha_1 = 1$ $\alpha_{2l} = 0$ for all $l \in \mathbb{N}$. Then $\alpha_n^\times = \alpha_n$ for all $n \in \mathbb{N}$. If $r \in \{1, 2\}$, the calculation of $\det(B_{2m+1}[2r])$ is very straightforward:

$$\det(B_{2m+1}[2](0, \alpha_3)) = -2(2m+1)\alpha_3 = 2!(-1)^1 \binom{2m+1}{1} \alpha_3$$

$$= 2!(-1)^1 \binom{2m+1}{1} B_{1,1}^\circ(\alpha_3),$$

9.1 Open Problem 1: Grothendieck Constant Versus Taylor Series Inversion

and

$$\det(B_{2m+1}[4](0, \alpha_3, 0, \alpha_5)) \stackrel{(9.1.13)}{=} -12\left(2(2m+1)\alpha_5 + \frac{2m+2}{2}\det(B_{2m+1}[2])\alpha_3\right)$$

$$= -12\left(2(2m+1)\alpha_5 + \frac{2m+2}{2}(-2(2m+1)\alpha_3)\alpha_3\right)$$

$$= 4!\left(-\binom{2m+1}{1}\alpha_5 + \binom{2m+2}{2}\alpha_3^2\right)$$

$$= 4!\left((-1)^1\binom{2m+1}{1}B_{2,1}^\circ(\alpha_3, \alpha_5)\right.$$

$$\left. + (-1)^2\binom{2m+2}{2}B_{2,2}^\circ(\alpha_3)\right).$$

If $r = 3$, a little more calculation effort is required, also triggered by a significant transformation step:

$$\det(B_{2m+1}[6](0, \alpha_3, 0, \alpha_5, 0, \alpha_7)) \stackrel{(9.1.13)}{=} -240\left(3(2m+1)\alpha_7 + \frac{4m+3}{2}\right.$$

$$\det(B_{2m+1}[2])\alpha_5 + \frac{2m+3}{4!}\det(B_{2m+1}[4])\alpha_3\right)$$

$$= -240\left(3\binom{2m+1}{1}\alpha_7\right.$$

$$- (4m+3)\binom{2m+1}{1}\alpha_3\alpha_5$$

$$+ (2m+3)\left(-\binom{2m+1}{1}\alpha_5\right.$$

$$\left. + \binom{2m+2}{2}\alpha_3^2\right)\alpha_3\right)$$

$$= -240\left(3\binom{2m+1}{1}\alpha_7 - 3\binom{2m+2}{2}(2\alpha_3\alpha_5)\right.$$

$$\left. + (2m+3)\binom{2m+2}{2}\alpha_3^3\right)$$

$$= 6!\left(-\binom{2m+1}{1}\alpha_7 + \binom{2m+2}{2}(2\alpha_3\alpha_5)\right.$$

$$\left. - \binom{2m+3}{3}\alpha_3^3\right)$$

$$= 6!\left((-1)^1\binom{2m+1}{1}B_{3,1}^\circ(\alpha_3, \alpha_5, \alpha_7)\right.$$

$$+ (-1)^2 \binom{2m+2}{2} B^\circ_{3,2}(\alpha_3, \alpha_5)$$

$$+ (-1)^3 \binom{2m+3}{3} B^\circ_{3,3}(\alpha_3) \Big).$$

In fact, the emerging structure can be kept in the case of $r = 4$, since

$$\det(B_{2m+1}[8])(0, \alpha_3, 0, \alpha_5, 0, \alpha_7, 0, \alpha_9)$$

$$\stackrel{(9.1.13)}{=} -2 \cdot 7! \Big(4(2m+1)\alpha_9 + \frac{6m+4}{2} \det(B_{2m+1}[2])\alpha_7$$

$$+ \frac{4m+4}{4!} \det(B_{2m+1}[4])\alpha_5 + \frac{2m+4}{6!} \det(B_{2m+1}[6])\alpha_3 \Big)$$

$$= 8! \Big(-\binom{2m+1}{1}\alpha_9 + \binom{2m+2}{2}(2\alpha_3\alpha_7 + \alpha_5^2) - \binom{2m+3}{3}(3\alpha_3^2\alpha_5)$$

$$+ \binom{2m+4}{4}\alpha_3^4 \Big)$$

$$= 8! \Big((-1)^1 \binom{2m+1}{1} B^\circ_{4,1}(\alpha_3, \alpha_5, \alpha_7, \alpha_9) + (-1)^2 \binom{2m+2}{2} B^\circ_{4,2}(\alpha_3, \alpha_5, \alpha_7)$$

$$+ (-1)^3 \binom{2m+3}{3} B^\circ_{4,3}(\alpha_3, \alpha_5) + (-1)^4 \binom{2m+4}{4} B^\circ_{4,4}(\alpha_3) \Big).$$

A relentless focus on Example 9.1 therefore leads to a non-obvious simplification of (9.1.7) which reduces the analysis of complex partition sets $P(2m, k)$ and related non-trivial ordinary partial Bell polynomials $B^\circ_{2m,k}(0, \alpha_3, 0, \alpha_5, 0, \alpha_7, 0, \ldots)$ to that one of partition sets $P(m, l)$ and related "fully occupied" ordinary partial Bell polynomials $B^\circ_{m,l}(\alpha_3, \alpha_5, \alpha_7, \ldots)$. In fact, the following result holds:

Proposition 9.1 *Let $(x_n)_{n \in \mathbb{N}}$ be an arbitrary sequence of complex numbers. Let $m \in \mathbb{N}$ and $r \in [m]$. Then*

$$\det(B_{2m+1}[2r])(0, x_1, 0, x_2, \ldots, 0, x_r))$$

$$= (2r)! \sum_{l=1}^{r} (-1)^l \binom{2m+l}{l} B^\circ_{r,l}(x_1, x_2, \ldots, x_{r-l+1}).$$

Proof If $m \in \mathbb{N}$ and $r \in [m]$, put $\alpha_{2i+1} := x_i$, where $i \in [r]$. We make use of a version of the principle of transfinite induction (which generalises induction over natural numbers), known as the principle of Noetherian induction. To this end, we consider the set

$$\mathcal{M} := \{(n, \nu) \in \mathbb{N} \times \mathbb{N} : n + 1 > \nu\} = \{(n, \nu) \in \mathbb{N} \times \mathbb{N} : n \geq \nu\}.$$

9.1 Open Problem 1: Grothendieck Constant Versus Taylor Series Inversion

Then $(\mathcal{M}, <_{\text{lex}})$ is a well-founded totally ordered set, where $<_{\text{lex}}$ denotes the lexicographic order on $(\mathbb{N}, <) \times (\mathbb{N}, <)$; i.e., $(l, \kappa) <_{\text{lex}} (m, \mu)$ if and only if $l < m$ or ($l = m$ and $\kappa < \mu$) (cf. [43, Chapter 6.3 including the table on p. 87]). Obviously, $(1, 1)$ is the minimal element of $\mathcal{M}$ which satisfies the claim, so that the induction basis is fulfilled. Let $(m, r) \in \mathcal{M}$ be non-minimal. Then $m \geq 2$. Clearly, if $r = 1$, then $(m, r) = (m, 1) \in \mathcal{M}$ satisfies the claim. So, we have to consider the case $r \geq 2$. Let $k \in [r - 1]$. Then $(m, k) <_{\text{lex}} (m, r)$. Hence, because of the induction assumption, it follows that

$$\det(B_{2m+1}[2k]) = (2k)! \sum_{i=1}^{k} (-1)^i \binom{2m+i}{i} B_{k,i}^{\circ}(\alpha_3, \alpha_5, \ldots, \alpha_{2(k-i)+3}).$$

(9.1.14)

If we insert (9.1.14) into (9.1.13), we obtain

$$\det(B_{2m+1}[2r]) = -2(2r-1)! \left((2m+1)r\, \alpha_{2r+1} + \sum_{k=1}^{r-1} p_k(r, m)\, \alpha_{2(r-k)+1} \right),$$

where

$$p_k(r, m) = \frac{2m(r-k) + r}{(2k)!} \det(B_{2m+1}[2k])$$

$$= \left((2m+1)r - 2mk \right) \sum_{i=1}^{k} (-1)^i \binom{2m+i}{i} B_{k,i}^{\circ}(\alpha_3, \alpha_5, \ldots, \alpha_{2k+3-2i}).$$

Consequently,

$$-\frac{\det(B_{2m+1}[2r])}{2(2r-1)!} - (2m+1)r\, \alpha_{2r+1}$$

$$= \sum_{k=1}^{r-1} \sum_{i=1}^{k} \left(2m(r-k) + r \right)(-1)^i \binom{2m+i}{i} \alpha_{2(r-k)+1}$$

$$B_{k,i}^{\circ}(\alpha_3, \alpha_5, \ldots, \alpha_{2k+3-2i})$$

$$= \sum_{l=1}^{r-1} (-1)^l \binom{2m+l}{l} \sum_{j=l}^{r-1} \left(2m(r-j) + r \right) \alpha_{2(r-j)+1}$$

$$B_{j,l}^{\circ}(\alpha_3, \alpha_5, \ldots, \alpha_{2(j-l+1)+1}).$$

Thereby, the last equality is a special case of the double sum representation

$$\sum_{k=1}^{r-1}\sum_{i=1}^{k} b_{ki}\, c_{ik} = \sum_{k=1}^{r-1}\sum_{i=k}^{r-1} b_{ik}\, c_{ki} = \sum_{l=1}^{r-1}\sum_{j=l}^{r-1} b_{jl}\, c_{lj}, \qquad (9.1.15)$$

which originates from the calculation of the trace of the matrix product of the lower triangular matrix $(b_{ki}\, \mathbb{1}_{\{(\mu,\nu):\mu\geq\nu\}}(k,i))$ and the upper triangular matrix $(c_{lj}\, \mathbb{1}_{\{(\mu,\nu):\mu\leq\nu\}}(l,j))$ and the trace of its transpose, which both are equal. Finally, since

$$\sum_{j=l}^{r-1}(2m(r-j)+r)\,\alpha_{2(r-j)+1}\, B^{\circ}_{j,l}(\alpha_3,\alpha_5,\ldots,\alpha_{2(j-l)+3})$$

$$= \frac{r(2m+l+1)}{l+1}\, B^{\circ}_{r,l+1}(\alpha_3,\ldots,\alpha_{2(r-l)+1})$$

(due to Lemma 9.1, applied to $n := r$, $k := l+1$, $\zeta := 2m$ and $i := j$), it follows that

$$r\left(\frac{\det(B_{2m+1}[2r])}{(2r)!} + (2m+1)\,\alpha_{2r+1}\right) = \frac{\det(B_{2m+1}[2r])}{2(2r-1)!} + (2m+1)r\,\alpha_{2r+1}$$

$$= -r\sum_{l=1}^{r-1}(-1)^l \binom{2m+l+1}{l+1}$$

$$B^{\circ}_{r,l+1}(\alpha_3,\ldots,\alpha_{2(r-l)+1})$$

$$= r\sum_{l=2}^{r}(-1)^l \binom{2m+l}{l}$$

$$B^{\circ}_{r,l}(\alpha_3,\ldots,\alpha_{2(r-l)+3}),$$

which concludes the Noetherian induction step, and the claim follows. □

Altogether, Lemma 9.1, Proposition 9.1 and (9.1.15), together with our previously mentioned analysis of the structure of β_{2m+1}, imply the following two fundamental results:

Theorem 9.1 *Let $m \in \mathbb{N}_2$ and $x_1,\ldots,x_{m-1} \in \mathbb{C}$. Then*

$$\sum_{r=1}^{m-1} \frac{2(m-r)+1}{(2r)!}\, x_{m-r}\, \det\left(B_{2m+1}[2r](0, x_1, 0, x_2, \ldots, 0, x_r)\right)$$

$$= \frac{1}{(2m)!} \sum_{r=2}^{m}(-1)^{r-1} \frac{(2m+r)!}{r!}\, B^{\circ}_{m,r}(x_1, x_2, \ldots, x_{m-r+1}).$$

9.1 Open Problem 1: Grothendieck Constant Versus Taylor Series Inversion

Proof By consecutive application of Proposition 9.1 and (9.1.15), it follows that

$$\sum_{r=1}^{m-1} \frac{2(m-r)+1}{(2r)!} x_{m-r} \det\left(B_{2m+1}[2r](0, x_1, 0, x_2, \ldots, 0, x_r)\right)$$

$$= \sum_{r=1}^{m-1} \sum_{l=1}^{r} (-1)^l (2(m-r)+1) \binom{2m+l}{l} x_{m-r} B_{r,l}^\circ(x_1, x_2, \ldots, x_{r-l+1})$$

$$= \sum_{r=1}^{m-1} (-1)^r \binom{2m+r}{r} \sum_{l=r}^{m-1} (2(m-l)+1) x_{m-l} B_{l,r}^\circ(x_1, x_2, \ldots, x_{l-r+1}).$$

However, the latter expression equals

$$\sum_{r=1}^{m-1} (-1)^r 2 \binom{2m+r}{r} \sum_{l=r}^{m-1} (m-l) x_{m-l} B_{l,r}^\circ(x_1, x_2, \ldots, x_{l-r+1})$$

$$+ \sum_{r=1}^{m-1} (-1)^r \binom{2m+r}{r} \sum_{l=r}^{m-1} x_{m-l} B_{l,r}^\circ(x_1, x_2, \ldots, x_{l-r+1}).$$

Thus, we can now apply Lemma 9.1 to both summands, and a final shift of the index r in the single remaining sum ($\sum_{r=1}^{m-1} a_{r+1} = \sum_{r=2}^{m} a_r$) clearly finishes the proof. □

Corollary 9.1 *Let $f(\rho) = \sum_{n=0}^{\infty} \alpha_{2n+1} \rho^{2n+1}$ be an odd real analytic function, convergent on $(-r, r) \subseteq \mathbb{R}$, where $r > 0$ denotes the radius of convergence of f. Assume that $f'(0) = \alpha_1 \neq 0$, implying that f is invertible around 0. Consider the real analytic odd inverse function $f^{-1} : V \longrightarrow \mathbb{R}$, where V is an open neighbourhood of $f(0) = 0$. If $f^{-1}(y) = \sum_{m=0}^{\infty} \beta_{2m+1} y^{2m+1}$ for all $y \in V$, then $\beta_1 = \frac{1}{\alpha_1}$ and*

$$\beta_{2m+1} = -\frac{1}{\alpha_1^{2m+1}} \left(\alpha_{2m+1}^\times + \frac{1}{2m+1} \sum_{r=1}^{m-1} \frac{2(m-r)+1}{(2r)!} \alpha_{2(m-r)+1}^\times \right.$$

$$\det\left(B_{2m+1}[2r](0, \alpha_3^\times, 0, \alpha_5^\times, \ldots, 0, \alpha_{2r+1}^\times)\right)\bigg)$$

$$= \frac{1}{(2m+1)! \alpha_1^{2m+1}} \sum_{r=1}^{m} (-1)^r \frac{(2m+r)!}{r!} B_{m,r}^\circ(\alpha_3^\times, \alpha_5^\times, \ldots, \alpha_{2(m-r+1)+1}^\times)$$

for all $m \in \mathbb{N}$, where $\alpha_{2\nu+1}^\times := \frac{\alpha_{2\nu+1}}{\alpha_1}$ ($\nu \in \mathbb{N}$).

As a little, yet illuminating exercise, we recommend the readers to perform the rather quick calculation of the first 3 Taylor coefficients of the inverse of the odd function $f := 3\sinh$, say, by applying Corollary 9.1 ! The outcome could be double-checked by means of https://en.wikipedia.org/wiki/Inverse_hyperbolic_functions#Series_expansions. Moreover, because of [146, Theorem 2] the coefficients β_{2m+1} satisfy the following, interesting recurrence relation:

$$\beta_{2m+1}\alpha_1^{2m+1}$$
$$= -\sum_{r=0}^{m-1}(\beta_{2r+1}\alpha_1^{2r+1})\, B^\circ_{2m+1,2r+1}(\alpha_1^\times, 0, \alpha_3^\times, 0, \alpha_5^\times, \ldots, 0, \alpha_{2(m-r)+1}^\times)$$
$$= -\sum_{r=0}^{m-1}\beta_{2r+1}\, B^\circ_{2m+1,2r+1}(\alpha_1, 0, \alpha_3, 0, \alpha_5, \ldots, 0, \alpha_{2(m-r)+1})$$

for all $m \in \mathbb{N}$.

In a nutshell, we recognise that already in the one-dimensional case, at least two hard open problems appear. On the one hand we need to know the explicit value of the Fourier-Hermite coefficients (cf. Proposition 6.1)

$$\sqrt{n!}\,\langle b, H_n\rangle_{\gamma_1} = \sqrt{n!}\,\mathbb{E}[b(X)H_n(X)] = \frac{d^n}{dt^n}\mathbb{E}[b(X+t)]\Big|_{t=0},$$

where $X \sim N_1(0,1)$. On the other hand, we have to look for a closed form expression of the coefficients β_n (if it were available at all), where the latter involves the complex recursive structure of ordinary partial Bell polynomials or related determinants. For example (keeping the Haagerup function in mind—cf. Example 7.1 and Remark 4.1), our question of the value of

$$\frac{\pi^k}{4^k}\, B^\circ_{n,k}\Big(\frac{1}{8}, \frac{3}{64}, \frac{25}{1024}, \ldots, \frac{((2(n-k)+1)!!)^2}{((2(n-k+1))!!)^2\,(n-k+2)}\Big)$$

very recently lead to an in depth-analysis, published in [40]. It appears to us that in general one cannot use proofs by standard induction on $n \in \mathbb{N}$ to verify statements about Bell polynomials. The Noetherian Induction Principle seems to be more appropriate here (as we have seen for example, in the proof of Proposition 9.1). In this context, we would like to draw attention to another recently published paper, where the authors point to similar difficulties including the formulation of related—open—problems (cf. [98]). Moreover, the solved examples in [98] show the large combinatorial barriers which we have to resolve while working with (partial) Bell polynomials.

Keeping these problems and barriers in mind, the following research topics and problems—which actually do not require any knowledge of the Grothendieck inequality—arise naturally:

(RP1) Continue to investigate the structure of partial Bell polynomials; possibly under inclusion of the use of supercomputers and related computer algebra systems.
(RP2) Develop a software package which puts Corollary 9.1 into practice.
(RP3) Look for an explicit analytic expression for the *inverse* function of the main building block of the Haagerup function; i.e., the inverse of the strictly increasing odd function

$$[-1, 1] \ni x \mapsto x\,_2F_1(\frac{1}{2}, \frac{1}{2}, 2; x^2)$$

(if available !), where as usual $_2F_1(a, b, c; \cdot)$ denotes the classic Gaussian hypergeometric function (cf. Example 7.1). Obviously, the inverse of $[-1, 1] \ni x \mapsto x \cdot \,_2F_1(\frac{1}{2}, \frac{1}{2}, \frac{3}{2}; x^2) = \arcsin(x) = \sin^{-1}(x)$ is the function sin. However, what about the inverses of (invertible) functions $_2F_1(a, b, c; \cdot)$ in general? Do we have to work with elliptic integrals here? What part does the Jacobi elliptic function play in this? A complex-analytic approach to a part of this problem using contour integration is given in [17, Chapter 5, including Theorem 5.6.18].

9.2 Open Problem 2: Interrelation Between the Grothendieck Inequality and Copulas

If we thoroughly overhaul the CCP function $[-1, 1] \ni \rho \mapsto \frac{2}{\pi} \arcsin(\rho)$ we recognise that some knowledge of Gaussian copulas (i.e., finite-dimensional multivariate distribution functions of univariate marginals generated by the distribution function of Gaussian random vectors—cf., e.g., [106, 116, 140, 150]) and (the probabilistic version) of the Hermite polynomials might become very fruitful regarding our indicated search for different " suitable" CCP functions. $[-1, 1] \ni \rho \mapsto \psi(\frac{1}{2}, \frac{1}{2}; t) = \frac{2}{\pi} \arcsin(\rho)$ namely reveals as a special case of the CCP function

$$[-1, 1] \ni \rho \mapsto \psi(p, p; \rho) = \frac{1}{c(p)} \sum_{n=1}^{\infty} \frac{1}{n} H_{n-1}^2(\Phi^{-1}(p)) \rho^n$$

$$= \frac{1}{2\pi\,p(1-p)} \exp(-(\Phi^{-1}(p))^2)$$

$$\rho \sum_{n=0}^{\infty} \frac{1}{n+1} H_n^2(\Phi^{-1}(p)) \rho^n ,$$

where $0 < p < 1$ and

$$c(p) := \frac{p(1-p)}{\varphi^2(\Phi^{-1}(p))} = 2\pi \, p(1-p) \exp((\Phi^{-1}(p))^2) = \sum_{n=0}^{\infty} \frac{1}{n+1} H_n^2(\Phi^{-1}(p)).$$

If we put

$$b_p(x) := \operatorname{sign}(xn\Phi^{-1}(p)) = 2\,\mathbb{1}_{[\Phi^{-1}(p),\infty)}(x) - 1$$
$$= 1 - 2\,\mathbb{1}_{(-\infty,\Phi^{-1}(p))}(x) \in \{-1, 1\},$$

where $x \in \mathbb{R}$, then the tetrachoric series expansion of the bivariate Gaussian copula (cf. [8, 58, 103]) implies the following generalisation of the Grothendieck equality:

$$h_p(\rho) := h_{b_p}(\rho) := \mathbb{E}[b_p(X) b_p(Y)] = (2p-1)^2$$
$$+ \frac{2}{\pi} \exp(-(\Phi^{-1}(p))^2) \sum_{n=1}^{\infty} \frac{1}{n} H_{n-1}^2(\Phi^{-1}(p))\rho^n$$
$$= (2p-1)^2 + 4p(1-p)\psi(p, p; \rho).$$

Due to our construction of $\psi(p, p; \cdot)$ the latter is clearly equivalent to

$$\rho(b_p(X), b_p(Y)) = \psi(p, p; \rho)$$

for all $p \in (0, 1)$, $\rho \in [-1, 1]$ and $(X, Y) \sim N_2(0, \Sigma_2(\rho))$, where $\rho(b_p(X), b_p(Y))$ denotes Pearson's correlation coefficient between the random variables $b_p(X)$ and $b_p(Y)$. Unfortunately,

$$h_p(\rho) = \psi(p, p; \rho) \text{ for all } \rho \in [-1, 1] \text{ if and only if } p = \frac{1}{2}.$$

These facts clearly lead to further research problems; namely:

(RP4) Prove whether there are $p \in (-1, 1) \setminus \{\frac{1}{2}\}$ and functions $\chi_p : \mathbb{R} \longrightarrow \{-1, 1\}$ such that $\psi(p, p; \rho) = h_{\chi_p}(\rho) = \mathbb{E}[\chi_p(X) \chi_p(Y)]$ for all $\rho \in [-1, 1]$ and $(X, Y) \sim N_2(0, \Sigma_2(\rho))$, so that the condition (SIGN) of our workflow is satisfied for h_{χ_p}.

(RP5) Generalise the above approach (which is built on the tetrachoric series of the bivariate Gaussian copula) to the n-variate case, where $n \in \mathbb{N}_3$.

(RP6) Verify whether the above approach can be transferred to the complex case. Could we then similarly generalise the Haagerup equality?

(RP7) If (RP4), respectively (RP5) holds, prove whether the condition (CRA) of the scheme holds. If this were the case, calculate (respectively approximate numerically) the related upper bound of $K_G^{\mathbb{R}}$. Include supercomputers and computer algebra systems if necessary.

9.3 Open Problem 3: Non-commutative Dependence Structures in Quantum Mechanics and the Grothendieck Inequality

Even a mathematical modelling of *non-commutative dependence* in quantum theory and its applications to quantum information and quantum computing is strongly linked with the existence of the real Grothendieck constant $K_G^{\mathbb{R}}$.

The latter can be very roughly adumbrated as follows: the experimentally proven non- Kolmogorovian (non-commutative) nature of the underlying probability theory of quantum physics leads to the well-known fact that in general a normal state of a composite quantum system cannot be represented as a convex combination of a product of normal states of the subsystems. This phenomenon is known as *entanglement* or *quantum correlation*. The Einstein-Podolsky-Rosen paradox, the violation of Bell's inequalities (limiting *spatial* correlation) and the Leggett-Garg inequalities (limiting *temporal* correlation) in quantum mechanics and related theoretical and experimental research implied a particular focus on a deeper understanding of this type of correlation—and hence to the *modelling of a specific type of dependence* of two (ore more) quantum observables in a composite quantum system, measured by two (or more) space-like separated instruments, each one having a classical parameter (such as the orientation of an instrument which measures the spin of a particle). In this context, a Leggett-Garg inequality (LGI) could be viewed as a "Bell inequality in time". The transition probability function, i.e., the *joint* probability distribution of observables in some fixed state of the system (considered as a function of the aforementioned parameters) may violate Bell's inequalities and is therefore not realisable in "classical" (commutative) physics. The surprising fact, firstly recognised by B. S. Tsirel'son (cf. [9, Ch. 11.2] and [119, 147, 149]), is that also this—experimentally verified—gap is an implication of the existence of the real Grothendieck constant $K_G^{\mathbb{R}} > 1$ (also known as *Tsirel'son bound*)! In other words, $K_G^{\mathbb{R}}$ indicates "how non-local quantum mechanics can be at most".

Already in the classical Kolmogorovian model, i.e., in the framework of probability space triples $(\Omega, \mathscr{F}, \mathbb{P})$, a rigorous description of tail dependence—which *exceeds* the standard dependence measure, given by Pearson's correlation coefficient, is a challenging task. To disclose (and simulate) the geometry of dependence one has to determine finite-dimensional multivariate distribution functions of univariate marginals, hence *copulas*. In the description of research problem 2 we have seen that Gaussian copulas are lurking in the Grothendieck equality. More precisely, we have (cf. [144]):

Example (Stieltjes, 1889) Let $\rho \in [-1, 1]$. Let $X, Y \sim N_1(0, 1)$ such that $\mathbb{E}[XY] = \rho$. Then

$$\mathbb{E}[\text{sign}(X)\text{sign}(Y)] = 4\, C^{\text{Ga}}(\tfrac{1}{2}, \tfrac{1}{2}; \rho) - 1 = \frac{2}{\pi} \arcsin(\rho) = \frac{2}{\pi} \arcsin(\mathbb{E}[XY]),$$

where $[-1, 1] \ni \rho \mapsto C^{\text{Ga}}(\frac{1}{2}, \frac{1}{2}; \rho)$ denotes the bivariate Gaussian copula with Pearson's correlation coefficient ρ as parameter, evaluated at $(\frac{1}{2}, \frac{1}{2})$.

Always keeping in mind a *non-commutative* version of the Grothendieck inequality (cf. [126, 147, 149]), our conjecture is that copulas in function spaces play a non-negligible role here. Unfortunately, compared to the finite-dimensional setting, the advent of the latter confronts us with hard problems. For example, by no means it is clear how marginals can be defined in an infinite-dimensional measurable vector space. If X is a random variable in a separable Hilbert space H, projections onto an orthonormal basis $(\langle X, e_n \rangle)_{n \in \mathbb{N}}$ are reasonable candidates. This case was studied in [65]. If in addition the space considered is a reproducing kernel Hilbert space of functions, over $[0, 1]$ say, an equally natural option for marginals would be function evaluations $\{X(t) : t \in [0, 1]\}$. Here, a new framework is required, including the preparation of a general concept of marginals for measurable vector spaces (cf. [13]). Consequently, we get the following problems:

(RP8) Look for objects like "non-commutative copulas", leading to a search for "non-commutative distribution functions in measurable vector spaces", including a non-commutative version of the famous result of Sklar (cf. [116] and the references therein).

(RP9) Define a "multivariate" spectral theory of *non-commuting* normal operator tuples and introduce non-commutative tail dependency measures in non-commutative C^*-algebras and operator spaces.

Let us close this chapter briefly with the following "blue-sky" research questions, which appear quite naturally and are completely unanswered. Can we improve the approximation results in the commutative case if we remove the underlying Gaussian structure in the Grothendieck inequality (for both fields, $\mathbb{R}$ and $\mathbb{C}$) and implement tail dependent distribution functions instead (such as the generalised extreme value (GEV) distribution)? What about infinitely divisible probability distributions in general? It is very likely that the use of correlation matrices and CCP functions, including linked *Gaussian* copula approaches, would no longer suffice (just as it is the case with Brownian motion which is in fact a particular case of a Lévy process, yet without jumps). So, could even general semimartingale techniques help to improve the approximations (cf., e.g., [44, 88])?

References

1. Abadir, K., Magnus, J.: Matrix Algebra. Econometric Exercises 1. Cambridge University Press, Cambridge (2005)
2. Acín, A., Gisin, N., Toner, B.: Grothendieck's constant and local models for noisy entangled quantum states. Phys. Rev. A **73**, 062105, 1–5 (2006)
3. Albiac, F., Kalton, N.J., (Godefroy, G.): Topics in Banach Space Theory, 2nd revised and updated edition. Graduate Texts in Mathematics, vol. 233. Springer, Cham (2016)
4. Aliprantis, C.D., Border, K.C.: Infinite Dimensional Analysis. A Hitchhiker's Guide, 3rd edn. Springer, Berlin (2006)
5. Andersen, H., Højbjerre, M., Sørensen, D., Eriksen, P.: Linear and Graphical Models for the Multivariate Complex Normal Distribution. Lecture Notes in Statistics, vol. 101. Springer, New York (1995)
6. Andrews, G.E., Askey, R., Roy, R.: Special Functions. Cambridge University Press, Cambridge (1999)
7. Apostol, T.M.: Calculating higher derivatives of inverses. Am. Math. Mon. **107**(8), 738–741 (2000)
8. Atiya, A.F., Fayed, H.A.: A novel series expansion for the multivariate normal probability integrals based on Fourier series. Math. Comput. **83**(289), 2385–2402 (2014)
9. Aubrun, G., Szarek, S.J.: Alice and Bob Meet Banach. The Interface of Asymptotic Geometric Analysis and Quantum Information Theory. Mathematical Surveys and Monographs, vol. 223. American Mathematical Society xxi, Providence (2017)
10. Bateman, H., Erdélyi, A., Magnus, A., Oberhettinger, W., Tricomi, F.G.: Higher Transcendental Functions, vol. I. McGraw-Hill Book Company, New York (1953—Reprint Ed. 1985)
11. Bauer, H.: Probability Theory. Translated from the German by Robert B. Burckel. de Gruyter Studies in Mathematics, vol. 23. Walter de Gruyter, Berlin-New York (1996)
12. Bauer, H.: Measure and Integration Theory. Translated from the German by Robert B. Burckel. de Gruyter Studies in Mathematics, vol. 26. Walter de Gruyter, Berlin-New York (2001)
13. Benth, F.E., Di Nunno, G., Schroers, D.: Copula measures and Sklar's theorem in arbitrary dimensions. Scand. J. Stat. **49**, 1144–1183 (2022)
14. Besançon, M., Designolle, S., Gelß, P., Iommazzo, G., Knebel, S., Pokutta, S.: Improved local models and new Bell inequalities via Frank-Wolfe algorithms. https://arxiv.org/abs/2302.04721 (2023)
15. Bhatia, R.: Matrix Analysis. Graduate Texts in Mathematics, vol. 169. Springer, New York (1997)

16. Bhatia, R.: Positive Definite Matrices. Princeton University Press, Princeton and Oxford (2007)
17. Bhattiprolu, V.: On the approximability of injective tensor norms. Ph.D. thesis, Carnegie Mellon University, Pittsburg (2019)
18. Bhattiprolu, V., Ghoshz, M., Guruswami, V., Lee, E., Tulsiani, M.: Approximating operator norms via generalized Krivine rounding. https://arxiv.org/abs/1804.03644 (2019)
19. Blekherman, G., Parrilo, P.A., Thomas, R.R. (eds.): Semidefinite Optimization and Convex Algebraic Geometry. MOS-SIAM Series on Optimization, vol. 13. Society for Industrial and Applied Mathematics (SIAM), Philadelphia (2012)
20. Bogachev, V.I.: Gaussian Measures. Transl. from the Russian by the author. Mathematical Surveys and Monographs, vol. 62. American Mathematical Society (AMS), xii, Providence (1998)
21. Boyd, S., Vandenberghe, L.: Convex Optimization. Cambridge University Press, Cambridge (2004)
22. Braverman, M., Makarychev, K., Makarychev, Y., Naor, A.: The Grothendieck constant is strictly smaller than Krivine's bound. Forum Math. Pi 1, Paper No. e4, 42 pp. (2013). https://arxiv.org/abs/1103.6161 (2011)
23. Briët, J., Buhrman, H., Toner, B.: A generalized Grothendieck inequality and nonlocal correlations that require high entanglement. Commun. Math. Phys. **305**, 827–843 (2011)
24. Briët, J., de Oliveira, F., Fernando, M., Vallentin, F.: Grothendieck inequalities for semidefinite programs with rank constraint. Theory Comput. **10**(4), 77–105 (2014)
25. Buckley, A.: The mathematics of quantum information theory: geometry of quantum states – an introduction to quantum entanglement. M.Sc. thesis, Faculty of Informatics of the Universitá della Svizzera Italiana, Lugano (2021). Available online via https://users.fmf.uni-lj.si/buckley/inf_master_thesis_inf_style_class%20(10).pdf
26. Cahill, N.D., D'Errico, J.R., Narayan, D.A., Narayan, J.Y.: Fibonacci determinants. College Math. J. **33**(3), 221–225 (2002)
27. Clauser, J.F., Horne, M.A., Shimony, A., Holt, R.A.: Proposed experiment to test local hidden-variable theories. Phys. Rev. Lett. **23**(15), 880–884 (1969)
28. Coine, C.: Schur multipliers on $\mathfrak{B}(L^p, L^q)$. J. Oper. Theory **79**(2), 301–326 (2018)
29. Comtet, L.: Advanced combinatorics. The art of finite and infinite expansions. Translated from the French by J. W. Nienhuys. Rev. and enlarged ed. D. Reidel Publishing Company, Dordrecht, Holland-Boston (1974)
30. Conway, J.B.: A Course in Functional Analysis, 2nd edn. Graduate Texts in Mathematics, vol. 96. Springer, New York (1990)
31. Cvijović, D.: New identities for the partial Bell polynomials. Appl. Math. Lett. **24**(9), 1544–1547 (2011)
32. Dattorro, J.: Convex Optimization & Euclidean Distance Geometry. Meboo Publishing, Palo Alto (2019). Available online via https://ccrma.stanford.edu/~dattorro/0976401304.pdf
33. Defant, A., Floret, K.: Tensor Norms and Operator Ideals. North-Holland Mathematics Studies, vol. 176. North-Holland, Amsterdam (1993)
34. Defant, M., Junge, M.: Grothendieck type inequalities and weak Hilbert spaces. In: Geometry of Banach Spaces, Proc. Conf., Strobl, 1989. London Mathematical Society Lecture Notes Series, vol. 158, pp. 71–88. Cambridge University Press, Cambridge (1991)
35. Deza, M.M., Laurent, M.: Geometry of Cuts and Metrics. Algorithms and Combinatorics, vol. 15. Springer, Berlin (1997)
36. Dick, J., Irrgeher, C., Leobacher, G., Pillichshammer, F.: On the optimal order of integration in Hermite spaces with finite smoothness. SIAM J. Numer. Anal. **56**(2), 684–707 (2018)
37. Diestel, J., Jarchow, H., Tonge, A.: Absolutely Summing Operators. Cambridge University Press, Cambridge (1995)
38. Diestel, J., Jarchow, H., Pietsch, A.: Operator ideals. In: Handbook of the Geometry of Banach Spaces, vol. 1, pp. 437–496. Elsevier, Amsterdam (2001)
39. Diestel, J., Fourie, J., Swart, J.: The Metric Theory of Tensor Products. Grothendieck's résumé revisited. American Mathematical Society. x, Providence (2008)

References

40. Du, W.-S., Lim, D., Qi, F.: Several recursive and closed-form formulas for some specific values of partial Bell polynomials. Adv. Theory Nonlinear Anal. Appl. **6**, 528–537 (2022)
41. Duistermaat, J.J., Kolk, J.A.C.: Multidimensional Real Analysis I: Differentiation. Transl. from the Dutch by J. P. van Braam Houckgeest. Cambridge Studies in Advanced Mathematics, vol. 86. Cambridge University Press, Cambridge (2004)
42. Dwork, C., Nikolov, A., Talwar, K.: Efficient algorithms for privately releasing marginals via convex relaxations. Discrete Comput. Geom. **53**(3), 650–673 (2015)
43. Eisinger, N., Ohlbach, H.J.: Design patterns for mathematical proofs. A guide, in particular for computer scientists. Design Patterns für mathematische Beweise. Ein Leitfaden insbesondere für Informatiker (in German). Springer Vieweg, Berlin (2017)
44. Eldan, R., Naor, A.: Krivine diffusions attain the Goemans-Williamson approximation ratio. https://arxiv.org/abs/1906.10615 (2019)
45. Fei, S.-M., Hua, B., Li, M., Li-Jost, X., Zhang, T., Zhou, C.: Towards Grothendieck constants and LHV models in quantum mechanics. J. Phys. A Math. Theor. **48**(6), 1–8 (2015)
46. Fishburn, P.C., Reeds, J.A.: Bell inequalities, Grothendieck's constant, and root two. SIAM J. Discrete Math. **7**(1), 48–56 (1994)
47. Fong, C.K., Radjavi, H., Rosenthal, P.: Norms for matrices and operators. J. Oper. Theory **18**(1), 99–113 (1987)
48. Friedland, S., Lim, L.-H.: Symmetric Grothendieck inequality. https://arxiv.org/abs/2003.07345 (2020)
49. Friedland, S., Lim, L.-H., Zhang, J.: An elementary and unified proof of Grothendieck's inequality. Enseign. Math. (2) **64**(3–4), 327–351 (2018)
50. Garnett, J.B.: Bounded Analytic Functions, revised 1st edn. Graduate Texts in Mathematics, vol. 236. Springer, New York (2006)
51. Ghanmi, A.: Operational formulae for the complex Hermite polynomials $H_{p,q}(z,\bar{z})$. Integral Transforms Spec. Funct. **24**(11), 884–895 (2013)
52. Gneiting, T.: Strictly and non-strictly positive definite functions on spheres. Bernoulli **19**(4), 1327–1349 (2013)
53. Goemans, M.X., Williamson, D.P.: Improved approximation algorithms for maximum cut and satisfiability problems using semidefinite programming. J. Assoc. Comp. Mach. **42**(6), 1115–1145 (1995)
54. González-Guillén, C.E., Lancien, C., Palazuelos, C., Villanueva, I.: Random quantum correlations are generically non-classical. Ann. Henri Poincaré **18**, 3793–3813 (2017)
55. Goodman, N.R.: Statistical analysis based on a certain multivariate complex Gaussian distribution (an introduction). Ann. Math. Stat. **34**, 152–177 (1963)
56. Grothendieck, A.: Produits tensoriels topologiques et espaces nucléaires. Memoirs of the American Mathematical Society, vol. 16 (1955)
57. Grothendieck, A.: Résumé de la Théorie Métrique des Produits Tensoriels Topologiques. Bol. Mat. Sao Paulo **8**, 1–79 (1953/1956). Reprinted in Resen. Inst. Mat. Estat. Univ. Sao Paulo **2**(4), 401–480 (1996)
58. Gupta, S.S.: Probability integrals of multivariate normal and multivariate t. Ann. Math. Stat. **34**, 792–828 (1963)
59. Gupta, V.P., Mandayam, P., Sunder, V.S.: The Functional Analysis of Quantum Information Theory. A Collection of Notes Based on Lectures by Gilles Pisier, K. R. Parthasarathy, Vern Paulsen and Andreas Winter. Lecture Notes in Physics, vol. 902. Springer, Cham (2015)
60. Haagerup, U.: A new upper bound for the complex Grothendieck constant. Isr. J. Math. **60**(2), 199–224 (1987)
61. Halliwell, L.J.: Complex Random Variables. Casualty Actuarial Society E-Forum, Fall 2015. Unpublished (2015). Cf. also: https://www.casact.org/sites/default/files/database/forum_15fforum_halliwell_complex.pdf
62. Hargé, G.: A particular case of correlation inequality for the Gaussian measure. Ann. Probab. **27**(4), 1939–1951 (1999)
63. Haroske, D.D., Triebel, H.: Distributions, Sobolev Spaces, Elliptic Equations. EMS Textbooks in Mathematics. European Mathematical Society, Zurich (2008)

64. Harris, B., Soms, A.P.: The use of the tetrachoric series for evaluating multivariate normal probabilities. J. Multivariate Anal. **10**, 252–267 (1980)
65. Hausenblas, E., Riedle, M.: Copulas in Hilbert spaces. Stochastics **89**, 222–239 (2017)
66. Helemskii, A.Ya.: Lectures and Exercises on Functional Analysis. American Mathematical Society (AMS), Providence (2006)
67. Hendrickx, J.M., Olshevsky, A.: Matrix p-norms are NP-hard to approximate if $p \neq 1, 2, \infty$. SIAM J. Matrix Anal. Appl. **31**(5), 2802–2812 (2010)
68. Hendy, M.D., Penny, D.: A framework for the quantitative study of evolutionary trees. Syst. Zool. **38**(4), 297–309 (1989)
69. Horn, R.A., Johnson, C.R.: Topics in Matrix Analysis. Cambridge University Press, Cambridge (1991)
70. Horn, R.A., Johnson, C.R.: Matrix Analysis, 2nd edn. Cambridge University Press, Cambridge (2013)
71. Hough, J.B., Krishnapur, M., Peres, Y., Virág, B.: Zeros of Gaussian Analytic Functions and Determinantal POint Processes. University Lecture Series, vol. 51. American Mathematical Society, Providence (2009)
72. Hu, Y.-Z.: Some operator inequalities. Séminaire de probabilités de Strasbourg, vol. 28, pp. 316–333 (1994). Available online via http://www.numdam.org/item/SPS_1994__28__316_0/
73. Hu, Y.: Analysis on Gaussian Spaces. World Scientific, Singapore (2017)
74. Hytönen, T., van Neerven, J., Veraar, M., Weis, L.: Analysis in Banach Spaces. Volume II: Probabilistic Methods and Operator Theory. Ergebnisse der Mathematik und ihrer Grenzgebiete. 3. Folge, vol. 67. Springer, Cham (2017)
75. Irrgeher, C., Leobacher, G.: High-dimensional integration on $\mathbb{R}^d$, weighted Hermite spaces, and orthogonal transforms. J. Complexity **31**(2), 174–205 (2015)
76. Ismail, M.E.H.: Analytic properties of complex Hermite polynomials. Trans. Am. Math. Soc. **368**(2), 1189–1210 (2016)
77. Itô, K.: Complex multiple Wiener integral. Jpn. J. Math. **22**, 63–86 (1953)
78. Jacob, N.: Pseudo Differential Operators and Markov Processes. Vol. I: Fourier Analysis and Semigroups. Imperial College Press, London (2001)
79. Jameson, G.J.O.: Summing and Nuclear Norms in Banach Space Theory. London Mathematical Society Student Texts, vol. 8. Cambridge University Press, Cambridge (1987)
80. Jarchow, H.: Locally Convex Spaces. Mathematische Leitfäden. B. G. Teubner, Stuttgart (1981)
81. Kaijser, S.: An application of Grothendieck's inequality to a problem in harmonic analysis. Sémin. Maurey-Schwartz 1975–1976, Espaces L^p. Appl. radonir., Géom. Espaces de Banach, Exposé **V**, 1–7, Polytechnic School, France (1975)
82. Katznelson, Y.: An Introduction to Harmonic Analysis, 3rd edn. Cambridge Mathematical Library. Cambridge University Press, Cambridge (2002)
83. Khare, A.: Matrix Analysis and Entrywise Positivity Preservers. London Mathematical Society Lecture Note Series. Cambridge University Press, Cambridge (2022)
84. Kinnewig, S.: Bell Inequalities and Grothendieck's Constant. B.Sc. thesis, Leibniz Universität Hannover, Institut für Theoretische Physik, Hannover (2017)
85. Knopp, K.: Theory and application of infinite series. Transl. from the 2nd ed. and revised in accordance with the Fourth by R. C. H. Young. Blackie & Son, Ltd. XII, London-Glasgow (1951)
86. König, H.: Some remarks on the Grothendieck inequality. In: General Inequalities 6, Proc. 6th Int. Conf., Oberwolfach, 1990, ISNM 103, pp. 201–206 (1992)
87. König, H.: On an extremal problem originating in questions of unconditional convergence. In: Recent Progress in Multivariate Approximation (Witten-Bommerholz, 2000), ISNM Internat. Ser. Numer. Math. Birkhäuser, Basel, vol. 137, pp. 185–192 (2001)
88. Kosmala, T., Riedle, M.: Stochastic Integration with Respect to Cylindrical Lévy Processes by p-Summing Operators. J. Theor. Probab. **34**, 477–497 (2020)

89. Krantz, S.G., Parks, H.R.: A Primer of Real Analytic Functions, 2nd edn. Birkhäuser Advanced Texts. Basler Lehrbücher, Boston (2002)
90. Krivine, J.-L.: Sur la complexification des opérateurs de L^∞ dans L^1 (in French). C. R. Acad. Sci. Paris Sér. A **284**, 377–379 (1977)
91. Krivine, J.-L.: Sur la constante de Grothendieck (in French). C. R. Acad. Sci. Paris Sér. A **284**, 445–446 (1977)
92. Krivine, J.-L.: Constantes de Grothendieck et fonctions de type positif sur les sphères (in French). Adv. Math. **31**, 16–30 (1979)
93. Krivine, J.-L.: A note about Grothendieck's constant. https://arxiv.org/abs/2306.00995 (2023)
94. Landsman, Z., Nešlehová, J.: Stein's lemma for elliptical random vectors. J. Multivariate Anal. **99**(5), 912–927 (2008)
95. Li, B.: Convex analysis and its application to quantum information theory. Ph.D. thesis, Case Western Reserve University, Cleveland (2018)
96. Li, B.: Exact values of quantum violations in low-dimensional Bell correlation inequalities. Linear Algebra Appl. **603**, 289–300 (2020)
97. Li, D., Queffélec, H.: Introduction to Banach Spaces: Analysis and Probability. Volume 1. Translated from the French by Danièle Gibbons and Greg Gibbons. Cambridge Studies in Advanced Mathematics, vol. 166. Cambridge University Press, Cambridge (2018)
98. Lim, D., Niu, D.-W., Qi, F., Yao, Y.-H.: Special values of the Bell polynomials of the second kind for some sequences and functions. J. Math. Anal. Appl. **491**(2), 124382 (2020)
99. Lindenstrauss, J., Pełczyński, A.: Absolutely summing operators in L_p-spaces and their applications. Stud. Math. **29**, 275–326 (1968)
100. Mamis, K.: Extension of Stein's lemma derived by using an integration by differentiation technique. Examples Counterexamples **2**, 100077 (2022)
101. Masjed-Jamei, M., Moalemi, Z., Koepf, W., Srivastava, H.M.: An extension of the Taylor series expansion by using the Bell polynomials. Rev. R. Acad. Cienc. Exactas Fís. Nat., Ser. A Mat., RACSAM **113**(2), 1445–1461 (2019)
102. Maurey, B.: Une nouvelle démonstration d'un théorème de Grothendieck (in French). Maurey-Schwartz seminar, Lecture 22, 1–7, Mathematics Center, Polytechnic School, May 9 (1973)
103. Meyer, C.: The bivariate normal copula. Commun. Stat. Theory Methods **42**(13), 2402–2422 (2013)
104. Miller, K.S., Samko, S.G.: Completely monotonic functions. Integral Transforms Spec. Funct. **12**(4), 389–402 (2001)
105. Mortini, R.: Ideals in the Wiener Algebra W^+. J. Austral. Math. Soc. (Ser. A) **46**, 220–228 (1989)
106. Nelsen, R.B.: An Introduction to Copulas, 2nd edn. Springer Series in Statistics. Springer, New York (2006)
107. Nesterov, Y.: Semidefinite relaxation and non-convex quadratic optimization. Optim. Methods Softw. **9**, 1–3, 141–160 (1998)
108. Nielsen, M.A., Chuang, I.L.: Quantum Computation and Quantum Information. Cambridge University Press, Cambridge (2000)
109. Niemi, H.: Grothendieck's inequality and minimal orthogonally scattered dilations. In: Probability Theory on Vector Spaces III, Proc. Conf., Lublin/Pol. 1983, Lect. Notes Math., vol. 1080, pp. 175–187 (1984)
110. Nualart, D.: The Malliavin Calculus and Related Topics, 2nd edn. Probability and Its Applications. Springer, Berlin (2006)
111. O'Donnell, R.: Analysis of Boolean functions. Cambridge University Press, Cambridge (2014)
112. Oertel, F.: Operator ideals and the principle of local reflexivity. Acta Univ. Carolinae - Math. Phys. **33**(2), 115–120 (1992)
113. Oertel, F.: Local properties of accessible injective operator ideals. Czech. Math. J. **48**(1), 119–133 (1998)

114. Oertel, F.: Extension of finite rank operators and operator ideals with the property (I). Math. Nachr. **238**, 144–159 (2002)
115. Oertel, F.: On normed products of operator ideals which contain $\mathcal{L}_2$ as a factor. Arch. Math. **80**, 61–70 (2003)
116. Oertel, F.: An analysis of the Rüschendorf transform - with a view towards Sklar's Theorem. Depend. Model. **3**(1), 113–125 (2015)
117. Owa, S., Srivastava, H.M.: Current Topics in Analytic Function Theory. World Scientific, Singapore (1992)
118. Pajot, P.: La revanche d'un théorème oublié (in French). https://www.larecherche.fr/la-revanche-dun-théorème-oublié (2015)
119. Palazuelos, C.: Random constructions in Bell inequalities: a survey. Found. Phys. **48**(8), 857–885 (2018)
120. Paulsen, V.I.: Completely Bounded Maps and Operator Algebras. Cambridge Studies in Advanced Mathematics, vol. 78. Cambridge University Press, Cambridge (2002)
121. Paulsen, V.I., Raghupathi, M.: An Introduction to the Theory of Reproducing Kernel Hilbert Spaces. Cambridge Studies in Advanced Mathematics, vol. 152. Cambridge University Press, Cambridge (2016)
122. Pełczynski, A.: Norms of classical operators in function spaces (Part V "Grothendieck's constants", 155–159). Colloque organisé en l'honneur du Professeur Laurent Schwartz, Polytechnic School, 1983, vol. 1, Astérisque 131, pp. 137–162 (1985)
123. Pietsch, A.: Operator Ideals. North-Holland Mathematical Library, vol. 20. North-Holland, Amsterdam (1980)
124. Pisier, G.: Grothendieck's theorem for noncommutative C*-algebras, with an appendix on Grothendieck's constants. J. Funct. Anal. **29**(3), 397–415 (1978)
125. Pisier, G.: Similarity Problems and Completely Bounded Maps. Second, expanded edition. Lecture Notes in Mathematics, vol. 1618. Springer, Berlin-Heidelberg (2001)
126. Pisier, G.: Grothendieck's theorem, past and present. Bull. Am. Math. Soc., New Ser. **49**(2), 237–323 (2012). Cf. also: http://www.math.tamu.edu/~pisier/grothendieck.UNCUT.pdf
127. Raghavendra, P., Steurer, D.: Towards computing the Grothendieck constant. In: Proceedings of the 20th Annual ACM-SIAM Symposium on Discrete Algorithms (SODA 09), Philadelphia, January 4–6, 2009, pp. 525–534. ACM, New York (2009)
128. Rietz, R.E.: A proof of the Grothendieck inequality. Isr. J. Math. **19**, 271–276 (1974)
129. Rohn, J.: Computing the norm $\|A\|_{\infty,1}$ is NP-hard. Linear Multilinear Algebra **47**, 195–204 (2000)
130. Rudin, W.: Real and Complex Analysis, 3rd edn. McGraw-Hill, New York (1987)
131. Rudin, W.: Functional Analysis, 2nd edn. International Series in Pure and Applied Mathematics. McGraw-Hill, New York (1991)
132. Rudin, W.: Function Theory in the Unit Ball of $\mathbb{C}^n$. Reprint of the 1980 original. Classics in Mathematics. Springer, Berlin (2008)
133. Ruess, W.M., Stegall, C.P.: Extreme points in duals of operator spaces. Math. Ann. **261**, 535–546 (1982)
134. Ryan, R.A.: Introduction to Tensor Products of Banach Spaces. Springer Monographs in Mathematics. Springer, London (2002)
135. Sasvári, Z.: Multivariate Characteristic and Correlation Functions. De Gruyter Studies in Mathematics, vol. 50. de Gruyter, Berlin (2013)
136. Schmeding, A.: An Introduction to Infinite-Dimensional Differential Geometry. Cambridge Studies in Advanced Mathematics, vol. 202. Cambridge University Press, Cambridge - open access (2023)
137. Schoenberg, I.J.: Positive definite functions on spheres. Duke Math. J. **9**, 96–108 (1942)
138. Schur I.: Bemerkungen zur Theorie der beschränkten Bilinearformen mit unendlich vielen xiii Veränderlichen. J. reine angew. Math. **140**, 1–28 (1911)
139. Silvester, J.R.: Determinants of block matrices. Math. Gazette **84**(501), 460–467 (2000)
140. Sklar, A.: Fonctions de répartition à n dimensions et leurs marges. Publications de l'Institut Statistique de l'Université de Paris (ISUP) **8**, 229–231 (1959)

References

141. Steeb, W.-H., Hardy, Y.: Problems and Solutions in Quantum Computing and Quantum Information, 4th edn. World Scientific, Singapore (2018)
142. Stein, E.M., Shakarchi, R.: Complex Analysis. Princeton University Press, Princeton (2003)
143. Steinberg, D.: Computation of matrix norms with applications to robust optimization. M.Sc. thesis, Technion - Israel Institute of Technology, Haifa (2005). Available online via https://www2.isye.gatech.edu/~nemirovs/Daureen.pdf
144. Stieltjes, T.S.: Extrait d'une lettre adressée à M. Hermite. Bull. Sci. Math. Ser. **2**(13), 170 (1889)
145. Szörényi, B.: Characterizing statistical query learning: simplified notions and proofs. In: Gavaldà, R., et al. (eds.) Algorithmic Learning Theory. 20th International Conference, ALT 2009, Porto, October 3–5, 2009. Lecture Notes in Artificial Intelligence, vol. 5809, pp. 186–200. Springer, Berlin (2009)
146. Todorov, P.G.: On the coefficients of the univalent functions of the Nevanlinna classes N_1 and N_2. In: Srivastava, H.M., et al. (eds.) Current Topics in Analytic Function Theory, pp. 363–370. World Scientific, Singapore (1992)
147. Tsirel'son, B.S.: Quantum generalizations of Bell's inequality. Lett. Math. Phys. **4**(2), 93–100 (1980)
148. Tsirel'son, B.S.: Quantum analogues of Bell's inequalities. The case of two spatially divided domains. J. Sov. Math. **36**(4), 557–570 (1987)
149. Tsirel'son, B.S.: Some results and problems on quantum Bell-type inequalities. Hadronic J. Suppl. **8**(4), 329–345 (1993)
150. Úbeda Flores, M., de Amo Artero, E., Durante, F., Fernández-Sánchez, J. (eds.) Copulas and Dependence Models with Applications. Contributions in Honor of Roger B. Nelsen. Springer, Berlin (2017)
151. Vapnik, V.N.: The Nature of Statistical Learning Theory, 2nd edn. Statistics for Engineering and Information Science. Springer, New York (2000)
152. Vourdas, A.: Grothendieck bound in a single quantum system. J. Phys. A: Math. Theor. **55**, 43 (2022)
153. Vourdas, A.: Corrigendum: Grothendieck bound in a single quantum system (2022 J. Phys. A: Math. Theor. **55**, 435206). J. Phys. A: Math. Theor. **56**(16) (2023)
154. Werner, D.: Functional Analysis, 8th revised edn. (in German). Springer, Berlin (2018)
155. Whittaker, E.T.: On the reversion of series. Gaz. Mat. **12**(50), 1–1 (1951)
156. Widder, D.V.: The Laplace Transform. Princeton University Press, Princeton (1941)
157. Wojtaszczyk, P.: Banach Spaces for Analysts. Cambridge Studies in Advanced Mathematics, vol. 25. Cambridge University Press, Cambridge (1991)
158. Zhu, L.: Grothendieck's Inequality. Unpublished (2018). Available online via http://www.cs.toronto.edu/~toni/Courses/Proofs-SOS-2018/Lectures/grothendieck.pdf

Index

n-correlation preserving (n-CCP) function, 113
2-dominated operator, 56
2-factorable operator, 50

absolute convex hull, 59
absolutely monotonic function, 108
absolutely p-summing operator ($1 \leq p < \infty$), 94

Bell inequality, 63

Catalan number, 121
characteristic function (complex case), 24
characteristic function (real case), 20
CHSH inequalities, 74
circularly symmetric function, 179
completely correlation preserving (CCP) function, 113
completely monotonic function, 115
Complex Gaussian random vector, 21
complex Hermite polynomial, 184
complex inner product rounding, 187
complex sign-function, 173
composition operator, 10
convex hull, 59

double factorial, 6

elementary matrix, 8
elliptope, 41

entanglement (quantum correlation), 217
entrywise functional calculus, 108
Euler-Mascheroni constant, 4
exponential partial Bell polynomial, 201

Frobenius inner product, 8

Gaussian copula, 144
Gaussian hypergeometric function, 85
Gaussian measure (complex case), 29
Gaussian measure (real case), 28
Gram matrix, 39
Grothendieck constant $K_G^{\mathbb{F}}$ ($\mathbb{F} \in \{\mathbb{R}, \mathbb{C}\}$), 3
Grothendieck equality, 86
Grothendieck inequality
    matrix form, 3
    tensor norm representation, 57

Haagerup equality, 86
Haagerup function, 191
Hadamard product (Schur product), 105
Hilbert-Schmidt operator, 8
hyperbolic CCP transform, 135

Krivine rounding scheme, 142
Kronecker product, 10

Leggett-Garg inequalities, 217
little Grothendieck inequality, 5

matrix vectorisation, 9
multivariate Gaussian law, 19
multivariate Hermite polynomial, 117

noise stability, 138
nuclear operator, 54

odd CCP transform, 147
ordinary partial Bell polynomial, 200

positive semidefinite, 14
principle of Noetherian induction, 210

quantum correlation matrix, 45
quantum state, 45

real inner product rounding, 159

subordinated matrix, 11
surface area of $\mathbb{S}^{n-1}$, 88

trace, 8
trace-class operator, 59

violation of a Bell inequality, 63

Walsh-Hadamard transform, 69
weak differentiation, 149
Werner state, 68
Wiener algebra, 111

LECTURE NOTES IN MATHEMATICS

Editors in Chief: J.-M. Morel, B. Teissier;

Editorial Policy

1. Lecture Notes aim to report new developments in all areas of mathematics and their applications – quickly, informally and at a high level. Mathematical texts analysing new developments in modelling and numerical simulation are welcome.

   Manuscripts should be reasonably self-contained and rounded off. Thus they may, and often will, present not only results of the author but also related work by other people. They may be based on specialised lecture courses. Furthermore, the manuscripts should provide sufficient motivation, examples and applications. This clearly distinguishes Lecture Notes from journal articles or technical reports which normally are very concise. Articles intended for a journal but too long to be accepted by most journals, usually do not have this "lecture notes" character. For similar reasons it is unusual for doctoral theses to be accepted for the Lecture Notes series, though habilitation theses may be appropriate.

2. Besides monographs, multi-author manuscripts resulting from SUMMER SCHOOLS or similar INTENSIVE COURSES are welcome, provided their objective was held to present an active mathematical topic to an audience at the beginning or intermediate graduate level (a list of participants should be provided).

   The resulting manuscript should not be just a collection of course notes, but should require advance planning and coordination among the main lecturers. The subject matter should dictate the structure of the book. This structure should be motivated and explained in a scientific introduction, and the notation, references, index and formulation of results should be, if possible, unified by the editors. Each contribution should have an abstract and an introduction referring to the other contributions. In other words, more preparatory work must go into a multi-authored volume than simply assembling a disparate collection of papers, communicated at the event.

3. Manuscripts should be submitted either online at www.editorialmanager.com/lnm to Springer's mathematics editorial in Heidelberg, or electronically to one of the series editors. Authors should be aware that incomplete or insufficiently close-to-final manuscripts almost always result in longer refereeing times and nevertheless unclear referees' recommendations, making further refereeing of a final draft necessary. The strict minimum amount of material that will be considered should include a detailed outline describing the planned contents of each chapter, a bibliography and several sample chapters. Parallel submission of a manuscript to another publisher while under consideration for LNM is not acceptable and can lead to rejection.

4. In general, **monographs** will be sent out to at least 2 external referees for evaluation.

   A final decision to publish can be made only on the basis of the complete manuscript, however a refereeing process leading to a preliminary decision can be based on a pre-final or incomplete manuscript.

   Volume Editors of **multi-author works** are expected to arrange for the refereeing, to the usual scientific standards, of the individual contributions. If the resulting reports can be forwarded to the LNM Editorial Board, this is very helpful. If no reports are forwarded or if other questions remain unclear in respect of homogeneity etc, the series editors may wish to consult external referees for an overall evaluation of the volume.

5. Manuscripts should in general be submitted in English. Final manuscripts should contain at least 100 pages of mathematical text and should always include

    – a table of contents;
    – an informative introduction, with adequate motivation and perhaps some historical remarks: it should be accessible to a reader not intimately familiar with the topic treated;
    – a subject index: as a rule this is genuinely helpful for the reader.
    – For evaluation purposes, manuscripts should be submitted as pdf files.

6. Careful preparation of the manuscripts will help keep production time short besides ensuring satisfactory appearance of the finished book in print and online. After acceptance of the manuscript authors will be asked to prepare the final LaTeX source files (see LaTeX templates online: https://www.springer.com/gb/authors-editors/book-authors-editors/manuscriptpreparation/5636) plus the corresponding pdf- or zipped ps-file. The LaTeX source files are essential for producing the full-text online version of the book, see http://link.springer.com/bookseries/304 for the existing online volumes of LNM). The technical production of a Lecture Notes volume takes approximately 12 weeks. Additional instructions, if necessary, are available on request from lnm@springer.com.

7. Authors receive a total of 30 free copies of their volume and free access to their book on SpringerLink, but no royalties. They are entitled to a discount of 33.3 % on the price of Springer books purchased for their personal use, if ordering directly from Springer.

8. Commitment to publish is made by a *Publishing Agreement*; contributing authors of multiauthor books are requested to sign a *Consent to Publish form*. Springer-Verlag registers the copyright for each volume. Authors are free to reuse material contained in their LNM volumes in later publications: a brief written (or e-mail) request for formal permission is sufficient.

Addresses:
Professor Jean-Michel Morel, CMLA, École Normale Supérieure de Cachan, France
E-mail: moreljeanmichel@gmail.com

Professor Bernard Teissier, Equipe Géométrie et Dynamique,
Institut de Mathématiques de Jussieu – Paris Rive Gauche, Paris, France
E-mail: bernard.teissier@imj-prg.fr

Springer: Ute McCrory, Mathematics, Heidelberg, Germany,
E-mail: lnm@springer.com

SPRINGER NATURE

GPSR Compliance

The European Union's (EU) General Product Safety Regulation (GPSR) is a set of rules that requires consumer products to be safe and our obligations to ensure this.

If you have any concerns about our products, you can contact us on ProductSafety@springernature.com

In case Publisher is established outside the EU, the EU authorized representative is:

Springer Nature Customer Service Center GmbH
Europaplatz 3
69115 Heidelberg, Germany

The manufacturer's authorised representative in the EU is Springer Nature Customer Service Centre GmbH, Europaplatz 3, 69115 Heidelberg, Germany. If you have any concerns regarding our products, please contact ProductSafety@springernature.com

Printed and bound by CPI Group (UK) Ltd, Croydon, CR0 4YY

25/03/2026

02078187-0011